Mathematical Systems: Techniques and Applications

Mathematical Systems: Techniques and Applications

Edited by
Calanthia Wright

Willford Press
www.willfordpress.com

Published by Willford Press,
118-35 Queens Blvd., Suite 400,
Forest Hills, NY 11375, USA

ISBN: 978-1-68285-578-2

Cataloging-in-Publication Data

Mathematical systems : techniques and applications / edited by Calanthia Wright.
p. cm.
Includes bibliographical references and index.
ISBN 978-1-68285-578-2
1. Mathematics. 2. Mathematics--Technological innovations. I. Wright, Calanthia.
QA36 .M38 2019
510--dc21

For information on all Willford Press publications
visit our website at www.willfordpress.com

Willford Press

Contents

Preface

The purpose of the book is to provide a glimpse into the dynamics and to present opinions and studies of some of the scientists engaged in the development of new ideas in the field from very different standpoints. This book will prove useful to students and researchers owing to its high content quality.

Mathematical systems are models of analysis and measurement of dynamic variables for applications in natural sciences and engineering disciplines. A mathematical model is significant to the mathematical study of a system and its different components. Prediction about system change and behavior is also studied in this discipline. Mathematical systems are involved in the characterization of abstract structures, statistical models and game theoretic models among many others. All mathematical models are built on the dynamics between different variables. Accordingly, these models are classified as linear or nonlinear, static or dynamic, deterministic or probabilistic, etc. This book present researches performed by experts across globe. It covers varied aspects of innovation and presents the ideas and concepts in a comprehensive manner. This book will assist researchers and students alike.

At the end, I would like to appreciate all the efforts made by the authors in completing their chapters professionally. I express my deepest gratitude to all of them for contributing to this book by sharing their valuable works. A special thanks to my family and friends for their constant support in this journey.

Editor

Preface

1

On the Caginalp phase-field system based on the Cattaneo law with nonlinear coupling

Armel Andami Ovono[1] **and Alain Miranville**[2,*]

[1] Université des Sciences et techniques de Masuku, Franceville, Gabon

[2] Université de Poitiers, Laboratoire de Mathématiques et Applications, UMR CNRS 7348 - SP2MI, Boulevard Marie et Pierre Curie - Téléport 2, F-86962 Chasseneuil Futuroscope Cedex, France

[*] **Correspondence:** Email:Alain.Miranville@math.univ-poitiers.fr

Abstract: We focus in this paper on a Caginalp phase-field system based on the Cattaneo law with nonlinear coupling. We start our analysis by establishing existence, uniqueness and regularity based on Moser's iterations. We finish with the study of the spatial behavior of the solutions in a semi-infinite cylinder, assuming the existence of such solutions.

Keywords: Caginalp phase field system; Cattaneo law; nonlinear coupling; well-posedness; regularity; spatial behavior

1. Introduction

The Caginalp phase-field model

$$\frac{\partial u}{\partial t} - \Delta u + f(u) = \theta \tag{1.1}$$

$$\frac{\partial \theta}{\partial t} - \Delta \theta = -\frac{\partial u}{\partial t} \tag{1.2}$$

as described in [1] has been the subject of numerous studies in recent years [2, 3, 4, 7, 11, 13, 35]. This model describes the behavior of certain materials in their stages of melting and solidification. In this case θ and u can represent respectively the temperature and the order parameter.

Using Fourier's law to the aforementioned model, one can observe a disparity between the observed results and the expected outcome. One of them is known as "paradox of heat conduction" [9]. In order to make the model more realistic by adjusting the latter, some alternative laws have been proposed, the

Maxwell-Cattaneo law [25] or the Gurtin-Pipkin law [21, 22]. Furthermore, in [18, 19, 20] Green and Naghdi proposed an alternative theory based on a thermomechanical theory of deformable media to obtain very rational models.

In recent years, the study of models derived from these new laws have been the subject of particular attention, especially with regard to the qualitative study of solutions [14, 15, 16, 17, 23, 27, 29, 30, 31, 32].

The purpose of our study is the following model

$$\frac{\partial u}{\partial t} - \Delta u + f(u) = g(u)\frac{\partial \alpha}{\partial t} \tag{1.3}$$

$$\eta\frac{\partial^2 \alpha}{\partial t^2} + \frac{\partial \alpha}{\partial t} - \Delta\alpha = -\eta g(u)\frac{\partial u}{\partial t} - G(u) + h, \quad \eta > 0 \tag{1.4}$$

$$u|_{\partial\Omega} = \alpha|_{\partial\Omega} = 0 \tag{1.5}$$

$$u|_{t=0} = u_0, \quad \alpha|_{t=0} = \alpha_0, \quad \left.\frac{\partial \alpha}{\partial t}\right|_{t=0} = \alpha_1, \tag{1.6}$$

where Ω is a bounded and regular domain of $\mathbb{R}^n$ with $n = 2$ or 3. This model is motivated by the recent works of Miranville and Quintanilla [24, 25, 26].

This paper is organized as follows. In Section 2, we give a rigorous derivation of our model using Cattaneo's law and a nonlinear coupling. Then, in Section 3 we prove existence, uniqueness and regularity results. We finish, in Section 4, by the study of the spatial behavior of the solutions in a semi-infinite cylinder, assuming that such solutions exist.

Throughout this paper, the same letters c, c' and c'' denote constants which may change from line to line.

2. Derivation of the model

Our equations (1.3)-(1.6) modeling phase transition are derived as follows.

Let Ψ be the total energy of the system defined as

$$\Psi(u,\theta) = \int_\Omega \left(\frac{1}{2}|\nabla u|^2 + F(u) - G(u)\theta - \frac{1}{2}\theta^2\right)dx,$$

with $G' = g$ and $F' = f$. Let H be the enthalpy satisfying

$$H = \partial_\theta \Psi = G(u) + \theta. \tag{2.1}$$

Furthermore,

$$\tfrac{\partial u}{\partial t} = -\partial_u \Psi \tag{2.2}$$

$$\tfrac{\partial H}{\partial t} + \mathrm{div} q = 0. \tag{2.3}$$

In particular, considering the Maxwell-Cattaneo law

$$(1 + \eta\frac{\partial}{\partial t})q = -\nabla\theta, \quad \eta > 0, \tag{2.4}$$

we get (using (2.1), (2.2) and (2.4))

$$(1+\eta\frac{\partial}{\partial t})\frac{\partial\theta}{\partial t}-\Delta\theta=(1+\eta\frac{\partial}{\partial t})\frac{\partial G(u)}{\partial t}. \tag{2.5}$$

Setting

$$\alpha=\int_0^t\theta(\tau)d\tau+\alpha_0,\quad \theta=\frac{\partial\alpha}{\partial t}, \tag{2.6}$$

we have, after integrating (2.5) over $[0,t]$,

$$\eta\frac{\partial^2\alpha}{\partial t^2}+\frac{\partial\alpha}{\partial t}-\Delta\alpha=-\eta g(u)\frac{\partial u}{\partial t}-G(u)+h, \tag{2.7}$$

with

$$h=\eta\frac{\partial^2\alpha}{\partial t^2}(0)+\frac{\partial\alpha}{\partial t}(0)-\Delta\alpha(0)-\eta g(u)\frac{\partial u}{\partial t}(0)-G(u)(0) \tag{2.8}$$

and

$$\frac{\partial G(u)}{\partial t}=g(u)\frac{\partial u}{\partial t}. \tag{2.9}$$

This leads to the above system (1.3)-(1.6).

3. Existence and uniqueness of solutions

We start by giving an existence result, the assumptions for the proof being the following: f and g are of class C^1 and

$$|G(s)|^2\le c_1\,F(s)+c_2,\ c_1,c_2\ge 0, \tag{3.1}$$

$$|g(s)s|\le c_3(|G(s)|+1),\ c_3\ge 0, \tag{3.2}$$

$$c_4\,s^{k+2}-c_5\le c_0F(s)-c_0'\le f(s)s\le c_6\,s^{k+2}-c_7,\ c_0,c_4,c_6>0,\ c_0',c_5,c_7\ge 0, \tag{3.3}$$

$$|g(s)|\le c_8(|s|+1),\quad |g'(s)|\le c_9,\ c_8,c_9\ge 0, \tag{3.4}$$

$$|f'(s)|\le c_{10}(|s|^k+1),\ c_{10}\ge 0, \tag{3.5}$$

where k is an integer.

We have the

Theorem 3.1. *We assume that (3.1)-(3.3) hold true. If in addition* $(u_0,\alpha_0,\alpha_1)\in H_0^1(\Omega)\cap L^{k+2}(\Omega)\times H_0^1(\Omega)\times L^2(\Omega)$, *then (1.3)-(1.6) admits a solution* (u,α) *such that* $u\in L^\infty(0,T;H_0^1(\Omega)\cap L^{k+2}(\Omega))$, $\frac{\partial u}{\partial t}\in L^2(0,T;L^2(\Omega))$, $\alpha\in L^\infty(0,T;H_0^1(\Omega))$ *and* $\frac{\partial\alpha}{\partial t}\in L^\infty(0,T;L^2(\Omega))$, $\forall T>0$.

Proof. We will focus on the priori estimates. The proof of existence follows from these estimates and a proper Galerkin scheme [12] and [34].

Multiplying (1.3) by $\frac{\partial u}{\partial t}$ and integrating over Ω, we have

$$\left\|\frac{\partial u}{\partial t}\right\|_2^2+\frac{1}{2}\frac{d}{dt}\|\nabla u\|_2^2+\frac{d}{dt}\int_\Omega F(u)dx=\int_\Omega g(u)\frac{\partial\alpha}{\partial t}\frac{\partial u}{\partial t}dx, \tag{3.6}$$

where $||.||_p$ denotes the usual L^p norm and $(.,.)$ the usual L^2 scalar product; more generally, we denote by $||.||_X$ the norm in the Banach space X.

Similarly, multiplying (1.4) by $\frac{\partial\alpha}{\partial t}$, we obtain

$$\begin{aligned}\frac{\eta}{2}\frac{d}{dt}\left\|\frac{\partial\alpha}{\partial t}\right\|_2^2 + \left\|\frac{\partial\alpha}{\partial t}\right\|_2^2 + \frac{1}{2}\frac{d}{dt}||\nabla\alpha||_2^2 = -\eta\int_\Omega g(u)\frac{\partial u}{\partial t}\frac{\partial\alpha}{\partial t}dx \\ - \int_\Omega G(u)\frac{\partial\alpha}{\partial t}dx + \int_\Omega h\frac{\partial\alpha}{\partial t}dx.\end{aligned} \tag{3.7}$$

Summing η(3.6) and (3.7), we find

$$\begin{aligned}\eta\left\|\frac{\partial u}{\partial t}\right\|_2^2 + \frac{\eta}{2}\frac{d}{dt}||\nabla u||_2^2 + \eta\frac{d}{dt}\int_\Omega F(u)dx + \frac{\eta}{2}\frac{d}{dt}\left\|\frac{\partial\alpha}{\partial t}\right\|_2^2 + \left\|\frac{\partial\alpha}{\partial t}\right\|_2^2 \\ + \frac{1}{2}\frac{d}{dt}||\nabla\alpha||_2^2 = -\int_\Omega G(u)\frac{\partial\alpha}{\partial t}dx + \int_\Omega h\frac{\partial\alpha}{\partial t}dx.\end{aligned} \tag{3.8}$$

We thus obtain a differential inequality of the form

$$\frac{d}{dt}E_1 + \eta\left\|\frac{\partial u}{\partial t}\right\|_2^2 + \left\|\frac{\partial\alpha}{\partial t}\right\|_2^2 = -\int_\Omega G(u)\frac{\partial\alpha}{\partial t}dx + \int_\Omega h\frac{\partial\alpha}{\partial t}dx, \tag{3.9}$$

with $E_1 = \frac{\eta}{2}||\nabla u||_2^2 + \eta\int_\Omega F(u)dx + \frac{\eta}{2}\left\|\frac{\partial\alpha}{\partial t}\right\|_2^2 + \frac{1}{2}||\nabla\alpha||_2^2$.

Multiplying (1.3) by u, we find

$$\frac{1}{2}\frac{d}{dt}||u||_2^2 + ||\nabla u||_2^2 + (f(u),u) = \int_\Omega g(u)\frac{\partial\alpha}{\partial t}udx. \tag{3.10}$$

We have, owing to (3.2), (3.3) and (3.10),

$$\frac{1}{2}\frac{d}{dt}||u||_2^2 + ||\nabla u||_2^2 + c_0\int_\Omega F(u)dx \le c\int_\Omega |G(u)|^2dx + \frac{1}{2}\left\|\frac{\partial\alpha}{\partial t}\right\|_2^2 + c''. \tag{3.11}$$

From (3.9) and (3.11), we obtain

$$\begin{aligned}\frac{d}{dt}\left(E_1 + \frac{1}{2}||u||_2^2\right) + \eta\left\|\frac{\partial u}{\partial t}\right\|_2^2 + \frac{1}{2}\left\|\frac{\partial\alpha}{\partial t}\right\|_2^2 + ||\nabla u||_2^2 + c_0\int_\Omega F(u)dx \le \\ -\int_\Omega G(u)\frac{\partial\alpha}{\partial t}dx + \int_\Omega h\frac{\partial\alpha}{\partial t}dx + c\int_\Omega |G(u)|^2dx + c''.\end{aligned} \tag{3.12}$$

Multiplying (1.4) by α, we get

$$\begin{aligned}\eta\frac{d}{dt}\left(\frac{\partial\alpha}{\partial t},\alpha\right) + \left(\frac{\partial\alpha}{\partial t},\alpha\right) + ||\nabla\alpha||_2^2 = -\eta\int_\Omega g(u)\frac{\partial u}{\partial t}\alpha dx \\ + \int_\Omega (h - G(u))\alpha dx + \eta\left\|\frac{\partial\alpha}{\partial t}\right\|_2^2.\end{aligned} \tag{3.13}$$

Adding δ(3.13) and (3.12) with $\delta > 0$, we find

$$\begin{aligned}
&\frac{d}{dt}\Big(E_1 + \frac{1}{2}\|u\|_2^2\Big) + \eta\Big\|\frac{\partial u}{\partial t}\Big\|_2^2 + \frac{1}{2}\Big\|\frac{\partial \alpha}{\partial t}\Big\|_2^2 + \|\nabla u\|_2^2 + c_0\int_\Omega F(u)dx \\
&+ \eta\delta\frac{d}{dt}\Big(\frac{\partial \alpha}{\partial t}, \alpha\Big) + \delta\Big(\frac{\partial \alpha}{\partial t}, \alpha\Big) + \delta\|\nabla \alpha\|_2^2 \leq \\
&-\int_\Omega G(u)\frac{\partial \alpha}{\partial t}dx + \int_\Omega h\frac{\partial \alpha}{\partial t}dx + c\int_\Omega |G(u)|^2 dx + c'' \\
&-\delta\eta\int_\Omega g(u)\frac{\partial u}{\partial t}\alpha dx + \delta\int_\Omega (h - G(u))\alpha dx + \delta\eta\Big\|\frac{\partial \alpha}{\partial t}\Big\|_2^2 .
\end{aligned} \tag{3.14}$$

Since

$$\int_\Omega g(u)\frac{\partial u}{\partial t}\alpha dx = \frac{d}{dt}\int_\Omega G(u)\alpha dx - \int_\Omega G(u)\frac{\partial \alpha}{\partial t}dx, \tag{3.15}$$

i.e.,

$$-\eta\delta\int_\Omega g(u)\frac{\partial u}{\partial t}\alpha dx = -\eta\delta\frac{d}{dt}\int_\Omega G(u)\alpha dx + \eta\delta\int_\Omega G(u)\frac{\partial \alpha}{\partial t}dx, \tag{3.16}$$

we get, owing to (3.14) and (3.16),

$$\begin{aligned}
&\frac{d}{dt}E_2 + \eta\Big\|\frac{\partial u}{\partial t}\Big\|_2^2 + \frac{1}{2}\Big\|\frac{\partial \alpha}{\partial t}\Big\|_2^2 + \|\nabla u\|_2^2 + c_0\int_\Omega F(u)dx + \delta\|\nabla \alpha\|_2^2 \leq \\
&\int_\Omega (h - G(u))\frac{\partial \alpha}{\partial t}dx + c\int_\Omega |G(u)|^2 dx + c'' \\
&+ \delta\eta\int_\Omega G(u)\frac{\partial \alpha}{\partial t}dx + \delta\int_\Omega (h - G(u))\alpha dx + \delta\eta\Big\|\frac{\partial \alpha}{\partial t}\Big\|_2^2,
\end{aligned} \tag{3.17}$$

with

$$\begin{aligned}
E_2 =& \frac{1}{2}\|u\|_2^2 + \frac{\eta}{2}\|\nabla u\|_2^2 + \eta\int_\Omega F(u)dx + \frac{\eta}{2}\Big\|\frac{\partial \alpha}{\partial t}\Big\|_2^2 + \frac{1}{2}\|\nabla \alpha\|_2^2 \\
&+ \eta\delta\int_\Omega G(u)\alpha dx + \frac{\delta}{2}\|\alpha\|_2^2 + \eta\delta\Big(\frac{\partial \alpha}{\partial t}, \alpha\Big).
\end{aligned}$$

Furthermore,

$$\begin{aligned}
&\frac{d}{dt}E_2 + \eta\Big\|\frac{\partial u}{\partial t}\Big\|_2^2 + \frac{1}{2}\Big\|\frac{\partial \alpha}{\partial t}\Big\|_2^2 + \|\nabla u\|_2^2 + c_0\int_\Omega F(u)dx + \delta\|\nabla \alpha\|_2^2 \leq \\
&\int_\Omega h\frac{\partial \alpha}{\partial t}dx + c\int_\Omega |G(u)|^2 dx + c'' + \delta\eta\Big\|\frac{\partial \alpha}{\partial t}\Big\|_2^2 \\
&+ (\delta\eta - 1)\int_\Omega G(u)\frac{\partial \alpha}{\partial t}dx - \delta\int_\Omega G(u)\alpha dx + \delta\int_\Omega h\alpha dx.
\end{aligned} \tag{3.18}$$

Noting that

$$\begin{aligned}
(\delta\eta - 1)\int_\Omega G(u)\frac{\partial \alpha}{\partial t}dx &\leq \frac{1}{8}\Big\|\frac{\partial \alpha}{\partial t}\Big\|_2^2 + c\int_\Omega |G(u)|^2 dx, \\
\delta\int_\Omega G(u)\alpha dx &\leq c\int_\Omega |G(u)|^2 dx + \frac{\delta}{4}\|\nabla \alpha\|_2^2,
\end{aligned} \tag{3.19}$$

$$\begin{aligned}\delta \int_\Omega h\alpha dx &\leq c\|h\|_2^2 + \frac{\delta}{4}\|\nabla\alpha\|_2^2 \\ \int_\Omega h\frac{\partial\alpha}{\partial t}dx &\leq \frac{1}{8}\left\|\frac{\partial\alpha}{\partial t}\right\|_2^2 + 2\|h\|_2^2,\end{aligned} \tag{3.20}$$

we obtain, owing to (3.18), (3.19) and (3.20)

$$\begin{aligned}&\frac{d}{dt}E_2 + \eta\left\|\frac{\partial u}{\partial t}\right\|_2^2 + \frac{5}{8}\left\|\frac{\partial\alpha}{\partial t}\right\|_2^2 + \|\nabla u\|_2^2 + c_0\int_\Omega F(u)dx \\ &+ \frac{\delta}{2}\|\nabla\alpha\|_2^2 \leq c\int_\Omega |G(u)|^2dx + c''.\end{aligned} \tag{3.21}$$

Choosing δ such that

$$\frac{\eta}{2}\left\|\frac{\partial\alpha}{\partial t}\right\|_2^2 + \delta\left(\frac{\partial\alpha}{\partial t}, \alpha\right) + \frac{1}{2}\|\nabla\alpha\|_2^2 \geq c\left(\left\|\frac{\partial\alpha}{\partial t}\right\|_2^2 + \|\nabla\alpha\|_2^2\right), \tag{3.22}$$

and using (3.1), we have

$$\eta\int_\Omega F(u)dx + \frac{1}{2}\|\nabla\alpha\|_2^2 + \eta\delta\int_\Omega G(u)\alpha dx \geq c\left(\int_\Omega F(u)dx + \|\nabla\alpha\|_2^2\right) - \delta\, c_2. \tag{3.23}$$

We have, taking into account (3.3), (3.22) and (3.23),

$$E_2 \leq c\left(\|u\|_{H^1(\Omega)}^2 + \|u\|_{k+2}^{k+2} + \left\|\frac{\partial\alpha}{\partial t}\right\|_2^2 + \|\alpha\|_{H^1(\Omega)}^2\right) + k_1 \quad k_1 > 0. \tag{3.24}$$

Similarly

$$E_2 \geq c\left(\|u\|_{H^1(\Omega)}^2 + \|u\|_{k+2}^{k+2} + \left\|\frac{\partial\alpha}{\partial t}\right\|_2^2 + \|\alpha\|_{H^1(\Omega)}^2\right) - k_1 \quad k_1 > 0. \tag{3.25}$$

There holds owing to (3.1) and(3.21)

$$\frac{d}{dt}E_2 + c\left\|\frac{\partial u}{\partial t}\right\|_2^2 \leq c'\, E_2 + c''. \tag{3.26}$$

Finally the proof is deduced from (3.24)-(3.26). □

Let us consider a more restrictive assumption on G as follows:

$$\forall\epsilon > 0, \quad |G(s)|^2 \leq \epsilon F(s) + c_\epsilon, \quad s \in \mathbb{R}. \tag{3.27}$$

We also have the

Theorem 3.2. *We assume that (3.2),(3.3) hold true and* $(u_0, \alpha_0, \alpha_1) \in H_0^1(\Omega)\cap L^{k+2}(\Omega)\times H_0^1(\Omega)\times L^2(\Omega)$. *If in addition we consider (3.27), then* $u \in L^\infty(\mathbb{R}^+; H_0^1(\Omega)$
$) \cap L^\infty(\mathbb{R}^+; L^{k+2}(\Omega))$, $\alpha \in L^\infty(\mathbb{R}^+; H_0^1(\Omega))$ *and* $\frac{\partial\alpha}{\partial t} \in L^\infty(\mathbb{R}^+; L^2(\Omega))$, $\forall T > 0$.

Proof. From (3.21), we had

$$\begin{aligned}&\frac{d}{dt}E_2+\eta\left\|\frac{\partial u}{\partial t}\right\|_2^2+\frac{5}{8}\left\|\frac{\partial \alpha}{\partial t}\right\|_2^2+\|\nabla u\|_2^2+c_0\int_\Omega F(u)dx\\&+\frac{\delta}{2}\|\nabla\alpha\|_2^2\le c\int_\Omega |G(u)|^2dx+c'',\end{aligned}\tag{3.28}$$

with

$$\begin{aligned}E_2=&\frac{1}{2}\|u\|_2^2+\frac{\eta}{2}\|\nabla u\|_2^2+\eta\int_\Omega F(u)dx+\frac{\eta}{2}\left\|\frac{\partial\alpha}{\partial t}\right\|_2^2+\frac{1}{2}\|\nabla\alpha\|_2^2\\&+\eta\delta\int_\Omega G(u)\alpha dx+\frac{\delta}{2}\|\alpha\|_2^2+\eta\delta\Big(\frac{\partial\alpha}{\partial t},\alpha\Big).\end{aligned}$$

Using (3.27), we obtain

$$\begin{aligned}&\frac{d}{dt}E_2+\eta\left\|\frac{\partial u}{\partial t}\right\|_2^2+\frac{5}{8}\left\|\frac{\partial \alpha}{\partial t}\right\|_2^2+\|\nabla u\|_2^2+(c_0-k_\epsilon)\int_\Omega F(u)dx\\&+\frac{\delta}{2}\|\nabla\alpha\|_2^2\le h_\epsilon,\end{aligned}\tag{3.29}$$

with $k_\epsilon=c.\epsilon$ and $h_\epsilon=c.c_\epsilon|\Omega|+c''$.

We also get by using Young's inequality ($|G(u)\alpha|\le\frac{1}{\delta}|G(u)|^2+\frac{\delta}{4}|\alpha|^2$) and (3.27)

$$\eta\int_\Omega F(u)dx+\eta\delta\int_\Omega G(u)\alpha dx\ge\eta(1-\epsilon)\int_\Omega F(u)dx-\frac{\eta\delta^2}{4}\|\alpha\|_2^2-p_\epsilon,\tag{3.30}$$

with $p_\epsilon=\eta|\Omega|c_\epsilon$.

In addition, choosing δ such that

$$\frac{\eta}{2}\left\|\frac{\partial\alpha}{\partial t}\right\|_2^2+\delta\Big(\frac{\partial\alpha}{\partial t},\alpha\Big)+\frac{1}{2}\|\nabla\alpha\|_2^2+\frac{\delta}{2}\|\alpha\|_2^2-\frac{\eta\delta^2}{4}\|\alpha\|_2^2\ge c\Big(\left\|\frac{\partial\alpha}{\partial t}\right\|_2^2+\|\nabla\alpha\|_2^2\Big),\ c>0,\tag{3.31}$$

and $\epsilon<1$ such that $c_0-k_\epsilon>0$, we deduce from (3.29), (3.30) and (3.31) that

$$\frac{d}{dt}E_2+c\Big(E_2+\left\|\frac{\partial u}{\partial t}\right\|_2^2\Big)\le c'',\ c>0.$$

The proof follows from Gronwall's lemma. □

Remark 3.3. *The previous theorem proves that the system is dissipative in* $L^{k+2}(\Omega)\cap H_0^1(\Omega)\times H_0^1(\Omega)\times L^2(\Omega)$.

We give in what follows a regularity result of the solution which is based on Moser's iterations. We will use a restriction on k, in particular k should be an even integer.

Theorem 3.4. *We assume that the assumptions of theorem 3.2 hold and that* $n=3$. *Let* u *be a classical solution to (1.3)-(1.6) defined in* $[0,T]$ *and* k *be an even integer. We consider for all* $q>1$, $U_q=\sup_{t\le T}\|u(t)\|_q<\infty$. *Then* $U_\infty<\infty$.

The proof is based on the following lemma.

Lemma 3.5. *Let u be a classical solution to (1.3)-(1.6) defined in* $[0, T]$ *and k be an even integer. Given* $r > 1$ *such that* $\tilde{U}_r = \max\{1, \|u_0\|_\infty, U_r = sup_{t\leq T}\|u(t)\|_r\}$, *then there exists a constant* $C_3 = C_3(\|\frac{\partial\alpha}{\partial t}\|_{L^\infty(0,\infty;\, L^2(\Omega))})$ *such that*

$$\tilde{U}_{2r} \leq (C_3)^{\sigma(r)} r^{\sigma(r)} \tilde{U}_r,$$

with

$$\sigma(r) = \frac{5q}{r(5q-6)}, \quad q > \frac{3}{2}. \tag{3.32}$$

Proof. Multiplying (1.3) by u^{2r-1} with (3.3) and (3.4), we get

$$\begin{aligned} &\frac{1}{2r}\frac{d}{dt}\int_\Omega u^{2r}dx + \frac{2r-1}{r^2}\int_\Omega |\nabla(u^r)|^2 dx + c_4\int_\Omega u^{k+2r}dx \\ &- c_5\int_\Omega u^{2r-2}dx \leq c_8\int_\Omega |u|^{2r}\frac{\partial\alpha}{\partial t}dx + c_8\int_\Omega u^{2r-1}\frac{\partial\alpha}{\partial t}dx, \end{aligned} \tag{3.33}$$

and using $c_4\int_\Omega u^{k+2r}dx \geq 0$, we have

$$\begin{aligned} &\frac{1}{2r}\frac{d}{dt}\int_\Omega u^{2r}dx + \frac{2r-1}{r^2}\int_\Omega |\nabla(u^r)|^2 dx \\ &- c_5\int_\Omega u^{2r-2}dx \leq c_8\int_\Omega |u|^{2r}\frac{\partial\alpha}{\partial t}dx + c_8\int_\Omega u^{2r-1}\frac{\partial\alpha}{\partial t}dx. \end{aligned} \tag{3.34}$$

Let $p > 1$ be such that $\frac{1}{p} + \frac{1}{q} = 1$. It is clear that condition $q > \frac{3}{2}$ is equivalent to $p < 3$. Taking $w = u^r$ in (3.34), we obtain after some calculations

$$\frac{1}{2r}\frac{d}{dt}\|w\|_2^2 + \frac{2r-1}{r^2}\|\nabla w\|_2^2 \leq \lambda_1\|w\|_{\kappa p}^{\kappa} + \lambda_2\|w\|_{\kappa p}^{\kappa-\frac{1}{r}} \tag{3.35}$$

with $\kappa = (2r-1)/r$, $\lambda_1 = \lambda_1(\|\frac{\partial\alpha}{\partial t}\|_2)$ and $\lambda_2 = \lambda_2(\|\frac{\partial\alpha}{\partial t}\|_2)$.

Let β be such that

$$\frac{1}{\kappa p} = \beta + \frac{1-\beta}{6}. \tag{3.36}$$

Since

$$\beta = \frac{6r - p(2r-1)}{5p(2r-1)}$$

we claim that $\beta \in (0, 1)$. In fact, from $p < \frac{6r}{2r-1}$ it follows that $\beta > 0$. Moreover

$$6r < 6p(2r-1),$$

i.e.

$$6r - p(2r-1) < 5p(2r-1),$$

proves that $\beta < 1$. This leads to $\beta \in (0, 1)$. Using proper interpolation inequalities, there holds

$$\begin{aligned} &\frac{1}{2r}\frac{d}{dt}\|w\|_2^2 + \frac{2r-1}{r^2}\|\nabla w\|_2^2 \leq \\ &\lambda_1\left(\|w\|_1^\beta\|w\|_{2^\star}^{1-\beta}\right)^2 + \lambda_2\left(\|w\|_1^\beta\|w\|_{2^\star}^{1-\beta}\right)^{\kappa-\frac{1}{r}} \end{aligned} \tag{3.37}$$

and using proper Sobolev's injections, there holds

$$\begin{aligned} &\frac{1}{2r}\frac{d}{dt}\|w\|_2^2 + \frac{2r-1}{r^2}\|\nabla w\|_2^2 \leq \\ &\lambda_1\Big[\|w\|_1^{2\beta} C^{2(1-\beta)}(4r)^{(1-\beta)}\Big]\Big[(\frac{1}{4r})^{(1-\beta)}\|\nabla w\|_2^{2(1-\beta)}\Big] \\ &+\lambda_2\Big[\|w\|_1^{\beta(\kappa-\frac{1}{r})} C^{(1-\beta)(\kappa-\frac{1}{r})}(4r)^{\frac{(\kappa-\frac{1}{r})(1-\beta)}{2}}\Big] \\ &\Big[(\frac{1}{4r})^{\frac{(\kappa-\frac{1}{r})(1-\beta)}{2}}\|\nabla w\|_2^{(1-\beta)(\kappa-\frac{1}{r})}\Big]. \end{aligned} \tag{3.38}$$

Using Young's inequality, we find

$$\begin{aligned} &\frac{1}{2r}\frac{d}{dt}\|w\|_2^2 + \frac{2r-1}{r^2}\|\nabla w\|_2^2 \leq \beta\Big[\lambda_1^{1/\beta}\|w\|_1^2 C^{\frac{2(1-\beta)}{\beta}}(4r)^{\frac{(1-\beta)}{\beta}}\Big] \\ &+(1-\beta)\Big[(\frac{1}{4r})\|\nabla w\|_2^2\Big] + \frac{(\kappa-\frac{1}{r})(1-\beta)}{2}\Big[(\frac{1}{4r})\|\nabla w\|_2^2\Big] \\ &+\delta_1\Big[\lambda_2^{1/\delta_1}\|w\|_1^{\beta(\kappa-\frac{1}{r})/\delta_1} C^{\frac{(1-\beta)(\kappa-\frac{1}{r})}{\delta_1}}(4r)^{\frac{(\kappa-\frac{1}{r})(1-\beta)}{2\delta_1}}\Big] \end{aligned} \tag{3.39}$$

with $\delta_1 = 1 - \frac{(\kappa-\frac{1}{r})(1-\beta)}{2}$, $\delta_1 \in (0,1)$. Hence

$$\begin{aligned} &\frac{1}{2r}\frac{d}{dt}\|w\|_2^2 + \frac{6r-4}{4r^2}\|\nabla w\|_2^2 \leq \Big[\lambda_1^{1/\beta}\|w\|_1^2 C^{\frac{2(1-\beta)}{\beta}}(4r)^{\frac{(1-\beta)}{\beta}}\Big] \\ &+\Big[\lambda_2^{1/\delta_1}\|w\|_1^{\beta(\kappa-\frac{1}{r})/\delta_1} C^{\frac{(1-\beta)(\kappa-\frac{1}{r})}{\delta_1}}(4r)^{\frac{(\kappa-\frac{1}{r})(1-\beta)}{2\delta_1}}\Big]. \end{aligned} \tag{3.40}$$

Since $\frac{6r-4}{2r} > 1$,

$$\begin{aligned} &\frac{d}{dt}\|w\|_2^2 + \|\nabla w\|_2^2 \leq \lambda_1^{1/\beta}\|w\|_1^2 C^{\frac{2(1-\beta)}{\beta}}(4r)^{\frac{(1-\beta)}{\beta}+1} \\ &+\lambda_2^{1/\delta_1}\|w\|_1^{\beta(\kappa-\frac{1}{r})/\delta_1} C^{\frac{(1-\beta)(\kappa-\frac{1}{r})}{\delta_1}}(4r)^{\frac{(\kappa-\frac{1}{r})(1-\beta)}{2\delta_1}+1}. \end{aligned} \tag{3.41}$$

Setting

$$\begin{aligned} &2r\sigma_1(r) - 1 = \frac{(1-\beta)}{\beta},\ 2r\sigma_2(r) - 1 = \frac{(\kappa-\frac{1}{r})(1-\beta)}{2\delta_1} \\ &2\rho_2(r) = \frac{\beta(\kappa-\frac{1}{r})}{\delta_1}, \end{aligned} \tag{3.42}$$

we have owing to Poincaré's inequality

$$\begin{aligned} \frac{d}{dt}\|w\|_2^2 + C_0\|w\|_2^2 &\leq \lambda_1^{1/\beta}\|w\|_1^2 C^{4r\sigma_1(r)-2}(4r)^{2r\sigma_1(r)} \\ &+\lambda_2^{1/\delta_1}\|w\|_1^{2\rho_2(r)} C^{4r\sigma_2(r)-2}(4r)^{2r\sigma_2(r)}, \end{aligned} \tag{3.43}$$

with

$$\rho_2(r) = \frac{(r-1)[6r - p(2r-1)]}{5rp(2r-1) - 6(r-1)[p(2r-1) - r]}.$$

We claim that $\rho_2(r) \in (0, 1)$. In fact, since $\beta > 0$, $\kappa - \frac{1}{r} > 0$ and $\delta_1 > 0$ it follows that $\rho_2(r) > 0$. In addition, from

$$\begin{aligned} 5(r-1)p(2r-1) &< 5rp(2r-1), \\ (r-1)[6r - p(2r-1)] &< 5rp(2r-1) \\ &\quad - 6p(2r-1)(r-1) + 6r(r-1), \end{aligned}$$

we see that

$$\rho_2(r) = \frac{(r-1)[6r - p(2r-1)]}{5rp(2r-1) - 6(r-1)[p(2r-1) - r]} < 1.$$

Hence $\rho_2(r) \in (0, 1)$. Integrating (3.43) over $[0, t)$, we get

$$\begin{aligned} \|w(t)\|_2^2 \le \|w(0)\|_2^2 &+ C_1^{1/\beta} \|w\|_1^2 \, C^{4r\sigma_1(r)-2} (4r)^{2r\sigma_1(r)} \\ &+ C_2^{1/\delta_1} \|w\|_1^2 \, C^{4r\sigma_2(r)-2} (4r)^{2r\sigma_2(r)}, \end{aligned} \tag{3.44}$$

with

$$C_1 = C_1\left(\left\|\frac{\partial \alpha}{\partial t}\right\|_{L^\infty(0,\infty;\, L^2(\Omega))}\right) \quad C_2 = C_2\left(\left\|\frac{\partial \alpha}{\partial t}\right\|_{L^\infty(0,\infty;\, L^2(\Omega))}\right).$$

In addition, we note that

$$\|w(0)\|_2^2 = \int_\Omega w(0)^2 dx = \int_\Omega u(0)^{2r} dx \le |\Omega| \|u(0)\|_\infty^{2r} \le |\Omega| \tilde{U}_r^{2r}. \tag{3.45}$$

It follows from (3.44) and (3.45) that

$$\begin{aligned} \tilde{U}_{2r}^{2r} \le |\Omega| \tilde{U}_r^{2r} &+ \frac{1}{C^2} C_1^{1/\beta} \, C^{4r\sigma_1(r)} (4r)^{2r\sigma_1(r)} \tilde{U}_r^{2r} \\ &+ \frac{1}{C^2} C_2^{1/\delta_1} \, C^{4r\sigma_2(r)} (4r)^{2r\sigma_2(r)} + \tilde{U}_r^{2r}. \end{aligned} \tag{3.46}$$

We also get from (3.42)

$$\sigma_1(r) = \frac{1}{2r\beta}, \ \sigma_2(r) = \frac{1}{2r\delta_1}, \quad \delta_1 > \beta, \tag{3.47}$$

with

$$\sigma_1(r) = \frac{5p(2r-1)}{2r(6r - (2r-1)p)}.$$

This implies

$$\sigma_2(r) < \sigma_1(r) \quad \forall r > 1. \tag{3.48}$$

Setting $\sigma(r) = \frac{5q}{r(5q-6)} = \frac{5p}{r(6-p)}$, it is not difficult to see that

$$\sigma_2(r) < \sigma_1(r) \le \sigma(r). \tag{3.49}$$

We obtain from (3.46), (3.47) and (3.49) that

$$\tilde{U}_{2r} \leq C_3^{\sigma(r)} r^{\sigma(r)} \tilde{U}_r, \tag{3.50}$$

with

$$C_3 = C_3\Big(\Big\|\frac{\partial \alpha}{\partial t}\Big\|_{L^\infty(0,\infty;\, L^2(\Omega))}\Big).$$

This achieves the proof of the lemma.

We now turn to the proof of Theorem 3.2. By Lemma 3.5, we had

$$\tilde{U}_{2r} \leq C_3^{\sigma(r)} r^{\sigma(r)} \tilde{U}_r. \tag{3.51}$$

Using Moser's iterations with $r = h$, $r = 2h$, $r = 2^2h$, etc, we get

$$\tilde{U}_{2^{n+1}h} \leq (C_3)^{\kappa_1} 2^{\kappa_2} h^{\kappa_1} \tilde{U}_h, \tag{3.52}$$

with

$$\kappa_1 := \sigma(h) + \sigma(2h) + \sigma(2^2h) + \cdots + \sigma(2^{n-1}h) + \sigma(2^n r),$$
$$\kappa_2 := \sigma(2h) + 2\sigma(2^2h) + 3\sigma(2^3h) + \cdots + (n-1)\sigma(2^{n-1}h) + n\sigma(2^n r).$$

Since $\sigma(2^{n+1}h) = \frac{1}{2^{n+1}}\sigma(h)$, a direct computation gives

$$\kappa_1 := \sum_{k=0}^{n} \sigma(2^k h) \leq \sum_{k=0}^{+\infty} \frac{1}{2^k}\sigma(h) = 2\sigma(h),$$
$$\kappa_2 := \sum_{k=1}^{n} k\sigma(2^k h) \leq \sum_{k=1}^{+\infty} \frac{k}{2^k}\sigma(h) = 4\sigma(2h).$$

This proves that $\kappa_1, \kappa_2 < +\infty$ at infinity and achieves the proof of the theorem. □

Remark 3.6. *The case where k is an even integer is more relevant in the sense that it allows us to consider physically realistic problems. In fact, we can already take the usual cubic nonlinear term* $f(u) = u^3 - u$.

Remark 3.7. *It is also possible to treat in a similar way the case* $n = 2$ *by choosing* β *such that* $\frac{1}{\kappa p} = \beta + \frac{1-\beta}{6}$.

We finally give a uniqueness result.

Theorem 3.8. *Let* (u_1, α_1) *and* (u_2, α_2) *be two solutions to (1.3)-(1.6). We assume that (3.4), (3.5) and the assumptions of Theorem 3.1 and Theorem 3.2 are satisfied. Then problem (1.3)-(1.6) admits a unique solution.*

Proof. We have

$$\frac{\partial u}{\partial t} - \Delta u + f(u_1) - f(u_2) = g(u_1)\frac{\partial \alpha_1}{\partial t} - g(u_2)\frac{\partial \alpha_2}{\partial t}, \tag{3.53}$$

$$\eta\frac{\partial^2\alpha}{\partial t^2}+\frac{\partial\alpha}{\partial t}-\Delta\alpha=-\eta\Big(g(u_1)\frac{\partial u_1}{\partial t}-g(u_2)\frac{\partial u_2}{\partial t}\Big)-G(u_1)+G(u_2), \tag{3.54}$$

with $u=u_1-u_2$ and $\alpha=\alpha_1-\alpha_2$. We also write

$$\frac{\partial u}{\partial t}-\Delta u+f(u_1)-f(u_2)=\Big(g(u_1)-g(u_2)\Big)\frac{\partial\alpha_1}{\partial t}+g(u_2)\frac{\partial\alpha}{\partial t}, \tag{3.55}$$

$$\eta\frac{\partial^2\alpha}{\partial t^2}+\frac{\partial\alpha}{\partial t}-\Delta\alpha=-\eta\Big(g(u_1)-g(u_2)\Big)\frac{\partial u_1}{\partial t}-\eta g(u_2)\frac{\partial u}{\partial t}-G(u_1)+G(u_2). \tag{3.56}$$

Multiplying (3.55) by $\frac{\partial u}{\partial t}$ and integrating over Ω, we obtain

$$\begin{aligned}\left\|\frac{\partial u}{\partial t}\right\|_2^2+\frac{1}{2}\frac{d}{dt}\|\nabla u\|_2^2+\Big(f(u_1)-f(u_2),\frac{\partial u}{\partial t}\Big)=&\int_\Omega\Big(g(u_1)-g(u_2)\Big)\frac{\partial\alpha_1}{\partial t}\frac{\partial u}{\partial t}\,dx\\&+\int_\Omega g(u_2)\frac{\partial\alpha}{\partial t}\frac{\partial u}{\partial t}dx.\end{aligned} \tag{3.57}$$

Similarly, multiplying (3.56) by $\frac{\partial\alpha}{\partial t}$, we have

$$\begin{aligned}&\frac{\eta}{2}\frac{d}{dt}\left\|\frac{\partial\alpha}{\partial t}\right\|_2^2+\left\|\frac{\partial\alpha}{\partial t}\right\|_2^2+\frac{1}{2}\frac{d}{dt}\|\nabla\alpha\|_2^2=-\eta\int_\Omega(g(u_1)-g(u_2))\frac{\partial u_1}{\partial t}\frac{\partial\alpha}{\partial t}dx\\&-\eta\int_\Omega g(u_2)\frac{\partial u}{\partial t}\frac{\partial\alpha}{\partial t}dx-\int_\Omega(G(u_1)-G(u_2))\frac{\partial\alpha}{\partial t}dx,\end{aligned} \tag{3.58}$$

and, adding (3.57) multiplied by η and (3.58), we obtain

$$\begin{aligned}&\frac{dE}{dt}+\eta\left\|\frac{\partial u}{\partial t}\right\|_2^2+\left\|\frac{\partial\alpha}{\partial t}\right\|_2^2+\leq-\eta\Big(f(u_1)-f(u_2),\frac{\partial u}{\partial t}\Big)\\&-\eta\int_\Omega(g(u_1)-g(u_2))\frac{\partial u_1}{\partial t}\frac{\partial\alpha}{\partial t}dx-\int_\Omega(G(u_1)-G(u_2))\frac{\partial\alpha}{\partial t}dx\\&+\eta\int_\Omega(g(u_1)-g(u_2))\frac{\partial\alpha_1}{\partial t}\frac{\partial u}{\partial t}\,dx\end{aligned} \tag{3.59}$$

with

$$E=\frac{\eta}{2}\|\nabla u\|_2^2+\frac{\eta}{2}\left\|\frac{\partial\alpha}{\partial t}\right\|_2^2+\frac{1}{2}\|\nabla\alpha\|_2^2.$$

Considering (3.1)-(3.3), we see that

$$\begin{aligned}\eta\int_\Omega|f(u_1)-f(u_2)|\left|\frac{\partial u}{\partial t}\right|dx&\leq c_{10}\eta\int_\Omega(|u_2|^k+1)\,|u|\left|\frac{\partial u}{\partial t}\right|dx\\&\leq c(\|u_2\|_{H^1(\Omega)}^{2k}+1)\|\nabla u\|_2^2+\frac{\eta}{2}\left\|\frac{\partial u}{\partial t}\right\|_2^2,\end{aligned} \tag{3.60}$$

$$\begin{aligned}\eta\int_\Omega(g(u_1)-g(u_2))\frac{\partial\alpha_1}{\partial t}\frac{\partial u}{\partial t}\,dx&\leq\eta c_9\int_\Omega|u|\left|\frac{\partial\alpha_1}{\partial t}\right|\left|\frac{\partial u}{\partial t}\right|dx\\&\leq c\Big(\left\|\nabla\frac{\partial\alpha_1}{\partial t}\right\|_2^2\|\nabla u\|_2^2\Big)+\frac{\eta}{2}\left\|\frac{\partial u}{\partial t}\right\|_2^2,\end{aligned} \tag{3.61}$$

$$\int_\Omega |G(u_1) - G(u_2)| \left|\frac{\partial \alpha}{\partial t}\right| dx \leq c(\|u_2\|^2_{H^1(\Omega)} + 1)\|\nabla u\|_2^2 + \frac{1}{2}\left\|\frac{\partial \alpha}{\partial t}\right\|_2^2, \tag{3.62}$$

and

$$\begin{aligned} \eta \int_\Omega |g(u_1) - g(u_2)| \left|\frac{\partial u_1}{\partial t}\right| \left|\frac{\partial \alpha}{\partial t}\right| dx &\leq \eta c_9 \|u\|_\infty \int_\Omega \left|\frac{\partial u_1}{\partial t}\right| \left|\frac{\partial \alpha}{\partial t}\right| dx \\ &\leq c\left\|\frac{\partial u_1}{\partial t}\right\|_2^2 \|\nabla u\|_2^2 + \frac{1}{2}\left\|\frac{\partial \alpha}{\partial t}\right\|_2^2. \end{aligned} \tag{3.63}$$

From (3.59)-(3.63), we deduce that

$$\frac{dE}{dt} \leq c\|\nabla u\|_2^2 \left(\|u_2\|^{2k}_{H^1(\Omega)} + 2 + \left\|\nabla \frac{\partial \alpha_1}{\partial t}\right\|_2^2 + \|u_2\|^2_{H^1(\Omega)} + \left\|\frac{\partial u_1}{\partial t}\right\|_2^2\right). \tag{3.64}$$

The proof follows from Gronwall's lemma. □

4. Spatial behavior of the solutions

To study the spatial behavior of the solutions in a semi-infinite cylinder we need to add some assumptions. We first assume that such solutions exist. We then consider the boundary conditions

$$u = \alpha = 0 \quad \text{on} \quad (0, +\infty) \times \partial D \times (0, T), \tag{4.1}$$

$$u(0, x_2, x_3, t) = h_1(x_2, x_3, t), \tag{4.2}$$

$$\alpha(0, x_2, x_3, t) = h_2(x_2, x_3, t) \quad \text{on} \quad \{0\} \times D \times (0, T) \tag{4.3}$$

and the initial conditions

$$u|_{t=0} = \alpha|_{t=0} = \left.\frac{\partial \alpha}{\partial t}\right|_{t=0} = 0 \quad \text{on} \quad R. \tag{4.4}$$

Here D denotes a two dimensional bounded domain and R a semi-infinite cylinder $(0, +\infty) \times D$. We will sometimes use some assumptions on the functions F and G; these will be specified later on. We further assume that $h = 0$.

We consider the function

$$F_\omega(z, t) = \int_0^t \int_{D(z)} e^{-(w\,s)}[\alpha_s \alpha_{,1} + u_{,1}(\gamma u + \eta u_s)] da\, ds \tag{4.5}$$

where $D(z) = \{x \in R, \quad x_1 = z\}$, $u_{,1} = \frac{\partial u}{\partial x_1}$, $u_s = \frac{\partial u}{\partial s}$ and w is a positive constant. By a differentiation of F_ω, we get

$$\begin{aligned} \frac{\partial F_\omega(z, t)}{\partial z} = \int_0^t \int_{D(z)} e^{-(w\,s)} \Big(&\nabla \alpha \nabla \alpha_s + \eta \alpha_s \alpha_{ss} + |\alpha_s|^2 + \eta g(u) u_s \alpha_s \\ &+ G(u)\alpha_s + \eta \nabla u \nabla u_s + \eta |u_s|^2 + \eta f(u) u_s \\ &- \eta g(u) u_s \alpha_s + \gamma |\nabla u|^2 + \gamma u\, u_s + \gamma f(u) u - \gamma g(u)\, u \alpha_s \Big) da\, ds \end{aligned} \tag{4.6}$$

which yields after simplification

$$\frac{\partial F_\omega(z,t)}{\partial z} = \int_0^t \int_{D(z)} e^{-(w\,s)}\Big(|\alpha_s|^2 + \eta|u_s|^2 + \gamma|\nabla u|^2\Big) da\, ds$$
$$+ \int_0^t \int_{D(z)} e^{-(w\,s)}\Big(\nabla\alpha\nabla\alpha_s + \eta\alpha_s\alpha_{ss} + \eta\nabla u\nabla u_s + \eta f(u)u_s + \gamma u\, u_s\Big) da\, ds \tag{4.7}$$
$$+ \int_0^t \int_{D(z)} e^{-(w\,s)}\Big((G(u) - \gamma g(u)u)\alpha_s + \gamma f(u)\, u\Big) da\, ds.$$

We also have

$$\frac{d}{dt}\int_{D(z)} e^{-(\omega\,s)}\Big(|\nabla\alpha|^2 + \eta|\alpha_s|^2 + \eta|\nabla u|^2 + 2\eta F(u) + \gamma|u|^2\Big) da =$$
$$-\,\omega \int_{D(z)} e^{-(\omega\,s)}\Big(|\nabla\alpha|^2 + \eta|\alpha_s|^2 + \eta|\nabla u|^2 + 2\eta F(u) + \gamma|u|^2\Big) da \tag{4.8}$$
$$+ \int_{D(z)} e^{-(w\,s)}\Big(\nabla\alpha\nabla\alpha_s + \eta\alpha_s\alpha_{ss} + \eta\nabla u\nabla u_s + \eta f(u)u_s + \gamma u\, u_s\Big) da.$$

In other words

$$\int_{D(z)} e^{-(w\,s)}\Big(\nabla\alpha\nabla\alpha_s + \eta\alpha_s\alpha_{ss} + \eta\nabla u\nabla u_s + \eta f(u)u_s + \gamma u\, u_s\Big) da =$$
$$\frac{1}{2}\frac{d}{dt}\int_{D(z)} e^{-(\omega\,s)}\Big(|\nabla\alpha|^2 + \eta|\alpha_s|^2 + \eta|\nabla u|^2 + 2\eta F(u) + \gamma|u|^2\Big) da \tag{4.9}$$
$$+ \frac{\omega}{2}\int_{D(z)} e^{-(\omega\,s)}\Big(|\nabla\alpha|^2 + \eta|\alpha_s|^2 + \eta|\nabla u|^2 + 2\eta F(u) + \gamma|u|^2\Big) da.$$

We deduce from (4.7) and (4.9) that

$$\frac{\partial F_\omega(z,t)}{\partial z} = \int_0^t \int_{D(z)} e^{-(\omega\,s)}\Big((|\alpha_s|^2 + \eta|u_s|^2 + \gamma|\nabla u|^2\Big) da\, ds$$
$$+ e^{-(\omega\,t)}\int_{D(z)} (|\nabla\alpha|^2 + \eta|\alpha_s|^2 + \eta|\nabla u|^2 + 2\eta F(u) + \gamma|u|^2) da$$
$$+ \int_0^t \int_{D(z)} e^{-(w\,s)}\Big[(G(u) - \gamma g(u)u)\alpha_s + \gamma f(u)\, u \tag{4.10}$$
$$+ \frac{\omega}{2}(|\nabla\alpha|^2 + \eta|\alpha_s|^2 + \eta|\nabla u|^2 + 2\eta F(u) + \gamma|u|^2)\Big] da\, ds.$$

We assume that, for γ large enough, $2\eta F(s) + \gamma|s|^2 \geq K_1(|s|^2 + |s|^{k+2})$, k integer, $K_1 > 0$. Then, there exists a constant $K_2 > 0$ such that

$$|\nabla\alpha|^2 + \eta|\alpha_s|^2 + \eta|\nabla u|^2 + 2\eta F(u) + \gamma|u|^2 \geq K_2(|\nabla\alpha|^2 + |\alpha_s|^2 + |\nabla u|^2 + |u|^2 + |u|^{k+2}). \tag{4.11}$$

We further assume that $(G(s)-\gamma g(s)s)^2 \leq K_3(|s|^2+|s|^{k+2})$, $K_3 > 0$, and that there exist positive constants κ_1 and κ_2 such that $f(s)s + \kappa_1|s|^2 \geq \kappa_2|s|^2$. Then, for ω large enough (here, ω depends on γ), there holds

$$
\begin{aligned}
&(G(u) - \gamma g(u)u)\alpha_s + \gamma f(u)\,u + \frac{\omega}{2}(|\nabla\alpha|^2 + \eta|\alpha_s|^2 \\
&+ \eta|\nabla u|^2 + 2\eta F(u) + \gamma|u|^2) \geq K_4(|\nabla\alpha|^2 + |\alpha_s|^2 + |\nabla u|^2 + |u|^2),
\end{aligned}
\tag{4.12}
$$

where K_4 is a positive constant. Note that g having at most a linear growth, $|g(s)| \leq c|s|$, $G(0) = 0$, and $F(s) = c's^4 + c''s^2$ (having in mind the usual cubic nonlinear term $f(s) = s^3 - s$), $c' > 0$, satisfy the above assumptions. We finally deduce from (4.10)-(4.12) the existence of $K_5 > 0$ such that

$$
\frac{\partial F_\omega(z,t)}{\partial z} \geq K_5 \int_0^t \int_{D(z)} e^{-(\omega s)}\Big[|\nabla\alpha|^2 + |\alpha_s|^2 + |\nabla u|^2 + |u|^2 + |u_s|^2\Big] da\, ds. \tag{4.13}
$$

We now give a spatial derivative estimate on $|F_\omega|$. Using Cauchy-Schwarz's inequality in (4.5), we obtain

$$
\begin{aligned}
|F_\omega| \leq &\Big(\int_0^t \int_{D(z)} e^{-(w s)}\alpha_s^2\, da\, ds\Big)^{\frac{1}{2}}\Big(\int_0^t \int_{D(z)} e^{-(w s)}\alpha_{,1}^2\, da\, ds\Big)^{\frac{1}{2}} \\
&+\Big(\int_0^t \int_{D(z)} e^{-(w s)}u_{,1}^2\, da\, ds\Big)^{\frac{1}{2}}\Big(\int_0^t \int_{D(z)} \gamma^2 e^{-(w s)}u^2\, da\, ds\Big)^{\frac{1}{2}} \\
&+\Big(\int_0^t \int_{D(z)} e^{-(w s)}u_{,1}^2\, da\, ds\Big)^{\frac{1}{2}}\Big(\int_0^t \int_{D(z)} \eta^2 e^{-(w s)}u_s^2\, da\, ds\Big)^{\frac{1}{2}}.
\end{aligned}
\tag{4.14}
$$

Hence

$$
|F_\omega| \leq K_6 \int_0^t \int_{D(z)} e^{-(\omega s)}\Big[|\nabla\alpha|^2 + |\alpha_s|^2 + |\nabla u|^2 + |u|^2 + |u_s|^2\Big] da\, ds. \tag{4.15}
$$

Choosing $K^\star = \frac{K_6}{K_5}$, there holds

$$
|F_\omega| \leq K^\star \frac{\partial F_\omega}{\partial z}. \tag{4.16}
$$

Due to (4.16), we arrive at a Phragmén-Lindelöf alternative (see [10], [33]) namely, either there exists $z_0 \geq 0$ such that $F(z_0,t) > 0$ or $F(z_0,t) \leq 0$ for all $z \geq 0$. In the first case our solution satisfies

$$
F_\omega(z,t) \geq e^{\left(K^{\star^{-1}}(z-z_0)\right)} F_\omega(z_0,t), \quad z \geq z_0, \tag{4.17}
$$

and, in the latter one $F(z_0,t) \leq 0$ for all $z \geq 0$, in which case our solution satisfies

$$
-F_\omega(z,t) \leq -e^{\left(-K^{\star^{-1}}z\right)} F_\omega(0,t), \quad z \geq 0. \tag{4.18}
$$

Inequality (4.17) shows that $F_\omega(z,t)$ tends exponentially fast to infinity.

On the contrary inequality (4.18) shows that $F_\omega(z,t)$ tends to 0 and

$$
G_\omega(z,t) \leq e^{\left(-K^{\star^{-1}}z\right)} G_\omega(0,t), \quad z \geq 0, \tag{4.19}
$$

where

$$
\begin{aligned}
G_\omega(z,t) = & \int_0^t \int_{R(z)} e^{-(\omega s)}\Big((|\alpha_s|^2 + \eta|u_s|^2 + \gamma|\nabla u|^2\Big)da\,ds \\
& + e^{-(\omega s)} \int_{R(z)} (|\nabla\alpha|^2 + \eta|\alpha_s|^2 + \eta|\nabla u|^2 + 2\eta F(u) + \gamma|u|^2)da \\
& + \int_0^t \int_{R(z)} e^{-(w s)}\Big[(G(u) - \gamma g(u)u)\alpha_s + \gamma f(u)\,u \\
& + \frac{\omega}{2}(|\nabla\alpha|^2 + \eta|\alpha_s|^2 + \eta|\nabla u|^2 + 2\eta F(u) + \gamma|u|^2)\Big]da\,ds
\end{aligned}
\tag{4.20}
$$

where $R(z) = \{x \in R, \quad z < x_1\}$. Setting

$$
\begin{aligned}
\mathcal{E}_\omega(z,t) = & \int_0^t \int_{R(z)} \Big((|\alpha_s|^2 + \eta|u_s|^2 + \gamma|\nabla u|^2\Big)da\,ds \\
& + \int_{R(z)} (|\nabla\alpha|^2 + \eta|\alpha_s|^2 + \eta|\nabla u|^2 + 2\eta F(u) + \gamma|u|^2)da \\
& + \int_0^t \int_{R(z)} \Big[(G(u) - \gamma g(u)u)\alpha_s + \gamma f(u)\,u \\
& + \frac{\omega}{2}(|\nabla\alpha|^2 + \eta|\alpha_s|^2 + \eta|\nabla u|^2 + 2\eta F(u) + \gamma|u|^2)\Big]da\,ds
\end{aligned}
\tag{4.21}
$$

we get

$$
\mathcal{E}_\omega(z,t) \le e^{\left(\omega t - K^{\star -1} z\right)} G_\omega(0,t), \quad z \ge 0. \tag{4.22}
$$

We give in what follows the main result of this section

Theorem 4.1. *Let (u,α) be a solution to problem (1.3)-(1.6) with the boundary conditions (4.1)-(4.4). Then, either this solution satisfies (4.17) or it satisfies (4.22).*

Remark 4.2. *Estimates (4.17) and (4.22) are known respectively as growth and decay estimates.*

Remark 4.3. *It is possible due to (4.17) to specify the rate of growth of our solutions to infinity. In fact, if (4.17) is satisfied, then*

$$
\begin{aligned}
& \int_0^t \int_{R(0,z)} e^{-(\omega s)}\Big((|\alpha_s|^2 + \eta|u_s|^2 + \gamma|\nabla u|^2\Big)da\,ds \\
& + e^{-(\omega s)} \int_{R(0,z)} (|\nabla\alpha|^2 + \eta|\alpha_s|^2 + \eta|\nabla u|^2 + 2\eta F(u) + \gamma|u|^2)da \\
& + \int_0^t \int_{R(0,z)} e^{-(w s)}\Big[(G(u) - \gamma g(u)u)\alpha_s + \gamma f(u)\,u \\
& + \frac{\omega}{2}(|\nabla\alpha|^2 + \eta|\alpha_s|^2 + \eta|\nabla u|^2 + 2\eta F(u) + \gamma|u|^2)\Big]da\,ds
\end{aligned}
\tag{4.23}
$$

where $R(0,z) = \{x \in R, \quad 0 < x_1 < z\}$, tends exponentially fast to infinity.

Acknowledgments

The authors wish to thank the referee for his careful reading of the paper.

References

1. G. Caginalp, *An analysis of a phase field model of a free boundary*, Arch. Ration. Mech. Anal., **92** (1986): 205-245.
2. S. Aizicovici, E. Feireisl, *Long-time stabilization of solutions to a phase-field model with memory.* J. Evol. Equ., **1** (2001), 69-84.
3. S. Aizicovici, E. Feireisl, *Long-time convergence of solutions to a phase-field system*, Math. Methods Appl. Sci., **24** (2001), 277-287.
4. D. Brochet, X. Chen, and D. Hilhorst, *Finite dimensional exponential attractors for the phase-field model*, Appl. Anal., **49** (1993), 197-212.
5. M. Brokate, J. Sprekels, Hysteresis and Phase Transitions, Springer, New York, 1996.
6. L. Cherfils, A. Miranville, *Some results on the asymptotic behavior of the Caginalp system with singular potentials*, Adv. Math. Sci. Appl.,(2007), 17107-129.
7. L. Cherfils, A. Miranville, *On the Caginalp system with dynamic boundary conditions and singular potentials*, Appl. Math., **54** (2009), 89-115.
8. R. Chill, E. Fasangová, and J. Prüss, *Convergence to steady states of solutions of the Cahn-Hilliard equation with dynamic boundary conditions*, Math. Nachr., **279** (2006), 1448-1462.
9. C.I. Christov, P.M. Jordan, *Heat conduction paradox involving second-sound propagation in moving media*, Phys. Rev. Lett., **94** (2005), 154-301.
10. J.N. Flavin, R.J. Knops, and L.E. Payne, *Decay estimates for the constrained elastic cylinder of variable cross-section*, Quart. Appl. Math., **47** (1989), 325-350.
11. S. Gatti, A. Miranville, *Asymptotic behavior of a phase-field system with dynamic boundary conditions*, in: Differential Equations: Inverse and Direct Problems (Proceedings of the workshop "Evolution Equations: Inverse and Direct Problems ", Cortona, June 21-25, 2004), in A. Favini, A. Lorenzi (Eds), A Series of Lecture Notes in Pure and Applied Mathematics, **251** (2006), 149-170.
12. C. Giorgi, M. Grasselli, and V. Pata, *Uniform attractors for a phase-field model with memory and quadratic nonlinearity*, Indiana Univ. Math. J., **48** (1999), 1395-1446.
13. M. Grasseli, A. Miranville, V. Pata, and S. Zelik, *Well-posedness and long time behavior of a parabolic-hyperbolic phase-field system with singular potentials*, Math. Nachr.. **280** (2007), 1475-1509.
14. M. Grasselli, *On the large time behavior of a phase-field system with memory*, Asymptot. Anal., **56** (2008), 229-249.
15. M. Grasselli, V. Pata, *Robust exponential attractors for a phase-field system with memory* J. Evol. Equ., **5** (2005), 465-483.
16. M. Grasselli, H. Petzeltová, and G. Schimperna, *Long time behavior of solutions to the Caginalp system with singular potentials*, Z. Anal. Anwend., **25** (2006), 51-73.

17. M. Grasselli, H. Wu, and S. Zheng, *Asymptotic behavior of a non-isothermal Ginzburg-Landau model*, Quart. Appl. Math., **66** (2008), 743-770.

18. A.E. Green, P.M. Naghdi, *A new thermoviscous theory for fluids*, J. Non-Newtonian Fluid Mech., **56** (1995), 289-306.

19. A.E. Green, P.M. Naghdi, *A re-examination of the basic postulates of thermomechanics*, Proc. Roy. Soc. Lond. A., **432** (1991), 171-194.

20. A.E. Green, P.M. Naghdi, *On undamped heat waves in an elastic solid*, J. Thermal. Stresses, **15** (1992), 253-264.

21. J. Jiang, *Convergence to equilibrium for a parabolic-hyperbolic phase-field model with Cattaneo heat flux law*, J. Math. Anal. Appl., **341** (2008), 149-169.

22. J. Jiang, *Convergence to equilibrium for a fully hyperbolic phase field model with Cattaneo heat flux law*, Math. Methods Appl. Sci., **32** (2009), 1156-1182.

23. Ph. Laurençot, *Long-time behaviour for a model of phase-field type*, Proc. Roy. Soc. Edinburgh Sect. A, **126** (1996), 167-185.

24. A. Miranville, R. Quintanilla, *Some generalizations of the Caginalp phase-field system*, Appl. Anal., **88** (2009), 877-894.

25. A. Miranville, R. Quintanilla, *A generalization of the Caginalp phase-field system based on the Cattaneo law*, Nonlinear Anal. TMA., **71** (2009), 2278-2290.

26. A. Miranville, R. Quintanilla, *A Caginalp phase-field system with a nonlinear coupling*. Nonlinear Anal.: Real World Applications, **11** (2010), 2849-2861.

27. A. Miranville, S. Zelik, *Robust exponential attractors for singularly perturbed phase-field type equations*, Electron. J. Diff. Equ., (2002), 1-28.

28. A. Miranville, S. Zelik, *Attractors for dissipative partial differential equations in bounded and unbounded domains*, in: C.M. Dafermos, M. Pokorny (Eds.) Handbook of Differential Equations, Evolutionary Partial Differential Equations. Elsevier, Amsterdam, 2008.

29. A. Novick-Cohen, *A phase field system with memory: Global existence*, J. Int. Equ. Appl. **14** (2002), 73-107.

30. R. Quintanilla, *On existence in thermoelasticity without energy dissipation*, J. Thermal. Stresses, **25** (2002), 195-202.

31. R. Quintanilla, *End effects in thermoelasticity*, Math. Methods Appl. Sci.. **24** (2001), 93-102.

32. R. Quintanilla, R. Racke, *Stability in thermoelasticity of type III*, Discrete Contin. Dyn. Syst. B, **3** (2003), 383-400.

33. R. Quintanilla, *Phragmén-Lindelöf alternative for linear equations of the anti-plane shear dynamic problem in viscoelasticity*, Dynam. Contin. Discrete Impuls. Systems, **2** (1996), 423-435.

34. R. Temam, *Infinite-dimensional Dynamical Systems in Mechanics and Physics, second edition*, Applied Mathematical Sciences, vol. **68**, Springer-Verlag, New York, 1997.

35. Z. Zhang, *Asymptotic behavior of solutions to the phase-field equations with Neumann boundary conditions*, Comm. Pure Appl. Anal., **4** (2005), 683-693.

2

Applications of the lichnerowicz Laplacian to stress energy tensors

Paul Bracken*

Department of Mathematics, University of Texas, TX 78539-2999 Edinburg, USA

* **Correspondence:** paul.bracken@utrgv.edu

Abstract: A generalization of the Laplacian for p-forms to arbitrary tensors due to Lichnerowicz will be applied to a 2-tensor which has physical applications. It is natural to associate a divergence-free symmetric 2-tensor to a critical point of a specific variational problem and it is this 2-tensor that is studied. Numerous results are obtained for the stress-energy tensor, such as its divergence and Laplacian. A remarkable integral formula involving a symmetric 2-tensor and a conformal vector field is obtained as well.

Keywords: Basis; tensor; connection; differential system; Laplacian; bundle; harmonic map
Mathematics Subject Classification: 53C20, 58E30

---

1. Introduction

Variational problems arise in various areas of mathematics and physics. Suppose (M, g) is a Riemannian manifold with volume form dv_M, it is the case that functionals of the form

$$I(\varphi, g) = \int_M \sigma(\varphi, g)\, dv_M \tag{1.1}$$

very often occur [1, 2]. Here φ could be a mapping between Riemannian manifolds of a vector bundle valued differential form. Given a variational problem starting from (1.1), the stress-energy tensor S can be derived by considering variations of the metric on M. If I has a critical point with respect to variations of φ, then the stress-energy tensor is divergence free, and there are conservation laws.

To provide some motivation, let (M, g) and (N, h) be two smooth Riemannian manifolds which are connected, compact, orientable and without boundary, and $\varphi : (M, g) \to (N, h)$ a smooth map. The differential of φ which is $d\varphi$ can be thought of as a section of the bundle $T^*M \otimes \varphi^{-1}TN$, with norm $|d\varphi|$. If $\{x^i\}$ and $\{u^a\}$ constitute local coordinate systems around x and $\varphi(x)$, respectively, then in terms of coordinates, we can write

$$|d\varphi|^2 = g^{ij}\, h_{ab}(\varphi) \frac{\partial \varphi^a}{\partial x^i} \frac{\partial \varphi^b}{\partial x^j}, \tag{1.2}$$

where $(\partial\varphi^a/\partial x^i)$ is the local representation of $d\varphi$. Then the energy density of φ can be defined as $e(\varphi) = 1/2|d\varphi|^2$ and the energy density of the field is given by the positive functional $E(\varphi) = \int_M e(\varphi)\, dv_M$.

A large class of maps which come up in physics, especially in gravity, are called harmonic. A mapping $\varphi : M \to N$ is harmonic if and only if it is an extremal of the energy. Consequently, it is the case that a map φ is harmonic if and only if it satisifes the Euler-Lagrange equation

$$\tau(\varphi) = -d^* d\,\varphi = \mathrm{tr}\nabla\, d\varphi.$$

This defines the tension field of φ, and may be expressed in local coordinates on M and N as follows

$$\begin{aligned}\nabla_{\partial_i}(d\varphi) &= (\frac{\partial\varphi^a_j}{\partial x^i})\, dx^j \frac{\partial}{\partial u^a} + \varphi^a_j(\nabla_{\partial_i} dx^j)\frac{\partial}{\partial u^a} + \varphi^a_j\, dx^j\, \nabla_{\partial_i}\frac{\partial}{\partial u^a} \\ &= \varphi^a_{ij} - {}^M\Gamma^k_{ij}\varphi^a_k + {}^N\Gamma^b_{b\gamma}\varphi^\gamma_j. \end{aligned} \tag{1.3}$$

The tension field is the trace of (1.3),

$$\tau(\varphi)^a = g^{ij}(\nabla\, d\varphi)^a_{ij} = -\Delta\varphi^a + {}^N\Gamma^a_{b\gamma}\varphi^b_i\varphi^\gamma_j g^{ij}. \tag{1.4}$$

Thus (1.4) is a semilinear, elliptic, second-order system. If N is the space $\mathbb{R}$, a harmonic map is called a harmonic function [4, 7, 8].

2. Energy Functional and Critical Point

Now let us extend this idea to another object which may be defined on a manifold. Let M be a Riemannian manifold and E a Riemannian vector bundle over M, where each fiber carries a positive definite inner product denoted by $\langle\cdot,\cdot\rangle^E$. Let $\Omega^p(E)$ be the space of smooth p-forms which have values in E, where it is assumed throughout that $p \geq 1$. For $\omega \in \Omega^p(E)$, define the energy functional

$$I(\omega, g) = \int_M \langle\omega(e_{i_1},\ldots,e_{i_p}), \omega(e_{i_1},\ldots,e_{i_p})\rangle^E\, dv_m. \tag{2.1}$$

where $\{e_i\}$ is an orthonormal basis on M and repeated indices are summed for $1 \leq i_1,\ldots,i_p \leq m$ and $m = \dim M$. With respect to a local coordinate system $\{x^i\}$ on M and local frame $\{s_a\}$ of E, the norm of ω, which is the integrand of (2.1) can be written

$$|\omega|^2 = \langle\omega(e_{i_1},\ldots,e_{i_p}), \omega(e_{i_1},\ldots,e_{i_p})\rangle = g^{i_1 j_1}\cdots g^{i_p j_p}\omega^a_{i_1\cdots i_p}\omega^b_{j_1\cdots j_p}h_{ab}. \tag{2.2}$$

Suppose M is compact, then vary the integral (2.1) with respect to metric g. If $g(u)$ is a smooth, one-parameter family of metrics such that $g(0) = g$, then the variation $\delta g = \partial g/\partial u|_{u=0}$ is a smooth symmetric tensor on M.

Theorem 1. For $\omega \in \Omega^p(E)$ and $p \geq 1$,

$$\frac{dI}{du}|_{u=0} = \int_M \langle S(\omega), \frac{\partial g}{\partial u}|_{u=0}\rangle\, dv_M, \tag{2.3}$$

where $S(\omega)$ is the symmetric two-tensor defined by

$$S(\omega) = \frac{1}{2}|\omega|^2 g - p\sum_{i_2,\cdots,i_p}\langle\omega(\cdot,e_{i_2},\ldots,e_{i_p}), \omega(\cdot,e_{i_2},\ldots,e_{i_p})\rangle^E. \tag{2.4}$$

where $\{e_i\}$ is an orthonormal basis on M.

Proof: Let $\{x^i\}$ be a local coordinate system on M, and $\{s_a\}$ a local frame for E,

$$\frac{dI(\omega)}{du}|_{u=0} = \int_M \frac{\partial|\omega|^2}{\partial g_{ij}}\delta g_{ij}\, dv_M + \int_M |\omega|^2 \frac{\partial(dv_M)}{\partial g_{ij}}\delta g_{ij}. \tag{2.5}$$

The volume form on M is given by

$$dv_M = (\det g)^{1/2}\, dx^1 \wedge \ldots \wedge dx^m,$$

where $\det g$ is the determinant of the metric tensor g_{ij} and therefore,

$$\frac{\partial}{\partial g_{ij}} dv_M = \frac{1}{2}(\det g)^{-1/2}\frac{\partial}{\partial g_{ij}}(\det g)\, dx^1 \wedge \ldots \wedge dx^m = \frac{1}{2}g^{ij}\, dv_M.$$

Differentiating the expression for the metric tensor $g^{ij}g_{jk} = \delta^i_k$, the following relation holds

$$\frac{\partial g^{i_s j_s}}{\partial g_{ij}} = -g^{i_s i}g^{j_s j}.$$

The first term in (2.5) is

$$\begin{aligned}\frac{\partial|\omega|^2}{\partial g_{ij}} &= \frac{\partial}{\partial g_{ij}}(g^{i_1 j_1}\cdots g^{i_p j_p}\omega^a_{i_1\cdots i_p}\omega^b_{j_1\cdots j_p}h_{ab})\\ &= -\sum_{s=1}^{p} g^{i_1 j_1}\cdots g^{i_s i}g^{j_s j}\cdots g^{i_p j_p}\omega^a_{i_1\cdots i_s\cdots i_p}\omega^b_{j_1\cdots j_s\cdots j_p}\cdot h_{ab}.\end{aligned}$$

Therefore,

$$\begin{aligned}\frac{\partial|\omega|^2}{\partial g_{ij}}g_{ik}g_{jl} &= -\sum_{s=1}^{p} g^{i_1 j_1}\cdots \delta^{i_s}_k\delta^{j_s}_l\cdots g^{i_p j_p}\omega^a_{i_1\cdots i_s\cdots i_p}\omega^b_{j_1\cdots j_s\cdots j_p}h_{ab}\\ &= -pg^{i_2 j_2}\cdots g^{i_p j_p}\omega^a_{k i_2\cdots i_p}\omega^b_{l j_2\cdots j_p}\cdot h_{ab}.\end{aligned}$$

□

Definition 1. Let M be an arbitrary not necessarily compact Riemannian manifold, and let E be a Riemannian vector bundle over M. Let $\omega \in \Omega^p(E)$ and define the stress-energy tensor of the form ω to be the following symmetric 2-tensor

$$S(\omega) = \frac{1}{2}|\omega|^2 g - p\sum_{j_1,\cdots,j_p}\langle\omega(\cdot, e_{j_2}, \cdots, e_{j_p}), \omega(\cdot, e_{j_2}, \cdots, e_{j_p})\rangle^E \tag{2.6}$$

at each point x where there is an orthonormal basis $\{e_i\}$.

Let the vector bundle E be endowed with a Riemannian connection denoted by ∇^E so

$$X\langle s, t\rangle^E = \langle\nabla^E_X s, t\rangle^E + \langle s, \nabla^E_X t\rangle^E, \tag{2.7}$$

Theorem 2. Let $\omega \in \Omega^p(E)$ and $S(\omega)$ be the stress-energy tensor associated with ω, then for all $x \in M$ and each $X \in T_xM$,

$$\operatorname{div} S(\omega)(X) = \langle\omega, d\omega\rfloor X\rangle + p\langle d^*\omega, \omega\rfloor X\rangle \tag{2.8}$$

where the contraction of a p-form with a vector field X is given by

$$(\omega \rfloor X)(X_1, \ldots, X_{p-1}) = \omega(X, X_1, \ldots, X_{p-1}).$$

Proof: Let $\{e_i\}$ be an orthonormal basis at x and extend objects to a neighborhood of x. Suppose that $\nabla_{e_i} e_j = 0$ at x for all i, j. The tensorial property allows one to evaluate at $X = e_k$ without loss of generality. Consequently,

$$\operatorname{div} S(X) = \sum_j (\nabla_{e_j} S)(e_j, e_k) = \sum_j e_j S(e_j, e_k) = \sum_j e_j \{\frac{1}{2}|\omega| g(e_j, e_k) - p\langle \omega \rfloor e_j, \omega \rfloor e_k \rangle\}$$

$$= \langle \nabla_{e_k} \omega, \omega \rangle - p \sum_j \langle \nabla_{e_j} (\omega \rfloor e_j), \omega \rfloor e_k \rangle - p \sum_j \langle \omega \rfloor e_j, \nabla_{e_j} (\omega \rfloor e_k) \rangle. \tag{2.9}$$

At x, it is the case that

$$d^* \omega(e_{j_1}, \ldots, e_{j_{p-1}}) = -\nabla^E_{e_j} (\omega(e_j, e_{j_1}, \ldots, e_{j_{p-1}})),$$

and

$$d\omega(e_k, e_{i_1}, \ldots, e_{i_p}) = \nabla^E_{e_k} (\omega(e_{i_1}, \ldots, e_{i_p})) + \sum_{k=1}^{p} (-1)^k \nabla^E_{e_{i_k}} (\omega(e_k, e_{i_1}, \ldots, \hat{e}_{i_k}, \ldots, e_{i_p})).$$

Solving for $\nabla^E_{e_k}$ and substituting it into the divergence, we get

$$\operatorname{div} S(X) = \langle d\omega \rfloor e_k, \omega \rangle$$

$$- \sum_{i_1, \cdots, i_p} \sum_{k=1}^{p} (-1)^k \langle \nabla^E_{e_{i_k}} (\omega(e_k, e_{i_1}, \ldots, \hat{e}_{i_k}, \ldots, e_{i_p})), \omega(e_{i_1}, \ldots, e_{i_p}) \rangle^E$$

$$+ p\langle d^* \omega, \omega \rfloor X \rangle - p \sum_j \langle \omega \rfloor e_j, \nabla_{e_j} (\omega \rfloor e_k) \rangle.$$

The double sum in this can be simplified to the form

$$\sum_{i_1, \cdots, i_p} \sum_{k=1}^{p} (-1)^{k+1} \langle \nabla^E_{e_{i_k}} (\omega(e_k, e_{i_1}, \ldots, \hat{e}_{i_k}, \ldots, e_{i_p})), \omega(e_1, \ldots, e_{i_p}) \rangle$$

$$= p \sum_{j=1}^{n} \langle \nabla_{e_j} (\omega \rfloor e_k), \omega \rfloor e_j \rangle.$$

The claim follows as a result of these

$$\operatorname{div} S(\omega) = \langle \omega, d\omega \rfloor e_k \rangle + p\langle d^* \omega, \omega \rfloor e_k \rangle + p \sum_j \langle \nabla_{e_j} (\omega \rfloor e_k), \omega \rfloor e_j \rangle_j \langle \omega \rfloor e_j, \nabla_{e_j} (\omega \rfloor e_k) \rangle$$

$$= \langle \omega, d\omega \rfloor e_k \rangle + p\langle d^* \omega, , \omega \rfloor e_k \rangle.$$

□

The form $\omega \in \Omega^p(E)$ is called harmonic if and only if $d\omega = d^* \omega = 0$. Substituting these derivatives into the right-hand side of (2.8), we prove the following result.

Corollary 1. If $\omega \in \Omega^p(E)$ is a harmonic form, then $\operatorname{div} S(\omega) = 0$. □

Clearly, S is symmetric and so given a symmetric, divergence-free 2-tensor S, conservation laws can be formulated by contracting with a Killing vector field. If X is a Killing vector field, then $S\rfloor X$ is also divergence free. However, the converse of Corollary 1 is not true. If $\operatorname{div} S(\omega) = 0$, it may not be concluded that ω is harmonic. However, when ω is a differential of a submersive mapping, there is equivalence.

Corollary 2. Let $\varphi : M \to N$ be a smooth mapping between Riemannian manifolds and let $\omega = d\varphi$ be the corresponding $\varphi^{-1}TN$- valued one-form on M. Then for each vector field X,

$$\operatorname{div} S(d\varphi)(X) = \langle \tau(\varphi), d\varphi(X) \rangle. \tag{2.10}$$

In (2.10), $\tau(\varphi)$ is called the tension field of φ and is defined as

$$\tau(\varphi) = -d^* d\varphi. \tag{2.11}$$

If φ is a submersive almost everywhere, then $\tau(\varphi) = 0$ if and only if $\operatorname{div} S(d\varphi) = 0$.

Proof: Substitute $\omega = d\varphi$ into (2.8) and use the identity $d(d\varphi) = 0$ to obtain,

$$\operatorname{div} S(d\varphi)(X) = \langle d\varphi, dd\varphi\rfloor X \rangle + \langle d^* d\varphi, d\varphi\rfloor X \rangle = -\langle \tau(\varphi), d\varphi\rfloor X \rangle. \tag{2.12}$$

Clearly, if $\tau(\varphi) = 0$, then the right side of (2.12) vanishes, $\operatorname{div} S(d\varphi)(X) = 0$, and conversely. □

The form ω is a critical form with respect to variations of the metric, so the conditions under which the stress-energy tensor vanishes should be studied.

Definition 2. The form $\omega \in \Omega^p(E)$ is conformal if the map $X \to \omega\rfloor X$ is conformal for each $x \in X$, or equivalently, ω is conformal if and only if there is a real λ such that

$$\langle \omega\rfloor X, \omega\rfloor Y \rangle = \lambda^2 \langle X, Y \rangle, \qquad X, Y \in T_x M. \tag{2.13}$$

To obtain an expression for λ^2, evaluate $S(\omega)$ in (2.13) on the vectors $X = Y = e_i \in T_x M$ and carry out the trace on both sides

$$m\lambda^2 = |\omega|^2. \tag{2.14}$$

If $\varphi : M \to N$ is a smooth map between Riemannian manifolds, then $d\varphi \in \Omega^1(\varphi^{-1}TN)$ is a conformal form if and only if the map is conformal as well.

Lemma 1. For $\omega \in \Omega^p(E)$ which is not identically zero, the stress-energy tensor (2.6) vanishes identically if and only if $m = 2p$ and ω is conformal.

Proof: Evaluating $S(\omega)$ on the pair (e_i, e_i) we obtain,

$$S(\omega)(e_i, e_i) = \frac{1}{2}|\omega|^2 - p\langle \omega\rfloor e_i, \omega\rfloor e_i \rangle.$$

Tracing on both sides gives

$$\operatorname{tr} S(\omega) = (\frac{m}{2} - p)|\omega|^2.$$

This vanishes for $\omega \neq 0$ when $m = 2p$. The definition of conformal form (2.13) gives equivalence. □

3. Weizenböck Formulas and Applications to the Stress-Energy Tensor

Now consider the Laplacian of the tensor $S(\omega)$. The Laplacian on p-forms was extended to arbitrary tensors on a Riemannian manifold by Lichnerowicz [5,6]. For a 2-tensor Q, the Laplacian on Q is given by

$$\Delta Q(X,Y) = -\mathrm{tr}\nabla^2 Q(X,Y) + S(\mathrm{Ricci}^M(X),Y) + S(X,\mathrm{Ricci}^M(Y)) - 2\sum_{i,j}\langle R^M(X,e_i)Y,e_j\rangle S(e_i,e_j). \tag{3.1}$$

where $\{e_i\}$ is an orthonormal basis. The curvature tensor on M is represented by R^M with sign convention

$$R^M(X,Y)Z = -\nabla_X\nabla_Y Z + \nabla_Y\nabla_X Z + \nabla_{[X,Y]}Z. \tag{3.2}$$

In (3.2), Ricci^M is the Ricci tensor defined by $\mathrm{Ricci}^M(X,Y) = \mathrm{tr}(Z \to R^M(X,Z)Y)$.

The following theorem given first by Lichnorowicz [5, 6] is very useful and is presented without proof.

Proposition 1. (Lichnerowicz) Let M be a Riemannian manifold with $\nabla\mathrm{Ricci}^M = 0$. Let Q be a 2-tensor on M, then the divergence commutes with the Laplacian

$$\mathrm{div}(\Delta Q) = \Delta(\mathrm{div}Q), \tag{3.3}$$

where the right-hand side is now the standard Laplacian on one-forms.

Define the symmetric 2-tensor $L(Q)$ to be

$$L(Q) = Q(\mathrm{Ricci}^M(X),Y) + Q(X,\mathrm{Ricci}^M(Y)) - 2\sum_{i,j}\langle R^M(X,e_i)Y,e_j\rangle\, Q(e_i,e_j). \tag{3.4}$$

Suppose $E \to M$ is a vector bundle endowed with a metric and Riemannian connection and associated curvature R^E. The Ricci operator on a p-form ω written as $\mathrm{Ricci}^p = \mathrm{Ricci}$ is defined to be

$$(\mathrm{Ricci}(\omega))(X_1,\ldots,X_p) = \sum_{i,k}(R^p(e_i,X_k)\omega)(e_i,X_1,\ldots,\hat{X}_k,\ldots,X_p). \tag{3.5}$$

In (3.5), R^p is the canonical curvature and is defined to be [4]

$$(R^p(X,Y)\omega)(X_1,\ldots,X_p) = R^E(X,Y)(\omega(X_1,\ldots,X_p)) - \sum_k \omega(X_1,\ldots,R^M(X,Y)X_k,\ldots,X_p). \tag{3.6}$$

The following theorem is due to Weizenböck [3].

Proposition 2. (Weitzenböck) Let $E \to M$ be a Riemannian vector bundle over a Riemannian manifold M. Then for any $\omega \in \Omega^p(E)$,

$$\begin{aligned} &(i) && \Delta\omega = -\mathrm{tr}\nabla^2\omega + \mathrm{Ricci}(\omega). \\ &(ii) && \tfrac{1}{2}\Delta|\omega|^2 = \langle\Delta\omega,\omega\rangle - |\nabla\omega|^2 - \langle\mathrm{Ricci}(\omega),\omega\rangle. \end{aligned} \tag{3.7}$$

Proof: Let us prove (*ii*) by using (*i*) for $-\mathrm{tr}\nabla^2\omega$,

$$\frac{1}{2}\Delta|\omega|^2 = -\frac{1}{2}\mathrm{tr}\nabla\, d|\omega|^2 = -\mathrm{tr}\nabla\langle \nabla\omega, \omega\rangle = -\mathrm{tr}\langle\nabla\nabla\omega, \omega\rangle - \langle\nabla\omega, \nabla\omega\rangle$$

$$= \langle\Delta\omega, \omega\rangle - \langle\mathrm{Ricci}(\omega), \omega\rangle - |\nabla\omega|^2.$$

□

It is worth noting a few important applications of this proposition. If $\omega \in \Omega^0(E)$, then (3.7) reduces to

$$\Delta\omega = -\mathrm{tr}\nabla d\omega, \qquad \frac{1}{2}\Delta|\omega|^2 = \langle\Delta\omega, \omega\rangle - |\nabla\omega|^2. \tag{3.8}$$

If $\omega \in \Omega^1(E)$ and $X \in \Gamma(TM)$, then

$$\Delta\omega(X) = -\mathrm{tr}\nabla^2\omega(X) - \sum_s R^E(e_s, X)\omega(e_s) + \sum_s \omega(R^M(e_s, X)e_s), \tag{3.9}$$

where $\{e_s\}$ is an orthonormal basis for TM and

$$\frac{1}{2}\Delta|\omega|^2 = \langle\Delta\omega, \omega\rangle - |\nabla\omega|^2 + \sum_{i,j}\langle R^E(e_i, e_j)\omega(e_i), \omega(e_j)\rangle - \sum_k \langle\omega(\mathrm{Ricci}(e_k)), \omega(e_k)\rangle. \tag{3.10}$$

Theorem 3. Let $X, Y \in \Gamma(TM)$, it holds that

$$(\mathrm{tr}\nabla^2\langle\omega\lrcorner\cdot, \omega\lrcorner\cdot\rangle)(X, Y) = \langle(\mathrm{tr}\nabla^2\omega)\lrcorner X, \omega\lrcorner Y\rangle + \langle\omega\lrcorner X, (\mathrm{tr}\nabla^2\omega)\lrcorner Y\rangle$$

$$+ 2\sum_i\langle(\nabla_{e_i}\omega)\lrcorner X, (\nabla_{e_i}\omega)\lrcorner Y\rangle. \tag{3.11}$$

Proof: Differentiating once gives,

$$\nabla_{e_j}\langle\omega\lrcorner\cdot, \omega\lrcorner\cdot\rangle(X, Y) = \langle\nabla_{e_j}\omega\lrcorner X, \omega\lrcorner Y\rangle + \langle\omega\lrcorner X, \nabla_{e_j}\omega\lrcorner Y\rangle.$$

Differentiating a second time gives

$$\nabla_{e_i}\nabla_{e_j}\langle\omega\lrcorner\cdot, \omega\lrcorner\cdot\rangle(X, Y) = \langle\nabla_{e_i}\nabla_{e_j}\omega\lrcorner X, \omega\lrcorner Y\rangle + \langle\nabla_{e_j}\omega\lrcorner X, \nabla_{e_j}\omega\lrcorner Y\rangle + \langle\nabla_{e_i}\omega\lrcorner X, \nabla_{e_j}\omega\lrcorner Y\rangle$$

$$+\langle\omega\lrcorner X, \nabla_{e_i}\nabla_{e_j}\omega\lrcorner Y\rangle.$$

Tracing on both sides of this equation yields (3.11). □

Theorem 4. Let $\omega \in \Omega^p(E)$ be a vector bundle valued p-form with stress-energy tensor $S(\omega)$. The Laplacian of $S(\omega)$ can be written in the following way,

$$\Delta S(\omega) = 2S(\langle\Delta\omega\lrcorner\cdot, \omega\lrcorner\cdot\rangle) - 2S(\sum_j\langle\nabla_{e_j}\omega\lrcorner\cdot, \nabla_{e_j}\omega\lrcorner\cdot\rangle) - 2S(\langle\mathrm{Ricci}(\omega)\lrcorner\cdot, \omega\lrcorner\cdot\rangle) + L(S(\omega)). \tag{3.12}$$

where Ricci^M is the Ricci tensor and the symmetric 2-tensor $L(S)$ was introduced in (3.4). Substituting Weitzenböck formula (3.7) (*i*) into (3.11), it follows that

$$\mathrm{tr}\nabla^2\langle\omega\lrcorner\cdot, \omega\lrcorner\cdot\rangle(X, Y) = \langle(\Delta\omega - \mathrm{Ricci}(\omega))\lrcorner X, \omega\lrcorner Y\rangle + \langle\omega\lrcorner X, (\Delta\omega - \mathrm{Ricci}(\omega))\lrcorner Y\rangle$$

$$-2\sum_i \langle \nabla_{e_i}\omega\lrcorner X, \nabla_{e_i}\omega\lrcorner Y\rangle$$

$$= \langle \Delta\omega\lrcorner X, \omega\lrcorner Y\rangle - \langle \mathrm{Ricci}(\omega)\lrcorner X, \omega\lrcorner Y\rangle + \langle \omega\lrcorner X, \Delta\omega\lrcorner Y\rangle - \langle \omega\lrcorner X, \mathrm{Ricci}(\omega)\lrcorner Y\rangle$$
$$-2\sum_i \langle \nabla_{e_i}\omega\lrcorner X, \nabla_{e_i}\omega\lrcorner Y\rangle. \tag{3.13}$$

Substituting (3) into (3.12), it follows by symmetry and linearity that,

$$\Delta S(\omega)(X,Y) = 2S(\langle \Delta\omega\lrcorner X, \omega\lrcorner Y\rangle) - 2S(\sum_i \langle \nabla_{e_i}\omega\lrcorner X, \nabla_{e_i}\omega\lrcorner Y\rangle)$$

$$-2S(\langle \mathrm{Ricci}(\omega)\lrcorner X, Y\rangle) + L(S(\omega))(X,Y).$$

This is the required result. □

For any 2-tensor $Q \in \Gamma(\otimes^2 T^*M)$, define the p-th stress-energy tensor associated to Q to be the 2-tensor

$$\mathcal{S}_p(Q) = \frac{1}{2}(\mathrm{tr}_g\, Q)g - p\,\mathrm{sym}\, Q, \tag{3.14}$$

where $\mathrm{tr}Q = \sum_i Q(e_i, e_i)$, and $\{e_i\}$ is orthonormal with respect to the metric g. Moreover, define

$$(\mathrm{sym}\, Q)(X,Y) = \frac{1}{2}(Q(X,Y) + Q(Y,X)). \tag{3.15}$$

Then sym Q in (3.15) is called the symmetrization of Q. For a p-form $\omega \in \Omega^p(E)$, this is simply

$$S(\omega) = \mathcal{S}_p(\langle \omega\lrcorner\cdot, \omega\lrcorner\cdot\rangle). \tag{3.16}$$

The previous result can be written in completely symmetric form in terms of $\mathcal{S}_p$.

Corollary 3. Let $\omega \in \Omega^p(E)$ be a vector-bundle valued p-form with associated stress-energy tensor $S(\omega)$. The Laplacian of $S(\omega)$ is given by

$$\Delta S(\omega) = 2\mathcal{S}_p(\langle \Delta\omega\lrcorner\cdot, \omega\lrcorner\cdot\rangle) - 2\mathcal{S}_p(\sum_i \langle \nabla_{e_i}\omega\lrcorner \cdot \nabla_{e_i}\omega\lrcorner\cdot\rangle) - 2\mathcal{S}_p(\langle \mathrm{Ricci}(\omega)\lrcorner\cdot, \omega\lrcorner\cdot\rangle) + L(S(\omega)). \tag{3.17}$$

□

It is also worth writing this out for the special case of a 1-form.

Corollary 4. Let $\omega \in \Omega^1(E)$ be a vector bundle valued 1-form, then (3.17) takes the form

$$\Delta S(\omega) = 2\mathcal{S}(\langle \Delta\omega, \omega\rangle) - 2\mathcal{S}(\sum_i \langle \nabla_{e_i}\omega, \nabla_{e_i}\omega\rangle) + 2\mathcal{S}(\sum_i \langle R^E(\omega)\lrcorner\cdot, \omega\lrcorner\cdot\rangle)$$

$$- 2\mathcal{S}\Big(\langle \omega(\mathrm{Ricci}^M(\cdot)), \omega\rangle\Big) + L(S(\omega)). \tag{3.18}$$

Moreover, for $\omega \in \Omega^1(E)$, a closed vector bundle valued 1-form,

$$\Delta S(\omega) = 2\mathcal{S}(\langle \Delta\omega, \omega\rangle) - 2S(\nabla\omega) + 2\mathcal{S}(\sum_i \langle R^E(e_i, \cdot)\omega(e_i), \omega(\cdot)\rangle)$$

$$- 2\mathcal{S}(\langle \omega(\mathrm{Ricci}^M(\cdot), \omega\rangle) + L(S(\omega)). \tag{3.19}$$

where $S(\nabla\omega)$ denotes the stress-energy tensor of the $T^*M \otimes E$ valued one-form $\nabla_X\omega$.

Proof: First (3.18) is a consequence of (3.17). For (3.19), since ω is closed, $d\omega = 0$, and it follows that $(\nabla_{e_i}\omega)(X) = (\nabla_X\omega)(e_i)$, consequently $\sum_i \langle \nabla_{e_i}\omega, \nabla_{e_i}\omega\rangle = \langle \nabla\omega, \nabla\omega\rangle$. □

Theorem 5. Let $\varphi : (M, g) \to (N, h)$ be a smooth map between Riemannian manifolds and let $d\varphi \in \Omega^1(\varphi^{-1}TN)$ be its exterior derivative.

$$\Delta S(d\varphi) = -2S(\langle \nabla\tau(\varphi), d\varphi\rangle) - 2S(\nabla d\varphi) - 2S(\langle d\varphi(\text{Ricci}^M(\cdot)), d\varphi(\cdot)\rangle)$$
$$- 2S(\sum_i \langle R^N(d\varphi(e_i), d\varphi(\cdot))\, d\varphi(e_i), d\varphi(\cdot)\rangle) + L(S(d\varphi)), \tag{3.20}$$

where $\tau(\varphi) = -d^*d\varphi$ denotes the tension field of the map φ.

Proof: For the one-form ω, substitute $\omega = d\varphi$ into (3.19). Since it follows that $\Delta d\varphi = (d^*d + dd^*)d\varphi = -d\tau(\varphi) = -\nabla\tau(\varphi)$, and $R^N \circ d\varphi$ is the induced curvature in the bundle $E = \varphi^{-1}TN$. □

Theorem 6. Let $\varphi : (M^2, g) \to (N^2, h)$ be a harmonic map between surfaces where M^2 has Gaussian curvature c^M, then the folowing holds,

$$\Delta S(d\varphi) = -2S(\nabla d\varphi) + 2c^M S(d\varphi). \tag{3.21}$$

In particular, if M has constant curvature, then it follows that $\text{div}S(\nabla d\varphi) = 0$.

Proof: If φ is harmonic, then $\nabla\tau(\varphi) = 0$, so the first term in (3.20) vanishes. Suppose $\sigma(Q) = \frac{1}{2}(\text{tr}Q)g - \text{sym}\, Q$ is a stress-energy tensor associated to the 2-tensor Q on M. Suppose z is an isothermal coordinate, then $\sigma(Q)$ can be expressed in diagonal form $\sigma(Q) = a\, dz \otimes dz + \bar{a}\, d\bar{z} \otimes d\bar{z}$ for some function $a = a(z, \bar{z})$. If $\sigma(Q)$ is divergence free, then $a_{\bar{z}}$ vanishes, so $\sigma(Q) = a\, dz \otimes dz$ defines a quadratic differential form on M^2. When φ is a smooth map, then locally φ^*h can be diagonalized to the form $\varphi^*h = \lambda_1\, \omega_1 \otimes \omega_1 + \lambda_2 \omega_2 \otimes \omega_2$, where $\{\omega_1, \omega_2\}$ is an orthonormal basis of 1-forms. If M and N have Gaussian curvatures c^M and c^N, then letting $\{e_1, e_2\}$ be the basis dual to $\{\omega_1, \omega_2\}$, summing on $i = 1, 2$ we have $\langle R^N(d\varphi(e_i), d\varphi(\cdot))d\varphi(e_i), d\varphi(\cdot)\rangle = c^N\lambda_1\lambda_2 g$, and consequently $\sigma(\langle R^N(d\varphi(e_i), d\varphi(\cdot))d\varphi(e_i), d\varphi(\cdot)\rangle) = 0$. Moreover, $L(S(\varphi)) = 4c^M S(d\varphi)$ and $\sigma(\langle d\varphi(\text{Ricci}^M(\cdot)), d\varphi(\cdot)\rangle) = c^M S(d\varphi)$, so substituting these calculations into (3.20), equation (3.21) is the result.

Recall that if $\omega \in \Omega^p(E)$ is harmonic, then $\text{div}\, S(\omega) = 0$. If we use $\omega = d\varphi \in \Omega^1(E)$, then $\text{div}S(d\varphi) = 0$. Take the divergence of both sides of (3.21) and note that c^M is constant, then the result follows since divergence and Laplacian commute be Proposition 1. □

Now $\tau(\varphi)$ defines an energy functional which is called the biharmonic energy functional

$$E_2(\varphi) = \frac{1}{2}\int_M |\tau(\varphi)|^2\, dv_M, \tag{3.22}$$

and has the associated stress-energy tensor $S_2(d\varphi)$,

$$S_2(d\varphi) = \frac{1}{2}|\tau(\varphi)|^2 g + 2S(\langle\nabla\tau(\varphi), d\varphi\rangle). \tag{3.23}$$

Theorem 7. Let $\varphi : (M, g) \to (N, h)$ be a smooth map between Riemannian manifolds and let $d\varphi \in \Omega^1(\varphi^{-1}TN)$ be the derivative.

(i) The Laplacian of $S(d\varphi)$ is given as

$$\Delta S(d\varphi) = \frac{1}{2}|\tau(\varphi)|^2 g - S_2(d\varphi) - 2S(\nabla d\varphi) - 2S(\langle d\varphi(\text{Ricci}^M(\cdot), d\varphi(\cdot)\rangle)$$

$$+ 2\mathcal{S}(\sum_i \langle R^N(d\varphi(e_i), d\varphi(\cdot))\, d\varphi(e_i), d\varphi(\cdot)\rangle^{TN}) + L(S\,(d\varphi)). \tag{3.24}$$

(ii) If $\varphi : M \to N$ is an isometric immersion,

$$\frac{1}{2}|\tau(\varphi)|^2 g - S_2(d\varphi) - 2S\,(\nabla d\varphi) - 2\mathcal{S}(\mathrm{Ricci}^M)$$

$$+ 2\mathcal{S}(\sum_i \langle R^N(d\varphi(e_i), d\varphi(\cdot))\, d\varphi(e_i), d\varphi(\cdot)\rangle) = 0. \tag{3.25}$$

(iii) If N is a space form of constant curvature $K^N = -1, 0, +1$,

$$\frac{1}{2}|\tau(\varphi)|^2 g - S_2(d\varphi) - 2\mathcal{S}(d\gamma) - 2\mathcal{S}(\mathrm{Ricci}^M) + 2(m-1)K^N S\,(d\varphi) = 0. \tag{3.26}$$

Proof: (i) Let $\omega = d\varphi$ then solve for the second term in (3.23) and substitute it into the previous result (3.21) which gives (3.24). (ii) If $\varphi : (M, g) \to (N, h)$ is an isometric immersion, then

$$S\,(d\varphi) = \frac{1}{2}(m-2)\,g, \tag{3.27}$$

hence the left-hand side of (3.24) as well as $L(S\,d\varphi))$ both vanish. (iii) Finally, if we map into a space form, the second fundamental form $\nabla d\varphi$ can be identified with the derivative of the associated Gauss map Γ and (3.26) follows. □

As an application of this theorem, suppose φ is a minimal immersion of a surface into Euclidean space, then $\tau(\varphi) = 0$, $S_2(d\varphi) = 0$ and $\sigma(\mathrm{Ricci}^M) = 0$. Then equation (3.26) implies that $S\,(d\gamma) = 0$.

If the divergence of both sides of (3.26) is worked out, since $\mathrm{div}\mathcal{S}(\mathrm{Ricci}^M) = 0$ as a consequence of the Bianchi identity, (3.26) implies that if $\varphi : M \to N$ is an isometric immersion into a space form, it follows that

$$\frac{1}{2}d|\tau(\varphi)|^2 - \mathrm{div}S_2(d\varphi) - 2\mathrm{div}\,S\,(d\gamma) = 0. \tag{3.28}$$

4. Further Results Including an Integral Theorem

Let us examine the influence of conformal vector fields to finish [3]. A monotonicity formula describes the growth properties of extremals of the functional (2.1) and may be used to establish their regularity in appropriate situations.

Let (M, g) be a Riemannian manifold of dimension m. A vector field X on M is called conformal if it satisfies

$$\mathcal{L}_X\, g = \xi g \tag{4.1}$$

for some function $\xi : M \to \mathbb{R}$ and $\mathcal{L}$ is the Lie derivative in the direction of X. Equivalently, X is conformal if and only if

$$\mathrm{sym}\; g(\nabla_i) = \frac{1}{m}(\mathrm{div}X)g, \tag{4.2}$$

in which case, we have $\xi = (2/m)\mathrm{div}X$.

Theorem 8. Let T be a symmetric 2-tensor, and let X be a conformal vector field, then

$$\mathrm{div}(T \lrcorner X) = (\mathrm{div}\,T)(X) + \frac{1}{m}(\mathrm{div}X)\cdot \mathrm{tr}T. \tag{4.3}$$

Proof: Let $\{e_i\}_1^m$ be a locally defined frame field. If X is a conformal vector field, the divergence can be calculated as follows,

$$\begin{aligned}\mathrm{div}(T\lrcorner X) &= \nabla_{e_i}(T\lrcorner X)(e_i) = (\nabla_{e_i}T)(X, e_i) + T(\nabla_{e_i}X, e_i)\\ &= (\mathrm{div}T)(X) + T(\langle\nabla_{e_i}X, e_j\rangle e_j, e_i).\end{aligned}$$

Since T is a symmetric 2-tensor, the definition of conformal implies that

$$\mathrm{div}(T\lrcorner X) = (\mathrm{div}T)(X) + \frac{1}{2}(\langle\nabla_{e_i}X, e_j\rangle + \langle\nabla_{e_j}X, e_i\rangle)T(e_i, e_j)$$

$$= (\mathrm{div}T)(X) + \frac{1}{m}\mathrm{div}X\, g(e_i, e_j)\, T(e_i, e_j) = (\mathrm{div}T) + \frac{1}{m}\mathrm{div}X\,\mathrm{tr}\,T,$$

since we have $\mathrm{tr}T = g(e_i, e_j)\, T(e_i, e_j)$.

Integrate (4.3) over a compact region, U, to obtain the monotonicity formula. The divergence theorem permits us to write

$$\int_U \mathrm{div}(T\lrcorner X)\, dv_M = \int_{\partial U} T(X, \mathbf{n})\, da_M. \tag{4.4}$$

where da_M is the volume form on ∂U.

Theorem 9. Let (M, g) be an oriented Riemannian manifold and let X be a conformal vector field on M. Suppose that $U \subset M$ is a compact region with a smooth boundary ∂U. Then for any symmetric 2-tensor T on M,

$$\int_{\partial U} T(X, \mathbf{n})\, da_M = \int_U (\mathrm{div}T)(X)\, dv_M + \frac{1}{m}\int_M \mathrm{div}X\,\mathrm{tr}T\, dv_M, \tag{4.5}$$

where dv_M is the volume element on M and da_M is the volume element on ∂U, $\mathbf{n}$ the outward pointing normal on ∂U. □

This is a remarkable theorem as there are numerous applications of it, but only one will be presented here. Let $\varphi : B^m \to (N, h)$ be a harmonic map from the Euclidean m-ball of radius R and $S = e(\varphi)g - \varphi^*h$ the corresponding stress-energy tensor. Clearly, $\mathrm{div}S = 0$ and $\mathrm{tr}S = e\,\mathrm{tr}g - \mathrm{tr}\varphi^*h = e\,m - 2e = (m-2)\,e(\varphi)$. Take X to be the conformal vector field $X = r\partial_r$ where $r = |\mathbf{x}|$, $\mathbf{x} \in \mathbb{R}^m$. By direct calculation, $\mathrm{div}X = m$, hence substituting these facts into (4.5), the following result holds for $r < R$,

$$\int_{\partial B^m} S(r\partial_r, \partial_r)\, da = \int_{B^m} (m-2)\, e(\varphi)\, dv_M. \tag{4.6}$$

Substituting for S yields

$$(m-2)\int_{B^m} e(\varphi)\, dv_M = \int_{\partial B^m} r\{e(\varphi) - h(\frac{\partial\varphi}{\partial r}, \frac{\partial\varphi}{\partial r})\}\, da_M. \tag{4.7}$$

The following conclusion can be drawn from (4.7). If $m > 2$ and $\varphi|_{\partial B_r^m} = c$ for some $r < R$ and constant c, then φ must be constant, that is, for $e(\varphi)|_{\partial B^m} = \frac{1}{2}h(\partial_r\varphi, \partial_r\varphi)$ gives the upper bound

$$\int_{B^m} e(\varphi)\, dv_M \leq 0,$$

which in turn implies that $e(\varphi) \equiv 0$.

Conflict of Interest

The author declares no conflicts of interest in this paper.

References

1. P. Baird, *Stress-energy tensors and the Lichnerowicz Laplacian,* J. of Geometry and Physics, **58** (2008), 1329-1342.
2. P. Blanchard and E. Brüning, *Variational Methods in Mathematical Physics, Springer-Verlag,* Berlin-Heidelberg, 1992.
3. S. S. Chern, *Minimal surfaces in Euclidean Space of N dimensions: Symposium in Honor of Marston Morse,* Princeton Univ. Press, 1965, 187-198.
4. J Eells and L. Lemaine, *Selected Topics in Harmonic Maps: CBMS Regional Conference Series in Mathematics,* American Mathematical Society, Providence, RI, **50** (1983).
5. A. Lichnerowicz, *Géometrie des Groups de Transformations,* Dunod, Paris, 1958.
6. A. Lichnerowicz, *Propagateurs et commutateurs en relativité générale,* Publ. Math. Inst. Hautes Ëtudes Sci., **10** (1961), 293-344.
7. P. Peterson, *Riemannian Geometry,* Springer-Verlag, NY, 1998.
8. R. Schoen, *The existence of weak solutions with prescribed singular behavior for a conformally invariant scalar equation,* Comm. Pure and Applied Math. XLI, 317-392, 1988.

3

Monotone Dynamical Systems with Polyhedral Order Cones and Dense Periodic Points

Morris W. Hirsch*

Department of Mathematics, University of Wisconsin, Madison WI 53706, USA

* **Correspondence:** Email: mwhirsch@chorus.net

Abstract: Let $X \subset \mathbb{R}^n$ be a set whose interior is connected and dense in X, ordered by a closed convex cone $K \subset \mathbb{R}^n$ having nonempty interior. Let $T\colon X \approx X$ be an order-preserving homeomorphism. The following result is proved: Assume the set of periodic points of T is dense in X, and K is a polyhedron. Then T is periodic.

Keywords: Dynamical systems; ordered spaces; convex cones; periodic orbits

1. Introduction

The following postulates and notation are used throughout:

- $K \subset \mathbb{R}^n$ (Euclidean n-space) is a *solid order cone*: a closed convex cone that has nonempty interior $\mathsf{Int}\,(K)$ and contains no affine line.
- $\mathbb{R}^n$ has the (partial) order $\succeq$ determined by K:

$$y \succeq x \iff y - x \in K,$$

  referred to as the K-order.
- $X \subset \mathbb{R}^n$ is a nonempty set whose $\mathsf{Int}\,(X)$ is connected and dense in X.
- $T\colon X \approx X$ is homeomorphism that is *monotone* for the K-order:

$$x \succeq y \implies Tx \succeq Ty.$$

A point $x \in X$ has *period* k provided k is a positive integer and $T^k x = x$. The set of such points is $\mathcal{P}_k = \mathcal{P}_k(T)$, and the set of periodic points is $\mathcal{P} = \mathcal{P}(T) = \bigcup_k \mathcal{P}_k$. T is *periodic* if $X = \mathcal{P}_k$, and *pointwise periodic* if $X = \mathcal{P}$.

Our main concern is the following speculation:

Conjecture. *If $\mathcal{P}$ is dense in X, then T is periodic.*

The assumptions on X show that T is periodic iff $T|\,\mathsf{Int}\,(X)$ is periodic. Therefore we assume henceforth:

- X is connected and open $\mathbb{R}^n$.

We prove the conjecture under the additional assumption that K is a *polyhedron*, the intersection of finitely many closed affine halfspaces of $\mathbb{R}^n$:

Theorem 1 (Main). *Assume K is a polyhedron, $T\colon\ X \approx X$ is monotone for the K-order, and $\mathcal{P}$ is dense in X. Then T is periodic.*

For analytic maps there is an interesting contrapositive:

Theorem 2. *Assume K is a polyhedron and $T\colon\ X \approx X$ is monotone for the K-order. If T is analytic but not periodic, $\mathcal{P}$ is nowhere dense.*

Proof. As X is open and connected but not contained in any of the closed sets $\mathcal{P}_k$, analyticity implies each $\mathcal{P}_k$ is nowhere dense. Since $\mathcal{P} = \bigcup_{k=1}^{\infty} \mathcal{P}_k$, a well known theorem of Baire [1] implies $\mathcal{P}$ is nowhere dense. ∎

The following result of D. Montgomery [4]* is crucial for the proof of the Main Theorem:

Theorem 3 (Montgomery). *Every pointwise periodic homeomorphism of a connected manifold is periodic.*

Notation

i, j, k, l denote positive integers. Points of $\mathbb{R}^n$ are denoted by $a, b, p, q, u, v, w, x, y, z$.

$x \leq y$ is a synonym for $y \geq x$. If $x \leq y$ and $x \neq y$ we write $x <$ or $y > x$.

The relations $x \ll y$ and $y \gg x$ mean $y - x \in \mathsf{Int}\,(K)$.

A set S is *totally ordered* if $x, y \in S \implies x \leq y$ or $x \geq y$.

If $x \leq y$, the *order interval* $[x, y]$ is $\{z\colon\ x \leq z \leq y\} = K_x \cap -K_y$.

The translation of K by $x \in \mathbb{R}^n$ is $K_x := \{w + x, w \in K.\}$

The image of a set or point ξ under a map H is denoted by $H\xi$ or $H(\xi)$. A set S is *positively invariant* under H if $HS \subset S$, *invariant* if $H\xi = \xi$, and *periodically invariant* if $H^k\xi = \xi$.

2. Proof of the Main Theorem

The following four topological consequences of the standing assumptions are valid even if K is not polyhedral.

Proposition 4. *Assune $p, q \in \mathcal{P}_k$ are such that*

$$p \ll q, \qquad p, q \in \mathcal{P}_k. \qquad [p, q] \subset X.$$

Then $T^k([p, q] = [p, q]$.

---

*See also S. Kaul [3].

Proof. It suffices to take $k = 1$. Evidently $T\mathcal{P} = \mathcal{P}$, and $T[p,q] \subset [p,q]$ because T is monotone, whence $\mathsf{Int}([p,q]) \cap \mathcal{P}$ is positively invariant under T. The conclusion follows because $\mathsf{Int}([p,q]) \cap \mathcal{P}$ is dense in $[p,q]$ and T is continuous. ∎

Proposition 5. *Assume $a, b \in \mathcal{P}_k, a \ll b$, and $[a,b] \subset X$. There is a compact arc $J \subset \mathcal{P}_k \cap [a,b]$ that joins a to b, and is totally ordered by $\ll$.*[†]

Proof. An application of Zorn's Lemma yields a maximal set $J \subset [a,b] \cap P$ such that: J is totally ordered by $\ll$, $a = \max J$, $b = \min J$. Maximality implies J is compact and connected and $a, b \in J$, so J is an arc (Wilder [7], Theorem I.11.23). ∎

Proposition 6. *Let $M \subset X$ be a homeomorphically embedded topological manifold of dimension $n-1$, with empty boundary.*

(i) *$\mathcal{P}$ is dense in M.*

(ii) *If M is periodically invariant, it has a neighborhood base $\mathcal{B}$ of periodically invariant open sets.*

Proof. (i) M locally separates X, by Lefschetz duality [5] (or dimension theory [6]. Therefore we can choose a family $\mathcal{V}$ of nonempty open sets in X that the family of sets $\mathcal{V}_M := \{V \cap M\colon\ V \in \mathcal{V})$ satisfies:

- $\mathcal{V}_M$ is a neighborhood basis of M,
- each set $V \cap M$ separates V.

By Proposition 5, for each $V \in \mathcal{V}$ there is a compact arc $J_V \cap \mathcal{P} \cap V$ whose endpoints a_V, b_v lie in different components of $V \setminus M$. Since J_V is connected, it contains a point in $V \cap M \cap \mathcal{P}$. This proves (i).

(ii) With notation as above, let $B_V := [a_V, b_V] \setminus \partial[a_V, b_V]$. The desired neighborhood basis is $\mathcal{B} := \{B_V\colon\ V \in \mathcal{V}\}$. ∎

From Propositions 4 and 6 we infer:

Proposition 7. *Suppose $p, q \in \mathcal{P}$, $p \ll q$ and $[p,q] \subset X$. Then $\mathcal{P}$ is dense in $\partial[p,q]$.* ∎

Let $\mathcal{T}(m)$ stand for the statement of Theorem 1 for the case $n = m$. Then $\mathcal{T}(0)$ is trivial, and we use the following inductive hypothesis:

Hypothesis (Induction). $n \geq 1$ and $\mathcal{T}(n-1)$ holds.

Let $Q \subset \mathbb{R}^n$ be a compact n-dimensional polyhedron. Its boundary ∂Q is the union of finitely many convex compact $(n-1)$-cells, the *faces* of Q. Each face F is the intersection of $\partial[p,q]$ with a unique affine hyperplane E^{n-1}. The corresponding *open face* $F^\circ := F \setminus \partial F$ is an open $(n-1)$-cell in E^{n-1}. Distinct open faces are disjoint, and their union is dense and open in ∂Q.

Proposition 8. *Assume $p, q \in \mathcal{P}_k$, $p \ll q$, $[p,q] \subset X$. Then $T|\partial[p,q]$ is periodic.*

[†]This result is adapted from Hirsch & Smith [2], Theorems 5,11 & 5,15.

Proof. $[p, q]$ is a compact, convex n-dimensional polyhedron, invariant under T^k (Proposition 4). By Proposition 6 applied to $M := \partial[p, q]$, there is a neighborhood base $\mathcal{B}$ for $\partial[p, q]$ composed of periodically invariant open sets. Therefore if $F^\circ \subset \partial[p, q]$ is an open face of $[p, q]$, the family of sets

$$\mathcal{B}_{F^\circ} := \{W \in \mathcal{B} : \ W \subset F^\circ\}$$

is a neighborhood base for F°, and each $W \in \mathcal{B}_{F^\circ}$ is a periodically invariant open set in which $\mathcal{P}$ is dense.

For every face F of $[p, q]$ the Induction Hypothesis shows that $F^\circ \subset \mathcal{P}$. Therefore Montgomery's Theorem implies $T|F^\circ$ is periodic, so $T|F$ is periodic by continuity. Since $\partial[p, q]$ is the union of the finitely many faces, it follows that $T|\partial[p, q]$ is periodic. ∎

To complete the inductive proof of the Main Theorem, it suffices by Montgomery's theorem to prove that an arbitrary $x \in X$ is periodic. As X is open in $\mathbb{R}^n$ and $\mathcal{P}$ is dense in X, there is an order interval $[a, b] \subset X$ such that

$$a \ll x \ll b, \qquad a, b \in \mathcal{P}_k.$$

By Proposition 5, a and b are the endpoints of a compact arc $J \subset \mathcal{P}_k \cap [a, b]$, totally ordered by $\ll$. Define $p, q \in J$:

$$p := \sup\{y \in J : \ y \preceq x\}, \qquad q := \inf\{y \in J : \ y \succeq x\}.$$

If $p = q = x$ then $x \in \mathcal{P}_k$. Otherwise $p \ll q$, implying $x \in \partial[p, q]$, whence $x \in \mathcal{P}$ by Proposition 8. ∎

Conflict of Interest

The author declares no conflicts of interest in this paper.

References

1. R. Baire, *Sur les fonctions de variables réelles,* Ann. di Mat. **3** (1899), 1-123.
2. M. Hirsch and H. Smith, *Monotone Dynamical Systems*, Handbook of Differential Equations, volume 2, chapter 4. A. Cañada, P. Drábek & A. Fonda, editors. Elsevier North Holland, 2005.
3. S. Kaul, *On pointwise periodic transformation groups,* Proceedings of the American Mathematical Society **27** (1971), 391-394.
4. D. Montgomery, *Pointwise periodic homeomorphisms,* American Journal of Mathematics **59** (1937), 118-120.
5. E. Spanier, Algebraic Topology, McGraw Hill, 1966.
6. W. Hurewicz and H. Wallman, Dimension Theory, Princeton University Press, 1941.
7. R. Wilder, Topology of Manifolds, American Mathematical Society, 1949.

4

Steady states of elastically-coupled extensible double-beam systems

Filippo Dell'Oro[1] **Claudio Giorgi**[2] **and Vittorino Pata**[1,*]

[1] Dipartimento di Matematica, Politecnico di Milano, Via Bonardi 9, 20133 Milano, Italy
[2] DICATAM, Università degli Studi di Brescia, Via Valotti 9, 25133 Brescia, Italy

[*] **Correspondence:** Email: vittorino.pata@polimi.it

Abstract: Given $\beta \in \mathbb{R}$ and $\varrho, k > 0$, we analyze an abstract version of the nonlinear stationary model in dimensionless form

$$\begin{cases} u'''' - \Big(\beta + \varrho \int_0^1 |u'(s)|^2 \, \mathrm{d}s\Big) u'' + k(u - v) = 0 \\ v'''' - \Big(\beta + \varrho \int_0^1 |v'(s)|^2 \, \mathrm{d}s\Big) v'' - k(u - v) = 0 \end{cases}$$

describing the equilibria of an elastically-coupled extensible double-beam system subject to evenly compressive axial loads. Necessary and sufficient conditions in order to have nontrivial solutions are established, and their explicit closed-form expressions are found. In particular, the solutions are shown to exhibit at most three nonvanishing Fourier modes. In spite of the symmetry of the system, nonsymmetric solutions appear, as well as solutions for which the elastic energy fails to be evenly distributed. Such a feature turns out to be of some relevance in the analysis of the longterm dynamics, for it may lead up to nonsymmetric energy exchanges between the two beams, mimicking the transition from vertical to torsional oscillations.

Keywords: Coupled-beams structures; steady states; bifurcations; buckling

---

1. Introduction

1.1. Physical motivations

For engineering purposes, the mathematical modeling process can be viewed as the first step towards the analysis of both static and dynamic responses of actual mechanical structures. Nevertheless, it relies on an idealization of the physical world, and has limits of validity that must be specified. For a given system, different models can be constructed, the "best" being the simplest one able to capture all the essential features needed in the investigation. Among others, models of elastic sandwich-structured composites are experiencing an increasing interest in the literature, mainly due to their wide use in sandwich panels and their applications in many branches of modern civil, mechanical and aerospace engineering [30]. Sandwich structures are in general symmetric, and their variety depends on the configuration of the core. Such devices are designed to have high bending stiffness with overall low density [9, 18]. In particular, sandwich beams, plates and shells are flexible elastic structures built up by attaching two thin and stiff external layers (beams, plates or shells) to a homogeneously-distributed lightweight and thick elastic core [23]. Their interest, which is relevant in structural mechanics, has been recently extended even to nanostructures (see e.g. [6] and references therein).

Models of elastic sandwich structures can be obtained by applying either the Euler-Bernoulli theory for beams or the Kirchhoff-Love theory for thin plates. In this context, several papers have been devoted to the mechanical properties of elastically-connected double Euler-Bernoulli beams systems. For instance, free and forced transverse vibrations of simply supported double-beam systems have been studied in [17, 22, 26], while the articles [31, 32] are concerned with the effect of compressive axial load on free and forced oscillations. Within the framework of nanostructures, axial instability and buckling of double-nanobeam systems have been analyzed in [21, 27].

Once a model is established, the next step is to (possibly) solve the mathematical equations, in order to discover the nature of the system response. In fact, the main goal is to predict and control the actual dynamics. To this end, the analysis of the steady states, and in particular of their closed-form expressions, becomes crucial. This is even more urgent when dealing with nonlinear systems, where the longterm dynamics is strongly influenced by the occurrence of a rich set of stationary solutions.

1.2. The model

In this paper, we aim to classify the stationary solutions, finding their explicit closed-form expressions, to symmetric elastically-coupled extensible double-beam systems. For instance, a sandwich structure composed of two elastic beams bonded to an elastic core (Figure 1a), or the road bed of a girder bridge composed of an elastic rug connecting two lateral elastic beams (Figure 1b). In both cases, the mechanical structure can be described by means of two equal beams complying with the nonlinear model of Woinowsky-Krieger [29], which takes into account extensibility, so that large deformations are allowed. The beams are supposed to have the same natural length $\ell > 0$, constant mass density, and common thickness $0 < h \ll \ell$. At their ends, they are simply supported and subject to evenly distributed axial loads. A system of linear springs models the elastic filler connecting the beams: when the system lies in its natural configuration, the beams are straight and parallel. The distance between the beams is equal to the free lengths of the springs. Denoting by $\nu \in (-1, \frac{1}{2})$ the Poisson ratio of the beams, the dynamics of the resulting undamped model is ruled by the following nonlinear equations in dimensionless form (see the final Appendix for more details about the derivation of the model)

$$\begin{cases} \dfrac{\ell(1-\nu)}{h}\Big(\partial_{tt} - \dfrac{h^2}{12\ell^2}\partial_{ttxx}\Big)u + \delta\partial_{xxxx}u - (\chi + \|\partial_x u\|^2)\partial_{xx}u + \kappa(u-v) = 0, \\ \dfrac{\ell(1-\nu)}{h}\Big(\partial_{tt} - \dfrac{h^2}{12\ell^2}\partial_{ttxx}\Big)v + \delta\partial_{xxxx}v - (\chi + \|\partial_x v\|^2)\partial_{xx}v - \kappa(u-v) = 0, \end{cases} \tag{1.1}$$

having set

$$\|f\| = \Big(\int_0^1 |f(s)|^2\, \mathrm{d}s\Big)^{\frac{1}{2}}.$$

In the vertical plane (x-z), system (1.1) describes the in-plane downward rescaled deflections of the midline of the beams*

$$u, v : [0, 1] \times \mathbb{R}^+ \to \mathbb{R}$$

with respect to their natural configuration (see Figure 1a). It may be also used to describe out-of-plane rescaled deflections of the same double-beam structure, accounting for both vertical and torsional oscillations (see Figure 1b). In the latter situation, each beam is assumed to swing in a vertical plane and the lateral movements are neglected. The structural constants $\delta, \kappa > 0$ are related to the common flexural rigidity of the beams and the common stiffness of the inner elastic springs, respectively, whereas the parameter $\chi \in \mathbb{R}$ summarizes the effect of the axial force acting at the right ends of the beams: positive when the beams are stretched, negative when compressed.

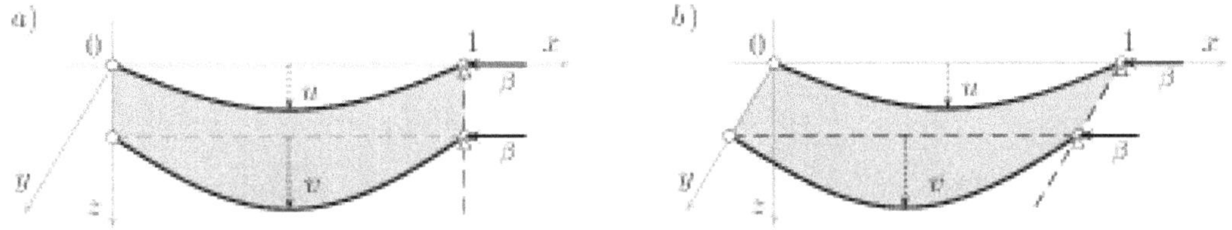

Figure 1. In-plane (a) and out-of-plane (b) deflections of a double-beam system under compressive axial loads $\beta = \chi/\delta$.

In this work, we are interested in the stationary solutions to the evolutionary problem (1.1), subject to the hinged boundary conditions. Namely, setting

$$\beta = \frac{\chi}{\delta} \in \mathbb{R}, \qquad \varrho = \frac{1}{\delta} > 0, \qquad k = \frac{\kappa}{\delta} > 0,$$

we consider the dimensionless system of ODEs

$$\begin{cases} u'''' - (\beta + \varrho\|u'\|^2)u'' + k(u - v) = 0, \\ v'''' - (\beta + \varrho\|v'\|^2)v'' - k(u - v) = 0, \end{cases} \tag{1.2}$$

supplemented with the boundary conditions

$$\begin{cases} u(0) = u(1) = u''(0) = u''(1) = 0, \\ v(0) = v(1) = v''(0) = v''(1) = 0. \end{cases} \tag{1.3}$$

It is apparent that problem (1.2)-(1.3) always admits the trivial solution $u = v = 0$, while the occurrence and the complexity of nontrivial solutions strongly depend on the values of the structural dimensionless parameters β, ϱ, k, all of which are allowed to be large (see the final comment in the Appendix).

*The functions u, v are appropriate rescaling of the original vertical deflections of the midline of the two beams

$$U, V : [0, \ell] \times \mathbb{R}^+ \to \mathbb{R},$$

in comply with the dimensionless character of system (1.1). See the Appendix for more details.

1.3. Earlier results on single-beam equations

When system (1.2) is uncoupled (i.e. in the limit situation when $k = 0$), the analysis reduces to the one of the single Woinowsky-Krieger beam

$$u'''' - (\beta + \varrho\|u'\|^2)u'' = 0.$$

In this case, it is well-known that an increasing compressive axial load leads to a series of fork bifurcations. The critical values of β at which bifurcations occur depend on the eigenvalues of the differential operator (see e.g. [2, 8]). After exceeding these values, the axial compression is sustained in one of two states of equilibrium: a purely compressed state with no lateral deviation (the trivial solution) or two symmetric laterally-deformed configurations (buckled solutions). This is why the phenomenon is usually referred to as *buckling*. Another interesting model, formally obtained by neglecting the second equation of system (1.2) and by taking $v \equiv 0$ in the first one, reads

$$u'''' - (\beta + \varrho\|u'\|^2)u'' + ku = 0,$$

namely, a single Woinowsky-Krieger beam which relies on an elastic foundation. In this case, bifurcations of the trivial solution split into two series, whose critical values depend also on the ratio k between the parameters κ and δ connected with the stiffness of the foundation and the flexural rigidity of the beam [3].

1.4. The goal of the present work

Clearly, when the double-beam system (1.2) is considered, the picture becomes much more difficult. To the best of our knowledge, in spite of the quite large number of papers about statics and dynamics of single Woinowsky-Krieger beams (e.g. [2, 3, 7, 8, 10, 11, 12, 14, 15, 24]), no analytic results concerning models with a coupling between two (or more) nonlinear beams of this type are available in the literature. This may be due to the fact that classifying and finding closed-form expressions for the solutions to equations of this kind is in general a very difficult, if not impossible, task. Indeed, it is usually unavoidable to replace distributed characteristics with discrete ones, so producing approximate solutions by resorting to some discretization procedures. Unfortunately, this strategy can be hardly applied when multiple stable states occur (see e.g. [18] and references therein).

Here, our aim is to fill this gap. To this end, we first recast (1.2)-(1.3) into an abstract nonlinear system involving an arbitrary strictly positive selfadjoint linear operator A with compact inverse. Then, we classify all the nontrivial solutions, finding also their explicit expressions. In particular, every solution is shown to exhibit at most three nonvanishing Fourier modes. According to our classification, the set of stationary solutions to nonlinear double-beam systems is very rich. The nonlinear terms accounting for extensibility substantially influence the instability (or buckling): the effects are higher with increasing values of (minus) the axial-load parameter β, and give rise to both in-phase (synchronous) buckling modes and out-of-phase (asynchronous) buckling modes. This feature becomes quite important in the study of the longterm behavior, as it may lead up to nonsymmetric energy exchanges between the two beams under small perturbations. In the asymptotic dynamics of a double-beam structure like the road bed of a girder bridge (Figure 1b), a nonsymmetric energy exchange of this kind is apt to mimic the transition from vertical to torsional oscillations, such as those occurred in the collapse of the Tacoma Narrows suspension bridge (see e.g. [20] and references therein). Another remarkable fact is that the model (1.2) has been derived under the assumption that the ratio h/ℓ between the thickness and the natural length of the beam is very small; the critical values at which bifurcations occur are consistent with such an assumption, namely, they are of order h/ℓ as well. We also stress that system (1.2) is dimensionless, and no physical parameters have been artificially set equal to one. Finally it is worth noting that, as a consequence of the abstract formulation, all the results are valid also for multidimensional structures. In particular, they are applicable to flexible double-plate sandwich structures with hinged boundaries, provided that the plates are modeled according to the Berger's approach [1, 16].

1.5. Plan of the paper

In the next §2 we introduce the aforementioned operator A, and we rewrite (1.2)-(1.3) in an abstract form. In §3 we prove that every solution can be expressed as a linear combination of at most three distinct eigenvectors of A. The subsequent §4 deals with the analysis of unimodal solutions (i.e. solutions with only one eigenvector involved). In particular, we show that not only a double series of fork bifurcations of the trivial solution occur, but also buckled solutions may suffer from a further bifurcation when $-\beta$ exceeds some greater critical value. In §5 we study the so-called equidistributed energy solutions (i.e. solutions with evenly distributed elastic energy), and we prove that bimodal and trimodal steady states pop up. In §6 we classify the general (not necessarily equidistributed) bimodal solutions, while in §7 we show that every trimodal solution is necessarily an equidistributed energy solution, The final §8 is devoted to a comparison with some single-beam equations previously studied in the literature. The derivation of the evolutionary physical model (1.1) is carried out in full detail in the concluding Appendix.

2. The Abstract Model

Let $(\mathrm{H}, \langle\cdot,\cdot\rangle, \|\cdot\|)$ be a separable real Hilbert space, and let

$$A : \mathfrak{D}(A) \Subset \mathrm{H} \to \mathrm{H}$$

be a strictly positive selfadjoint linear operator, where the (dense) embedding $\mathfrak{D}(A) \Subset \mathrm{H}$ is compact. In particular, the inverse A^{-1} of A turns out to be a compact operator on H. Accordingly, for $r \geq 0$, we introduce the compactly nested family of Hilbert spaces (the index r will be omitted whenever zero)

$$\mathrm{H}^r = \mathfrak{D}(A^{\frac{r}{2}}), \qquad \langle u, v\rangle_r = \langle A^{\frac{r}{2}}u, A^{\frac{r}{2}}v\rangle, \qquad \|u\|_r = \|A^{\frac{r}{2}}u\|.$$

Then, given $\beta \in \mathbb{R}$ and $\varrho, k > 0$, we consider the abstract nonlinear stationary problem in the unknown variables $(u, v) \in \mathrm{H}^2 \times \mathrm{H}^2$

$$\begin{cases} A^2u + C_uAu + k(u-v) = 0, \\ A^2v + C_vAv - k(u-v) = 0, \end{cases} \tag{2.1}$$

where

$$C_u = \beta + \varrho\|u\|_1^2 \qquad \text{and} \qquad C_v = \beta + \varrho\|v\|_1^2. \tag{2.2}$$

Definition 2.1. A couple $(u, v) \in \mathrm{H}^2 \times \mathrm{H}^2$ is called a *weak solution* to (2.1) if

$$\begin{cases} \langle u, \phi\rangle_2 + C_u\langle u, \phi\rangle_1 + k\langle(u-v), \phi\rangle = 0, \\ \langle v, \psi\rangle_2 + C_v\langle v, \psi\rangle_1 - k\langle(u-v), \psi\rangle = 0, \end{cases} \tag{2.3}$$

for every test $(\phi, \psi) \in \mathrm{H}^2 \times \mathrm{H}^2$.

It is apparent that the trivial solution $u = v = 0$ always exists.

Example 2.2. The concrete physical system (1.2) is recovered by setting $\mathrm{H} = L^2(0,1)$ and $A = L$, where

$$L = -\frac{\mathrm{d}^2}{\mathrm{d}x^2} \qquad \text{with} \qquad \mathfrak{D}(L) = H^2(0,1) \cap H_0^1(0,1).$$

Here $L^2(0,1)$, as well as $H_0^1(0,1)$ and $H^2(0,1)$, denote the usual Lebesgue and Sobolev spaces on the unit interval $(0,1)$. In particular

$$\mathrm{H}^2 = H^2(0,1) \cap H_0^1(0,1) \Subset \mathrm{H}^1 = H_0^1(0,1) \Subset \mathrm{H} = L^2(0,1).$$

Notation. For any $n \in \mathbb{N} = \{1, 2, 3, \ldots\}$ we denote by

$$0 < \lambda_n \to \infty$$

the increasing sequence of eigenvalues of A, and by $e_n \in \mathrm{H}$ the corresponding normalized eigenvectors, which form a complete orthonormal basis of H. In this work, all the eigenvalues λ_n are assumed to be *simple*, which is certainly true for the concrete realization $A = L$ arising in the considered physical models. Indeed, in such a case, the eigenvalues are equal to

$$\lambda_n = n^2\pi^2$$

with corresponding eigenvectors

$$e_n(x) = \sqrt{2}\sin(n\pi x).$$

3. General Structure of the Solutions

In this section we provide two general results on the solutions to system (2.1). To this end, we introduce the set of *effective modes*

$$\mathbb{E} = \{n : \lambda_n < -\beta\}.$$

Clearly,

$$\mathbb{E} \neq \emptyset \quad \Leftrightarrow \quad \beta < -\lambda_1. \tag{3.1}$$

Therefore, if $\mathbb{E} \neq \emptyset$,

$$\mathbb{E} = \{1, 2, \ldots, n_\star\},$$

where[†]

$$n_\star = \max\{n : \lambda_n < -\beta\} = |\mathbb{E}|.$$

Example 3.1. When $A = L$ (the Laplace-Dirichlet operator introduced in the previous section), we have

$$\mathbb{E} = \{n : n^2\pi^2 < -\beta\}.$$

Accordingly, in the nontrivial case $\beta < 0$,

$$|\mathbb{E}| = \left\lceil \sqrt{-\frac{\beta}{\pi^2}} \right\rceil - 1,$$

the symbol $\lceil a \rceil$ standing for the smallest integer greater than or equal to a.

[†]Here and in what follows $|\mathbb{S}|$ denotes the cardinality of a set $\mathbb{S} \subset \mathbb{N}$.

We begin to prove that the picture is trivial whenever the set $\mathbb{E}$ is empty.

Proposition 3.2. *If* $\mathbb{E} = \emptyset$ *system* (2.1) *admits only the trivial solution.*

Proof. Let (u, v) be a weak solution to (2.1). Choosing $(\phi, \psi) = (u, v)$ in the weak formulation (2.3), and adding the resulting expressions, we obtain the identity

$$\|u\|_2^2 + \|v\|_2^2 + (\beta + \varrho\|u\|_1^2)\|u\|_1^2 + (\beta + \varrho\|v\|_1^2)\|v\|_1^2 + k\|u - v\|^2 = 0.$$

Then, exploiting the Poincaré inequality

$$\lambda_1\|w\|_1^2 \leq \|w\|_2^2, \quad \forall w \in \mathrm{H}^2,$$

we infer that

$$(\lambda_1 + \beta)(\|u\|_1^2 + \|v\|_1^2) + \varrho\|u\|_1^4 + \varrho\|v\|_1^4 + k\|u - v\|^2 \leq 0,$$

and, since $\lambda_1 + \beta \geq 0$, we conclude that $u = v = 0$. □

Accordingly, from now on we will assume (often without explicit mention) that (3.1) be satisfied. As it will be clear from the subsequent analysis, this condition turns out to be sufficient as well in order to have nontrivial solutions. Hence, *a posteriori*, we can reformulate Proposition 3.2 by saying that system (2.1) admits nontrivial solutions if and only if the set $\mathbb{E}$ is nonempty.

The next result shows that every weak solution can be written as linear combination of at most three distinct eigenvectors of A.

Lemma 3.3. *Let* (u, v) *be a weak solution of system* (2.1). *Then*

$$u = \sum_n \alpha_n e_n \qquad \textit{and} \qquad v = \sum_n \gamma_n e_n$$

for some $\alpha_n, \gamma_n \in \mathbb{R}$, *where* $\alpha_n \neq 0$ *for at most three distinct values of* $n \in \mathbb{N}$. *Moreover,*

$$\alpha_n = 0 \quad \Leftrightarrow \quad \gamma_n = 0.$$

Proof. Let (u, v) be a weak solution to (2.1). Then, writing

$$u = \sum_n \alpha_n e_n \qquad \text{and} \qquad v = \sum_n \gamma_n e_n$$

for some $\alpha_n, \gamma_n \in \mathbb{R}$, and choosing $\phi = \psi = e_n$ in the weak formulation (2.3), we obtain for every $n \in \mathbb{N}$ the system

$$\begin{cases} \lambda_n^2\alpha_n + C_u\lambda_n\alpha_n + k(\alpha_n - \gamma_n) = 0, \\ \lambda_n^2\gamma_n + C_v\lambda_n\gamma_n - k(\alpha_n - \gamma_n) = 0. \end{cases} \tag{3.2}$$

It is apparent that

$$\alpha_n = 0 \quad \Leftrightarrow \quad \gamma_n = 0.$$

Substituting the first equation into the second one, we get

$$\gamma_n(\lambda_n^2 + C_v\lambda_n + k)(\lambda_n^2 + C_u\lambda_n + k) = k^2\gamma_n.$$

Hence, if $\gamma_n \neq 0$ (and so $\alpha_n \neq 0$), we end up with

$$\lambda_n^3 + (C_u + C_v)\lambda_n^2 + (C_uC_v + 2k)\lambda_n + k(C_u + C_v) = 0.$$

Since the equation above admits at most three distinct solutions λ_{n_i} we are done. □

Summarizing, every weak solution (u, v) can be written as

$$u = \sum_{i=1}^{3} \alpha_{n_i} e_{n_i} \qquad \text{and} \qquad v = \sum_{i=1}^{3} \gamma_{n_i} e_{n_i}, \tag{3.3}$$

for three distinct $n_i \in \mathbb{N}$ and some coefficients $\alpha_{ni}, \gamma_{n_i} \in \mathbb{R}$. In particular, from (2.2), we deduce the explicit expressions

$$C_u = \beta + \varrho \sum_{i=1}^{3} \lambda_{n_i} \alpha_{n_i}^2 \qquad \text{and} \qquad C_v = \beta + \varrho \sum_{i=1}^{3} \lambda_{n_i} \gamma_{n_i}^2. \tag{3.4}$$

In addition, when

$$\alpha_{n_i} \neq 0 \quad \Leftrightarrow \quad \gamma_{n_i} \neq 0,$$

the corresponding eigenvalue λ_{n_i} is a root of the cubic polynomial

$$P(\lambda) = \lambda^3 + (C_u + C_v)\lambda^2 + (C_u C_v + 2k)\lambda + k(C_u + C_v).$$

Notably, when the equality $C_u = C_v$ holds, the polynomial $P(\lambda)$ can be written in the simpler form

$$P(\lambda) = (\lambda + C_u)(\lambda^2 + C_u\lambda + 2k).$$

Remark 3.4. Adding the two equations of system (3.2), we infer that

$$\lambda_n = -\frac{C_u \alpha_n + C_v \gamma_n}{\alpha_n + \gamma_n} \tag{3.5}$$

whenever $\alpha_n + \gamma_n \neq 0$. This relation will be crucial for our purposes.

As an immediate consequence of Lemma 3.3, we also have

Corollary 3.5. *Every weak solution* (u, v) *is actually a strong solution. Namely,* $(u, v) \in \mathrm{H}^4 \times \mathrm{H}^4$ *and* (2.1) *holds. Even more so,* $(u, v) \in \mathrm{H}^r \times \mathrm{H}^r$ *for every* r*.*

Remark 3.6. In the concrete situation when $A = L$, every weak solution (u, v) is regular, that is, $(u, v) \in C^\infty([0, 1]) \times C^\infty([0, 1])$.

Finally, in the light of Lemma 3.3, we give the following definition.

Definition 3.7. We call a solution (u, v) *unimodal*, *bimodal* or *trimodal* if it involves one, two or three distinct eigenvectors, that is, if $\alpha_n \neq 0$ (and so $\gamma_n \neq 0$) for one, two or three indexes n, respectively.

4. Unimodal Solutions

We now focus on unimodal solutions. More precisely, we look for solutions (u, v) of the form

$$\begin{cases} u = \alpha_n e_n, \\ v = \gamma_n e_n, \end{cases} \tag{4.1}$$

for a fixed $n \in \mathbb{N}$ and some coefficients $\alpha_n, \gamma_n \neq 0$. In order to classify such solutions, we introduce the positive sequences[‡]

$$\mu_n = \frac{2k}{\lambda_n} + \lambda_n \qquad \text{and} \qquad \nu_n = \frac{3k}{\lambda_n} + \lambda_n,$$

along with the (disjoint) subsets of $\mathbb{E}$

$$\begin{aligned} \mathbb{E}_1 &= \{n : \lambda_n < -\beta \leq \mu_n\}, \\ \mathbb{E}_2 &= \{n : \mu_n < -\beta \leq \nu_n\}, \\ \mathbb{E}_3 &= \{n : \nu_n < -\beta\}. \end{aligned}$$

Clearly,

$$\mathbb{E}_1 \cup \mathbb{E}_2 \cup \mathbb{E}_3 = \mathbb{E}.$$

Then, we consider the real numbers (whenever defined)

$$\begin{cases} \alpha^{\pm}_{n,1} = \pm\sqrt{\dfrac{-\beta - \lambda_n}{\varrho\lambda_n}}, \\ \alpha^{\pm}_{n,2} = \pm\sqrt{\dfrac{-\beta - \mu_n}{\varrho\lambda_n}}, \\ \alpha^{\pm}_{n,3} = \pm\sqrt{\dfrac{-\beta + \mu_n - \nu_n - \lambda_n + \sqrt{(\beta + \lambda_n + \mu_n - \nu_n)(\beta + \nu_n)}}{2\varrho\lambda_n}}, \\ \alpha^{\pm}_{n,4} = \pm\sqrt{\dfrac{-\beta + \mu_n - \nu_n - \lambda_n - \sqrt{(\beta + \lambda_n + \mu_n - \nu_n)(\beta + \nu_n)}}{2\varrho\lambda_n}}, \end{cases} \tag{4.2}$$

hereafter called *unimodal amplitudes*, or U-*amplitudes* for brevity. By elementary calculations, one can easily verify that

$$\begin{aligned} \alpha^{\pm}_{n,1} \in \mathbb{R} \quad &\Leftrightarrow \quad \lambda_n \leq -\beta, \\ \alpha^{\pm}_{n,2} \in \mathbb{R} \quad &\Leftrightarrow \quad \mu_n \leq -\beta, \\ \alpha^{\pm}_{n,3} \in \mathbb{R} \quad &\Leftrightarrow \quad \nu_n \leq -\beta, \\ \alpha^{\pm}_{n,4} \in \mathbb{R} \quad &\Leftrightarrow \quad \nu_n \leq -\beta. \end{aligned}$$

Lemma 4.1. *For every fixed $n \in \mathbb{N}$, let us consider the set*

$$\Gamma_n = \{\alpha^{\pm}_{n,i} : i = 1, 2, 3, 4\}.$$

Then, Γ_n contains exactly

- *2 distinct nontrivial* U-*amplitudes $\{\alpha^{\pm}_{n,1}\}$ if $n \in \mathbb{E}_1$;*
- *4 distinct nontrivial* U-*amplitudes $\{\alpha^{\pm}_{n,1}, \alpha^{\pm}_{n,2}\}$ if $n \in \mathbb{E}_2$;*
- *8 distinct nontrivial* U-*amplitudes $\{\alpha^{\pm}_{n,1}, \alpha^{\pm}_{n,2}, \alpha^{\pm}_{n,3}, \alpha^{\pm}_{n,4}\}$ if $n \in \mathbb{E}_3$.*

[‡]Observe that $\lambda_n < \mu_n < \nu_n$.

If $n \notin \mathbb{E}$, the set Γ_n is either empty or it contains exactly the (trivial) u*-amplitudes* $\alpha^+_{n,1} = \alpha^-_{n,1} = 0$.

Proof. We analyze separately all the possible cases.

- If $n \in \mathbb{E}_1$, there are only two distinct nontrivial u-amplitudes, that is, $\alpha^\pm_{n,1}$. Indeed, when $\mu_n = -\beta$,

$$\alpha^\pm_{n,2} = 0.$$

- If $n \in \mathbb{E}_2$, there are only four distinct nontrivial u-amplitudes, that is, $\alpha^\pm_{n,1}$ and $\alpha^\pm_{n,2}$. Indeed, when $\nu_n = -\beta$,

$$\alpha^+_{n,3} = \alpha^+_{n,4} = \alpha^+_{n,2} \qquad \text{and} \qquad \alpha^-_{n,3} = \alpha^-_{n,4} = \alpha^-_{n,2}.$$

- If $n \in \mathbb{E}_3$, all the eight u-amplitudes $\alpha^\pm_{n,i}$ are distinct and nontrivial.

If $n \notin \mathbb{E}$, all the u-amplitudes $\alpha^\pm_{n,i}$, whenever defined, are trivial. In particular, the only two allowed amplitudes are $\alpha^+_{n,1} = \alpha^-_{n,1} = 0$. □

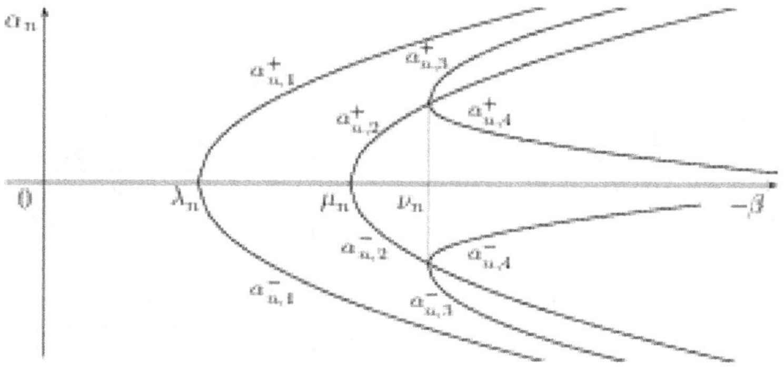

Figure 2. The u-amplitudes $\alpha^\pm_{n,i}$ for a fixed $n \in \mathbb{N}$.

We are now in a position to state our main result on unimodal solutions.

Theorem 4.2. *System* (2.1) *admits nontrivial unimodal solutions if and only if the set $\mathbb{E}$ is nonempty. More precisely, for every $n \in \mathbb{N}$, one of the following disjoint situations occurs.*

- *If $n \in \mathbb{E}_1$, we have exactly 2 nontrivial unimodal solutions of the form*

$$(u, v) = \begin{cases} (\alpha^+_{n,1}\, e_n,\ \alpha^+_{n,1}\, e_n) \\ (\alpha^-_{n,1}\, e_n,\ \alpha^-_{n,1}\, e_n). \end{cases}$$

- *If $n \in \mathbb{E}_2$, we have exactly 4 nontrivial unimodal solutions of the form*

$$(u, v) = \begin{cases} (\alpha^+_{n,1}\, e_n,\ \alpha^+_{n,1}\, e_n) \\ (\alpha^-_{n,1}\, e_n,\ \alpha^-_{n,1}\, e_n) \\ (\alpha^+_{n,2}\, e_n,\ \alpha^-_{n,2}\, e_n) \\ (\alpha^-_{n,2}\, e_n,\ \alpha^+_{n,2}\, e_n). \end{cases}$$

- *If $n \in \mathbb{E}_3$, we have exactly 8 nontrivial unimodal solutions of the form*

$$(u,v)=\begin{cases}(\alpha^+_{n,1}\,e_n,\ \alpha^+_{n,1}\,e_n)\\(\alpha^-_{n,1}\,e_n,\ \alpha^-_{n,1}\,e_n)\\(\alpha^+_{n,2}\,e_n,\ \alpha^-_{n,2}\,e_n)\\(\alpha^-_{n,2}\,e_n,\ \alpha^+_{n,2}\,e_n)\\(\alpha^+_{n,3}\,e_n,\ \alpha^-_{n,4}\,e_n)\\(\alpha^-_{n,3}\,e_n,\ \alpha^+_{n,4}\,e_n)\\(\alpha^+_{n,4}\,e_n,\ \alpha^-_{n,3}\,e_n)\\(\alpha^-_{n,4}\,e_n,\ \alpha^+_{n,3}\,e_n).\end{cases}$$

- *If $n \notin \mathbb{E}$, all the unimodal solutions involving the eigenvector e_n are trivial.*

In summary, system (2.1) *admits* $2|\mathbb{E}_1| + 4|\mathbb{E}_2| + 8|\mathbb{E}_3|$ *nontrivial unimodal solutions.*

Proof. Let us look for nontrivial solutions (u, v) of the form (4.1). Choosing $\phi = \psi = e_n$ in the weak formulation (2.3) and recalling (3.4), we obtain the system

$$\begin{cases}\lambda_n^2\alpha_n + (\beta + \varrho\lambda_n\alpha_n^2)\lambda_n\alpha_n + k(\alpha_n - \gamma_n) = 0,\\ \lambda_n^2\gamma_n + (\beta + \varrho\lambda_n\gamma_n^2)\lambda_n\gamma_n - k(\alpha_n - \gamma_n) = 0,\end{cases}$$

which, setting

$$\eta_n = 1 + \frac{\beta}{\lambda_n} + \frac{k}{\lambda_n^2} \qquad \text{and} \qquad \omega_n = \frac{\lambda_n^2}{k},$$

can be rewritten as

$$\begin{cases}\gamma_n = \omega_n\alpha_n(\eta_n + \varrho\alpha_n^2),\\ \alpha_n = \omega_n\gamma_n(\eta_n + \varrho\gamma_n^2).\end{cases} \tag{4.3}$$

Solving with respect to α_n, we arrive at the nine-order equation

$$\alpha_n(\varrho^4\alpha_n^8\omega_n^4 + 3\varrho^3\alpha_n^6\omega_n^4\eta_n + 3\varrho^2\alpha_n^4\omega_n^4\eta_n^2 + \varrho\alpha_n^2\omega_n^4\eta_n^3 + \varrho\alpha_n^2\omega_n^2\eta_n + \omega_n^2\eta_n^2 - 1) = 0.$$

If $\alpha_n = 0$ the solution is trivial (since in this case also γ_n is zero). Otherwise, introducing the auxiliary variable

$$x_n = \omega_n(\eta_n + \varrho\alpha_n^2),$$

we end up with

$$(x_n^2 - 1)(x_n^2 - x_n\omega_n\eta_n + 1) = 0.$$

Making use of the relations

$$\begin{cases}\omega_n\eta_n = -\dfrac{\lambda_n}{k}(-\beta + \mu_n - \nu_n - \lambda_n),\\[2mm] \omega_n^2\eta_n^2 - 4 = \dfrac{\lambda_n^2}{k^2}(\beta + \lambda_n + \mu_n - \nu_n)(\beta + \nu_n),\end{cases} \tag{4.4}$$

one can easily realize that the solutions are the u-amplitudes $\alpha^\pm_{n,i}$ given by (4.2). Hence, according to Lemma 4.1, we have exactly

- 2 distinct nontrivial solutions $\{\alpha^{\pm}_{n,1}\}$ for every $n \in \mathbb{E}_1$;
- 4 distinct nontrivial solutions $\{\alpha^{\pm}_{n,1}, \alpha^{\pm}_{n,2}\}$ for every $n \in \mathbb{E}_2$;
- 8 distinct nontrivial solutions $\{\alpha^{\pm}_{n,1}, \alpha^{\pm}_{n,2}, \alpha^{\pm}_{n,3}, \alpha^{\pm}_{n,4}\}$ for every $n \in \mathbb{E}_3$.

By the same token, when $n \notin \mathbb{E}$, we have only the trivial solution. We are left to find the explicit values $\gamma^{\pm}_{n,i}$, which can be obtained from (4.3). To this end, it is apparent to see that

$$\begin{cases} \gamma^{\pm}_{n,1} = \alpha^{\pm}_{n,1}, \\ \gamma^{\pm}_{n,2} = \alpha^{\mp}_{n,2}. \end{cases}$$

Moreover, invoking (4.4) and observing that the product $\omega_n\eta_n$ is negative when $n \in \mathbb{E}_3$,

$$\begin{aligned} \gamma^{\pm}_{n,3} &= \pm\frac{\sqrt{k}(\omega_n\eta_n + \sqrt{\omega_n^2\eta_n^2 - 4})}{2}\sqrt{\frac{-\omega_n\eta_n + \sqrt{\omega_n^2\eta_n^2 - 4}}{2\varrho\lambda_n^2}} \\ &= \mp\sqrt{k}\sqrt{\frac{-\omega_n\eta_n - \sqrt{\omega_n^2\eta_n^2 - 4}}{2\varrho\lambda_n^2}} = \alpha^{\mp}_{n,4}, \end{aligned}$$

and

$$\begin{aligned} \gamma^{\pm}_{n,4} = &= \pm\frac{\sqrt{k}(\omega_n\eta_n - \sqrt{\omega_n^2\eta_n^2 - 4})}{2}\sqrt{\frac{-\omega_n\eta_n - \sqrt{\omega_n^2\eta_n^2 - 4}}{2\varrho\lambda_n^2}} \\ &= \mp\sqrt{k}\sqrt{\frac{-\omega_n\eta_n + \sqrt{\omega_n^2\eta_n^2 - 4}}{2\varrho\lambda_n^2}} = \alpha^{\mp}_{n,3}. \end{aligned}$$

The theorem is proved. □

5. Equidistributed Energy Solutions

In order to investigate the existence of solutions to system (2.1) which are not necessarily unimodal, we begin to analyze a particular but still very interesting situation.

Definition 5.1. A nontrivial solution (u, v) is called an *equidistributed energy solution* (EE-*solution* for brevity) if

$$\|u\|_1 = \|v\|_1 \quad \Leftrightarrow \quad C_u = C_v. \tag{5.1}$$

At first glance, this condition might look restrictive. Though, as we will see in the next two lemmas, EE-solutions are in fact quite general. In particular, they pop up whenever a mode of u is equal or opposite to the corresponding mode of v.

Lemma 5.2. *With reference to* (3.3), *if*

$$\alpha_{n_i}\alpha_{n_j} = \pm\gamma_{n_i}\gamma_{n_j} \neq 0$$

for some (possibly coinciding) n_i, n_j, then (u, v) is an EE-*solution. In particular, this is the case when*[§]

$$|\alpha_{n_i}| = |\gamma_{n_i}| \neq 0$$

for some n_i.

[§]In fact, we will implicitly show in our analysis that the latter condition is necessary as well in order to have EE-solutions.

Proof. Let n_i, n_j be such that

$$\alpha_{n_i}\alpha_{n_j} = \pm\gamma_{n_i}\gamma_{n_j} \neq 0.$$

Choosing $\phi = \psi = e_{n_i}$ in the weak formulation (2.3), we obtain

$$\begin{cases} \lambda_{n_i}^2\alpha_{n_i} + C_u\lambda_{n_i}\alpha_{n_i} + k(\alpha_{n_i} - \gamma_{n_i}) = 0, \\ \lambda_{n_i}^2\gamma_{n_i} + C_v\lambda_{n_i}\gamma_{n_i} - k(\alpha_{n_i} - \gamma_{n_i}) = 0, \end{cases} \tag{5.2}$$

while, choosing $\phi = \psi = e_{n_j}$, we get

$$\begin{cases} \lambda_{n_j}^2\alpha_{n_j} + C_u\lambda_{n_j}\alpha_{n_j} + k(\alpha_{n_j} - \gamma_{n_j}) = 0, \\ \lambda_{n_j}^2\gamma_{n_j} + C_v\lambda_{n_j}\gamma_{n_j} - k(\alpha_{n_j} - \gamma_{n_j}) = 0. \end{cases} \tag{5.3}$$

Then, from (5.2),

$$\begin{cases} C_u = -\lambda_{n_i} - \dfrac{k(\alpha_{n_i} - \gamma_{n_i})}{\lambda_{n_i}\alpha_{n_i}}, \\ C_v = -\lambda_{n_i} + \dfrac{k(\alpha_{n_i} - \gamma_{n_i})}{\lambda_{n_i}\gamma_{n_i}}. \end{cases}$$

These expressions, substituted into (5.3), yield

$$\begin{cases} \lambda_{n_j}^2\lambda_{n_i}\alpha_{n_i}\alpha_{n_j} - \lambda_{n_i}^2\lambda_{n_j}\alpha_{n_i}\alpha_{n_j} - k\lambda_{n_j}\alpha_{n_j}(\alpha_{n_i} - \gamma_{n_i}) + k\lambda_{n_i}\alpha_{n_i}(\alpha_{n_j} - \gamma_{n_j}) = 0, \\ \lambda_{n_j}^2\lambda_{n_i}\gamma_{n_i}\gamma_{n_j} - \lambda_{n_i}^2\lambda_{n_j}\gamma_{n_i}\gamma_{n_j} + k\lambda_{n_j}\gamma_{n_j}(\alpha_{n_i} - \gamma_{n_i}) - k\lambda_{n_i}\gamma_{n_i}(\alpha_{n_j} - \gamma_{n_j}) = 0. \end{cases}$$

If

$$\alpha_{n_i}\alpha_{n_j} = \gamma_{n_i}\gamma_{n_j} \neq 0,$$

subtracting the two equations of the system above we readily find

$$|\alpha_{n_i}| = |\gamma_{n_i}|.$$

On the other hand, if

$$\alpha_{n_i}\alpha_{n_j} = -\gamma_{n_i}\gamma_{n_j} \neq 0,$$

(implying $n_i \neq n_j$), adding the two equations of the system we still conclude that

$$|\alpha_{n_i}| = |\gamma_{n_i}|.$$

At this point, an exploitation of (5.2) gives $C_u = C_v$. □

Lemma 5.3. *With reference to* (3.3), *if*

$$\alpha_{n_i}\gamma_{n_j} = \alpha_{n_j}\gamma_{n_i} \neq 0$$

for some $n_i \neq n_j$, then (u, v) is an EE-*solution.*

Proof. By assumption, there exists $\varpi \neq 0$ such that

$$\alpha_{n_i} = \varpi\gamma_{n_i} \qquad \text{and} \qquad \alpha_{n_j} = \varpi\gamma_{n_j}.$$

Due to Lemma 5.2, to reach the conclusion it is sufficient to show that $\varpi = -1$. If not, exploiting (3.5),

$$\lambda_{n_i} = -\frac{C_u\alpha_{n_i} + C_v\gamma_{n_i}}{\alpha_{n_i} + \gamma_{n_i}} = -\frac{C_u\varpi + C_v}{\varpi + 1} = -\frac{C_u\alpha_{n_j} + C_v\gamma_{n_j}}{\alpha_{n_j} + \gamma_{n_j}} = \lambda_{n_j},$$

yielding a contradiction. □

We now proceed with a detailed description of the class of EE-solutions.

5.1. The unimodal case

The unimodal solutions have been already classified in the previous section. In particular, from Theorem 4.2 we learn that all unimodal solutions, except the ones involving the U-amplitudes $\alpha^{\pm}_{n,3}$ and $\alpha^{\pm}_{n,4}$ arising from the further bifurcation at $\nu_n = -\beta$, are in fact EE-solutions. That is, system (2.1) admits

$$2|\mathbb{E}_1| + 4|\mathbb{E}_2| + 4|\mathbb{E}_3|$$

unimodal EE-solutions, explicitly computed.

5.2. The bimodal case

In order to classify the bimodal EE-solutions, we introduce the (disjoint and possibly empty) subsets of $\mathbb{E} \times \mathbb{E}$

$$\mathbb{B}_1 = \{(n_1, n_2) : n_1 < n_2,\ \lambda_{n_1} + \lambda_{n_2} < -\beta \text{ and } \lambda_{n_1}\lambda_{n_2} = 2k\}$$

and

$$\mathbb{B}_2 = \{(n_1, n_2) : n_1 < n_2,\ \lambda_{n_2} < -\beta \text{ and } \lambda_{n_1}(\lambda_{n_2} - \lambda_{n_1}) = 2k\}.$$

Then, setting

$$\mathbb{B} = \mathbb{B}_1 \cup \mathbb{B}_2,$$

we have the following result.

Theorem 5.4. *System* (2.1) *admits bimodal* EE*-solutions if and only if the set* $\mathbb{B}$ *is nonempty. More precisely, for every couple* $(n_1, n_2) \in \mathbb{N}\times\mathbb{N}$ *with* $n_1 < n_2$*, one of the following disjoint situations occurs.*

- *If* $(n_1, n_2) \in \mathbb{B}_1$*, we have exactly the (infinitely many) solutions of the form*

$$\begin{cases} u = xe_{n_1} + ye_{n_2}, \\ v = -xe_{n_1} - ye_{n_2}, \end{cases}$$

  for all $(x, y) \in \mathbb{R}^2$ *satisfying the equality*

$$\varrho x^2\lambda_{n_1} + \varrho y^2\lambda_{n_2} + \lambda_{n_1} + \lambda_{n_2} + \beta = 0 \qquad \textit{with} \qquad xy \neq 0.$$

- *If* $(n_1, n_2) \in \mathbb{B}_2$*, we have exactly the (infinitely many) solutions of the form*

$$\begin{cases} u = xe_{n_1} + ye_{n_2}, \\ v = -xe_{n_1} + ye_{n_2}, \end{cases}$$

  for all $(x, y) \in \mathbb{R}^2$ *satisfying the equality*

$$\varrho x^2\lambda_{n_1} + \varrho y^2\lambda_{n_2} + \lambda_{n_2} + \beta = 0 \qquad \textit{with} \qquad xy \neq 0.$$

- *If* $(n_1, n_2) \notin \mathbb{B}$*, there are no bimodal* EE*-solutions involving the eigenvectors* e_{n_1} *and* e_{n_2}*.*

Proof. Let us look for bimodal EE-solutions (u, v) of the form

$$\begin{cases} u = \alpha_{n_1} e_{n_1} + \alpha_{n_2} e_{n_2}, \\ v = \gamma_{n_1} e_{n_1} + \gamma_{n_2} e_{n_2}, \end{cases}$$

with $n_1 < n_2 \in \mathbb{N}$ and $\alpha_{n_i}, \gamma_{n_i} \in \mathbb{R} \setminus \{0\}$. Choosing $\phi = \psi = e_{n_1}$ in the weak formulation (2.3), we obtain

$$\begin{cases} \lambda_{n_1}^2 \alpha_{n_1} + C_u \lambda_{n_1} \alpha_{n_1} + k(\alpha_{n_1} - \gamma_{n_1}) = 0, \\ \lambda_{n_1}^2 \gamma_{n_1} + C_v \lambda_{n_1} \gamma_{n_1} - k(\alpha_{n_1} - \gamma_{n_1}) = 0, \end{cases}$$

while, choosing $\phi = \psi = e_{n_2}$, we get

$$\begin{cases} \lambda_{n_2}^2 \alpha_{n_2} + C_u \lambda_{n_2} \alpha_{n_2} + k(\alpha_{n_2} - \gamma_{n_2}) = 0, \\ \lambda_{n_2}^2 \gamma_{n_2} + C_v \lambda_{n_2} \gamma_{n_2} - k(\alpha_{n_2} - \gamma_{n_2}) = 0. \end{cases}$$

Since we require $C_u = C_v$, we infer that

$$C_u = -\lambda_{n_1} - \frac{k(\alpha_{n_1} - \gamma_{n_1})}{\lambda_{n_1} \alpha_{n_1}}, \tag{5.4}$$

$$C_u = -\lambda_{n_1} + \frac{k(\alpha_{n_1} - \gamma_{n_1})}{\lambda_{n_1} \gamma_{n_1}}, \tag{5.5}$$

$$C_u = -\lambda_{n_2} - \frac{k(\alpha_{n_2} - \gamma_{n_2})}{\lambda_{n_2} \alpha_{n_2}}, \tag{5.6}$$

$$C_u = -\lambda_{n_2} + \frac{k(\alpha_{n_2} - \gamma_{n_2})}{\lambda_{n_2} \gamma_{n_2}}. \tag{5.7}$$

At this point, we shall distinguish three cases.

◇ When

$$\begin{cases} \gamma_{n_1} + \alpha_{n_1} = 0, \\ \gamma_{n_2} + \alpha_{n_2} = 0, \end{cases}$$

equations (5.4)-(5.7) reduce to

$$\begin{cases} \lambda_{n_1} C_u = -\lambda_{n_1}^2 - 2k, \\ \lambda_{n_2} C_u = -\lambda_{n_2}^2 - 2k, \end{cases}$$

implying

$$\lambda_{n_1} \lambda_{n_2} = 2k.$$

Moreover, the value C_u is determined by (3.4), which provides the equality

$$\varrho \alpha_{n_1}^2 \lambda_{n_1} + \varrho \alpha_{n_2}^2 \lambda_{n_2} + \lambda_{n_1} + \lambda_{n_2} + \beta = 0.$$

Hence, there exist bimodal EE-solutions (explicitly computed) if and only if the pair $(n_1, n_2) \in \mathbb{B}_1$.

◇ When

$$\begin{cases} \gamma_{n_1} + \alpha_{n_1} = 0, \\ \gamma_{n_2} + \alpha_{n_2} \neq 0, \end{cases}$$

we take the difference of (5.7) and (5.6), establishing the identity

$$\gamma_{n_2} = \alpha_{n_2}.$$

Thus, equations (5.4)-(5.7) reduce to

$$\begin{cases} \lambda_{n_1} C_u = -\lambda_{n_1}^2 - 2k, \\ C_u = -\lambda_{n_2}, \end{cases}$$

implying

$$\lambda_{n_1}(\lambda_{n_2} - \lambda_{n_1}) = 2k.$$

Again, the value C_u is determined by (3.4), which gives

$$\varrho\alpha_{n_1}^2\lambda_{n_1} + \varrho\alpha_{n_2}^2\lambda_{n_2} + \lambda_{n_2} + \beta = 0.$$

Hence, there exist bimodal EE-solutions (explicitly computed) if and only if the pair $(n_1, n_2) \in \mathbb{B}_2$.

⋄ We show that the remaining case

$$\gamma_{n_1} + \alpha_{n_1} \neq 0$$

is impossible. Indeed, taking the difference of (5.5) and (5.4), we find

$$\gamma_{n_1} = \alpha_{n_1}.$$

If $\gamma_{n_2} + \alpha_{n_2} = 0$, from (5.4) and (5.6) we conclude that

$$0 < 2k = \lambda_{n_2}(\lambda_{n_1} - \lambda_{n_2}) < 0,$$

yielding a contradiction. On the other hand, if $\gamma_{n_2} + \alpha_{n_2} \neq 0$, we learn once more that

$$\gamma_{n_2} = \alpha_{n_2}.$$

But in this situation, equations (5.4) and (5.6) lead to $\lambda_{n_1} = \lambda_{n_2}$, and the sought contradiction follows. □

5.3. The trimodal case

Finally, we classify the trimodal EE-solutions. To this end, we consider the (possibly empty) subset of $\mathbb{E} \times \mathbb{E} \times \mathbb{E}$

$$\mathbb{T} = \{(n_1, n_2, n_3) : n_1 < n_2 < n_3,\ \lambda_{n_3} < -\beta \text{ and } \lambda_{n_1}(\lambda_{n_3} - \lambda_{n_1}) = \lambda_{n_2}(\lambda_{n_3} - \lambda_{n_2}) = 2k\}.$$

The result reads as follows.

Theorem 5.5. *System* (2.1) *admits trimodal* EE*-solutions if and only if the set* $\mathbb{T}$ *is nonempty. More precisely, for every triplet* $(n_1, n_2, n_3) \in \mathbb{N} \times \mathbb{N} \times \mathbb{N}$ *with* $n_1 < n_2 < n_3$*, one of the following disjoint situations occurs.*

- *If* $(n_1, n_2, n_3) \in \mathbb{T}$, *we have exactly the (infinitely many) solutions of the form*

$$\begin{cases} u = xe_{n_1} + ye_{n_2} + ze_{n_3}, \\ v = -xe_{n_1} - ye_{n_2} + ze_{n_3}, \end{cases}$$

for all $(x, y, z) \in \mathbb{R}^3$ *satisfying the equality*

$$\varrho x^2 \lambda_{n_1} + \varrho y^2 \lambda_{n_2} + \varrho z^2 \lambda_{n_3} + \lambda_{n_3} + \beta = 0 \qquad \text{with} \qquad xyz \neq 0.$$

- *If* $(n_1, n_2, n_3) \notin \mathbb{T}$, *there are no trimodal* EE-*solutions involving the eigenvectors* $e_{n_1}, e_{n_2}, e_{n_3}$.

Proof. The argument goes along the same lines of Theorem 5.4. For this reason, we limit ourselves to give a short (albeit complete) proof, leaving the verification of some calculations to the reader.

As customary, let us look for trimodal EE-solutions (u, v) of the form

$$\begin{cases} u = \alpha_{n_1} e_{n_1} + \alpha_{n_2} e_{n_2} + \alpha_{n_3} e_{n_3}, \\ v = \gamma_{n_1} e_{n_1} + \gamma_{n_2} e_{n_2} + \gamma_{n_3} e_{n_3}, \end{cases}$$

with $n_1 < n_2 < n_3 \in \mathbb{N}$ and $\alpha_{n_i}, \gamma_{n_i} \in \mathbb{R} \setminus \{0\}$. Accordingly, from the weak formulation (2.3), choosing first $\phi = \psi = e_{n_1}$, then $\phi = \psi = e_{n_2}$, and finally $\phi = \psi = e_{n_3}$, we obtain the six equations

$$\begin{cases} C_u = -\lambda_{n_1} - \dfrac{k(\alpha_{n_1} - \gamma_{n_1})}{\lambda_{n_1}\alpha_{n_1}}, \\ C_u = -\lambda_{n_1} + \dfrac{k(\alpha_{n_1} - \gamma_{n_1})}{\lambda_{n_1}\gamma_{n_1}}, \\ C_u = -\lambda_{n_2} - \dfrac{k(\alpha_{n_2} - \gamma_{n_2})}{\lambda_{n_2}\alpha_{n_2}}, \\ C_u = -\lambda_{n_2} + \dfrac{k(\alpha_{n_2} - \gamma_{n_2})}{\lambda_{n_2}\gamma_{n_2}}, \\ C_u = -\lambda_{n_3} - \dfrac{k(\alpha_{n_3} - \gamma_{n_3})}{\lambda_{n_3}\alpha_{n_3}}, \\ C_u = -\lambda_{n_3} + \dfrac{k(\alpha_{n_3} - \gamma_{n_3})}{\lambda_{n_3}\gamma_{n_3}}, \end{cases} \tag{5.8}$$

where the condition $C_u = C_v$ has been used. The next step is to show that

$$\begin{cases} \gamma_{n_1} + \alpha_{n_1} = 0, \\ \gamma_{n_2} + \alpha_{n_2} = 0, \\ \gamma_{n_3} + \alpha_{n_3} \neq 0, \end{cases} \tag{5.9}$$

being the remaining cases impossible. To prove the claim, the argument is similar to the one of Theorem 5.4. For instance, assuming

$$\begin{cases} \gamma_{n_1} + \alpha_{n_1} = 0, \\ \gamma_{n_2} + \alpha_{n_2} = 0, \\ \gamma_{n_3} + \alpha_{n_3} = 0, \end{cases}$$

system (5.8) reduces to

$$\begin{cases} \lambda_{n_1} C_u = -\lambda_{n_1}^2 - 2k, \\ \lambda_{n_2} C_u = -\lambda_{n_2}^2 - 2k, \\ \lambda_{n_3} C_u = -\lambda_{n_3}^2 - 2k, \end{cases}$$

forcing

$$2k = \lambda_{n_1}\lambda_{n_2} = \lambda_{n_2}\lambda_{n_3}$$

and yielding a contradiction. The other cases can be carried out analogously; the details are left to the reader. Within (5.9), we take the difference of the last two equations of (5.8), and we obtain

$$\gamma_{n_3} = \alpha_{n_3}.$$

Thus, system (5.8) turns into

$$\begin{cases} \lambda_{n_1} C_u = -\lambda_{n_1}^2 - 2k, \\ \lambda_{n_2} C_u = -\lambda_{n_2}^2 - 2k, \\ C_u = -\lambda_{n_3}, \end{cases}$$

implying

$$\lambda_{n_1}(\lambda_{n_3} - \lambda_{n_1}) = \lambda_{n_2}(\lambda_{n_3} - \lambda_{n_2}) = 2k.$$

Moreover, the value C_u is determined by (3.4), which provides the equality

$$\varrho\alpha_{n_1}^2\lambda_{n_1} + \varrho\alpha_{n_2}^2\lambda_{n_2} + \varrho\alpha_{n_3}^2\lambda_{n_3} + \lambda_{n_3} + \beta = 0.$$

Hence, there exist trimodal EE-solutions (explicitly computed) if and only if the triplet $(n_1, n_2, n_3) \in \mathbb{T}$. □

Corollary 5.6. *Let (u, v) be a trimodal* EE*-solution. Then, with reference to* (3.3), *if $n_1 < n_2 < n_3$ the eigenvalues $\lambda_{n_1}, \lambda_{n_2}, \lambda_{n_3}$ fulfill the relation*

$$\lambda_{n_1} + \lambda_{n_2} = \lambda_{n_3}.$$

Proof. In the light of Theorem 5.5, we know that $(n_1, n_2, n_3) \in \mathbb{T}$. In particular,

$$\lambda_{n_1}(\lambda_{n_3} - \lambda_{n_1}) = \lambda_{n_2}(\lambda_{n_3} - \lambda_{n_2}).$$

Since $\lambda_{n_1} \neq \lambda_{n_2}$, the conclusion follows. □

6. General Bimodal Solutions

In this section, we investigate the existence of general (not necessarily equidistributed) bimodal solutions to system (2.1). First, specializing Lemmas 5.2 and 5.3, we obtain

Theorem 6.1. *Let (u, v) be a bimodal solution. With reference to* (3.3), *if*

- $|\alpha_{n_1}| = |\gamma_{n_1}| \neq 0$, *or*

- $|\alpha_{n_2}| = |\gamma_{n_2}| \neq 0$, *or*
- $\alpha_{n_1}\alpha_{n_2} = \pm\gamma_{n_1}\gamma_{n_2} \neq 0$, *or*
- $\alpha_{n_1}\gamma_{n_2} = \alpha_{n_2}\gamma_{n_1} \neq 0$,

then (u, v) *is an* EE-*solution.*

Even if Theorem 6.1 somehow tells that a bimodal solution is likely to be an EE-solution, it is possible to have bimodal solutions of not equidistributed energy. Indeed, the complete picture will be given in the next Theorem 6.8 of §6.4. Some preparatory work is needed.

6.1. Technical lemmas

In what follows, $(n_1, n_2) \in \mathbb{N} \times \mathbb{N}$ is an arbitrary, but fixed, pair of natural numbers, with $n_1 < n_2$. We will introduce several quantities depending on (n_1, n_2). Setting

$$\zeta = \zeta(n_1, n_2) = \frac{\lambda_{n_2}}{\lambda_{n_1}} > 1, \tag{6.1}$$

and

$$\sigma = \sigma(n_1, n_2) = \frac{k - \lambda_{n_1}\lambda_{n_2}}{k} \in \mathbb{R}, \tag{6.2}$$

we consider the real numbers (defined whenever $\sigma \neq 0$)

$$\Phi = \Phi(n_1, n_2) = \frac{(\zeta + 1) + (\zeta - 1)\sigma^2}{\sigma\zeta},$$

and

$$\Psi = \Psi(n_1, n_2) = \frac{(\zeta + 1) - (\zeta - 1)\sigma^2}{\sigma}.$$

By direct computations, we have the identity

$$\Phi^2\zeta^2 - \Psi^2 = 4(\zeta^2 - 1),$$

which, in turn, yields

$$(\Phi^2 - 4)\zeta^2 = \Psi^2 - 4 = \frac{(\zeta - 1)^2\sigma^4 - 2(\zeta^2 + 1)\sigma^2 + (\zeta + 1)^2}{\sigma^2}. \tag{6.3}$$

This relation will be useful later. Then, we introduce the real numbers (whenever defined)

$$X = X(n_1, n_2) = \frac{\Phi + \sqrt{\Phi^2 - 4}}{2},$$

$$Y = Y(n_1, n_2) = \frac{\Phi - \sqrt{\Phi^2 - 4}}{2},$$

$$W = W(n_1, n_2) = \frac{\Psi + \sqrt{\Psi^2 - 4}}{2},$$

$$Z = Z(n_1, n_2) = \frac{\Psi - \sqrt{\Psi^2 - 4}}{2}.$$

Lemma 6.2. *The following are equivalent.*

- *At least one of the numbers X, Y, W, Z belongs to $\mathbb{R}$.*
- *All the numbers X, Y, W, Z belong to $\mathbb{R}$.*
- $\lambda_{n_1}\lambda_{n_2} \in (0, 2k] \setminus \{k\}$ *or* $\lambda_{n_1}(\lambda_{n_2} - \lambda_{n_1}) \in [2k, \infty)$.

Proof. It is apparent to see that

$$X \in \mathbb{R} \quad \Leftrightarrow \quad \Phi^2 \geq 4 \quad \Leftrightarrow \quad Y \in \mathbb{R},$$

and

$$W \in \mathbb{R} \quad \Leftrightarrow \quad \Psi^2 \geq 4 \quad \Leftrightarrow \quad Z \in \mathbb{R}.$$

Moreover, in the light of (6.3),

$$\Phi^2 \geq 4 \quad \Leftrightarrow \quad \Psi^2 \geq 4.$$

Therefore, in order to reach the conclusion, it is sufficient to show that

$$\Psi^2 \geq 4 \quad \Leftrightarrow \quad \lambda_{n_1}\lambda_{n_2} \in (0, 2k] \setminus \{k\} \quad \text{or} \quad \lambda_{n_1}(\lambda_{n_2} - \lambda_{n_1}) \in [2k, \infty).$$

To this end, exploiting (6.3),

$$\Psi^2 \geq 4 \quad \Leftrightarrow \quad \begin{cases} \lambda_{n_1}\lambda_{n_2} \neq k, \\ (\zeta - 1)^2\sigma^4 - 2(\zeta^2 + 1)\sigma^2 + (\zeta + 1)^2 \geq 0. \end{cases}$$

Making use of the trivial inequality $\sigma < 1$, one can verify by elementary calculations that

$$(\zeta - 1)^2\sigma^4 - 2(\zeta^2 + 1)\sigma^2 + (\zeta + 1)^2 \geq 0$$

if and only if

$$\sigma \in \left(-\infty, \frac{\zeta + 1}{1 - \zeta} \right] \cup [-1, 1).$$

Since

$$\sigma \in \left(-\infty, \frac{\zeta + 1}{1 - \zeta} \right] \quad \Leftrightarrow \quad \lambda_{n_1}(\lambda_{n_2} - \lambda_{n_1}) \in [2k, \infty),$$

and

$$\sigma \in [-1, 1) \quad \Leftrightarrow \quad \lambda_{n_1}\lambda_{n_2} \in (0, 2k] \setminus \{k\},$$

the proof is finished. □

Lemma 6.3. *The following are equivalent.*

- $X = Y$.
- $W = Z$.
- $\lambda_{n_1}\lambda_{n_2} = 2k$ *or* $\lambda_{n_1}(\lambda_{n_2} - \lambda_{n_1}) = 2k$.

The argument goes along the same lines of Lemma 6.2 (actually, it is even simpler). For this reason, the proof is omitted and left to the reader.

At this point, we state a simple but crucial identity, which follows immediately from (6.3) and the definitions of the numbers $\zeta, \Phi, \Psi, X, Y, W, Z$.

Lemma 6.4. *We have the equality*

$$\zeta X - W = \zeta Y - Z = (\zeta - 1)\sigma, \tag{6.4}$$

provided that the expressions above are well-defined.

6.2. The numbers $\mathfrak{m}$ *and* $\mathfrak{M}$

A crucial role in our analysis will be played by the following two real numbers (again, defined whenever $\sigma \neq 0$)

$$\mathfrak{m} = \mathfrak{m}(n_1, n_2) = \frac{k^2 + k\lambda_{n_2}(\lambda_{n_2} - \lambda_{n_1}) + \lambda_{n_1}^2\lambda_{n_2}^2}{(\lambda_{n_1}\lambda_{n_2} - k)\lambda_{n_2}}, \tag{6.5}$$

and

$$\mathfrak{M} = \mathfrak{M}(n_1, n_2) = \frac{k^2 - k\lambda_{n_1}(\lambda_{n_2} - \lambda_{n_1}) + \lambda_{n_1}^2\lambda_{n_2}^2}{(\lambda_{n_1}\lambda_{n_2} - k)\lambda_{n_1}}. \tag{6.6}$$

In particular, it is immediate to verify that

$$\sigma < 0 \quad \Rightarrow \quad \mathfrak{M} > \mathfrak{m} > 0.$$

Such numbers can be written in several different ways as functions of X, Y, W, Z. To see that, we will exploit the relations

$$\begin{cases} XY = 1, \\ X + Y = \Phi, \\ WZ = 1, \\ W + Z = \Psi, \end{cases} \tag{6.7}$$

valid whenever $X, Y, W, Z \in \mathbb{R}$. Then, setting

$$f = f(n_1, n_2) = \frac{kX - \lambda_{n_1}^2 - k}{\lambda_{n_1}},$$

$$g = g(n_1, n_2) = \frac{kY - \lambda_{n_1}^2 - k}{\lambda_{n_1}},$$

and making use of (6.4), it is easy to prove that

$$\begin{cases} f = \dfrac{kW - \lambda_{n_2}^2 - k}{\lambda_{n_2}}, \\ g = \dfrac{kZ - \lambda_{n_2}^2 - k}{\lambda_{n_2}}. \end{cases} \tag{6.8}$$

Lemma 6.5. *We have the equalities*

$$\mathfrak{m} = -g - \frac{kW^2(X - Y)}{\lambda_{n_1}(W^2 - 1)} = -g - \frac{k(X - Y)}{\lambda_{n_1}(1 - Z^2)},$$

and

$$\mathfrak{M} = -g - \frac{kX^2(X - Y)}{\lambda_{n_1}(X^2 - 1)} = -g - \frac{k(X - Y)}{\lambda_{n_1}(1 - Y^2)},$$

provided that the expressions above are well-defined.

Proof. Exploiting (6.7), we obtain the identities

$$\frac{W^2}{W^2-1} = \frac{W}{W-Z} = \frac{1}{1-Z^2},$$
$$\frac{X^2}{X^2-1} = \frac{X}{X-Y} = \frac{1}{1-Y^2}.$$

Thus, in order to complete the proof, it is sufficient to show that

$$-\mathfrak{m} = g + \frac{kW^2(X-Y)}{\lambda_{n_1}(W^2-1)},$$

and

$$-\mathfrak{M} = g + \frac{kX^2(X-Y)}{\lambda_{n_1}(X^2-1)}.$$

To this end, in the light of (6.4), (6.7), (6.8) and the definitions of ζ, σ, Ψ, g, we compute

$$\begin{aligned}
g + \frac{kW^2(X-Y)}{\lambda_{n_1}(W^2-1)} &= \frac{kY - \lambda_{n_1}^2 - k}{\lambda_{n_1}} + \frac{kW^2(X-Y)}{\lambda_{n_1}(W^2-1)} \\
&= \frac{kZ - \lambda_{n_2}^2 - k}{\lambda_{n_2}} + \frac{kW^2(W-Z)}{\lambda_{n_2}(W^2-1)} \\
&= \frac{kZ - \lambda_{n_2}^2 - k}{\lambda_{n_2}} + \frac{kW}{\lambda_{n_2}} \\
&= \frac{k\Psi - \lambda_{n_2}^2 - k}{\lambda_{n_2}} \\
&= \frac{k\zeta - k\sigma^2\zeta + k\sigma^2 - \sigma\lambda_{n_2}^2 + \lambda_{n_1}\lambda_{n_2}}{\sigma\lambda_{n_2}} \\
&= \frac{(k - \lambda_{n_1}\lambda_{n_2})^2 + k\lambda_{n_2}^2 + k\lambda_{n_1}\lambda_{n_2}}{\sigma k\lambda_{n_2}} \\
&= -\mathfrak{m},
\end{aligned}$$

while, making use of (6.7), along with the definitions of ζ, σ, Φ, g, we have

$$\begin{aligned}
g + \frac{kX^2(X-Y)}{\lambda_{n_1}(X^2-1)} &= \frac{kY - \lambda_{n_1}^2 - k}{\lambda_{n_1}} + \frac{kX}{\lambda_{n_1}} \\
&= \frac{k\Phi - \lambda_{n_1}^2 - k}{\lambda_{n_1}} \\
&= \frac{k + k\sigma^2\zeta - k\sigma^2 - \sigma\lambda_{n_1}\lambda_{n_2} + \lambda_{n_2}^2}{\sigma\lambda_{n_1}\zeta} \\
&= \frac{(k - \lambda_{n_1}\lambda_{n_2})^2 + k\lambda_{n_1}^2 + k\lambda_{n_2}\lambda_{n_1}}{\sigma k\lambda_{n_1}} \\
&= -\mathfrak{M}.
\end{aligned}$$

The lemma is proved. □

6.3. The circle-ellipse systems

We need to investigate the solvability of the circle-ellipse systems

$$\begin{cases} \varrho r^2 \lambda_{n_1} + \varrho t^2 \lambda_{n_2} + \beta = f, \\ \varrho r^2 \lambda_{n_1} X^2 + \varrho t^2 \lambda_{n_2} W^2 + \beta = g, \end{cases} \tag{6.9}$$

and

$$\begin{cases} \varrho r^2 \lambda_{n_1} + \varrho t^2 \lambda_{n_2} + \beta = g, \\ \varrho r^2 \lambda_{n_1} Y^2 + \varrho t^2 \lambda_{n_2} Z^2 + \beta = f, \end{cases} \tag{6.10}$$

in the unknowns r and t.

Lemma 6.6. *The following hold.*

- *Let* $\lambda_{n_1}\lambda_{n_2} \in (0, k)$*. Then neither system* (6.9) *nor* (6.10) *admit real solutions.*
- *Let* $\lambda_{n_1}\lambda_{n_2} \in (k, 2k)$*. Then system* (6.9) *admits real solutions* (r, t) *with* $rt \neq 0$ *if and only if the same does* (6.10)*, if and only if*
$$\mathfrak{m} < -\beta < \mathfrak{M}.$$
In which case, system (6.9) *admits exactly four distinct real solutions, and the same does* (6.10)*. Besides, they do not share any solution.*
- *Let* $\lambda_{n_1}(\lambda_{n_2} - \lambda_{n_1}) \in (2k, \infty)$*. Then system* (6.9) *admits real solutions* (r, t) *with* $rt \neq 0$ *if and only if the same does* (6.10)*, if and only if*
$$\mathfrak{M} < -\beta.$$
In which case, system (6.9) *admits exactly four distinct real solutions, and the same does* (6.10)*. Besides, they do not share any solution.*

Proof. We first observe that systems (6.9) and (6.10) do not share any solution. Indeed, if it were so, we would have $f = g$ (meaning that $X = Y$) and therefore, in the light of Lemma 6.3,

$$\lambda_{n_1}\lambda_{n_2} = 2k \qquad \text{or} \qquad \lambda_{n_1}(\lambda_{n_2} - \lambda_{n_1}) = 2k.$$

Then, setting $s = \sqrt{\zeta}t$, we can rewrite (6.9) and (6.10) as

$$\begin{cases} r^2 + s^2 = F, \\ X^2 r^2 + W^2 s^2 = G, \end{cases} \tag{6.11}$$

and

$$\begin{cases} r^2 + s^2 = G, \\ Y^2 r^2 + Z^2 s^2 = F, \end{cases} \tag{6.12}$$

where

$$F = \frac{f - \beta}{\varrho \lambda_{n_1}} \qquad \text{and} \qquad G = \frac{g - \beta}{\varrho \lambda_{n_1}}.$$

In particular, calling

$$\nu = \frac{k(X - Y)}{\varrho \lambda_{n_1}^2} \geq 0,$$

we have the equality

$$F = G + \nu. \tag{6.13}$$

Systems (6.11) and (6.12) represent the intersection between a circle and an ellipse, both centered at the origin. Therefore, real solutions (r, s) with $rs \neq 0$ exist if and only if the radius of the circle is strictly greater than the minor semi-axis of the ellipse and strictly smaller than the major semi-axis of the ellipse. In such a case, there are exactly four distinct solutions. We shall distinguish three cases.

⋄ *Case 1:* $\lambda_{n_1}\lambda_{n_2} \in (0, k)$. By direct computations, one can easily see that

$$\Psi > \Phi > 2,$$

implying

$$W > X > 1 > Y > Z > 0.$$

In particular, the number ν is strictly positive. As a consequence, in the light of the discussion above and (6.13), system (6.11) admits real solutions (r, s) with $rs \neq 0$ if and only if

$$\frac{G}{W^2} < G + \nu < \frac{G}{X^2}.$$

Being $X^2 > 1$, it is apparent to see that the relation above is impossible. Analogously, system (6.12) admits real solutions (r, s) with $rs \neq 0$ if and only if

$$\frac{G + \nu}{Y^2} < G < \frac{G + \nu}{Z^2}.$$

Again, being $Y^2 < 1$, the relation is impossible. In conclusion, neither system (6.11) nor (6.12) admit real solutions.

⋄ *Case 2:* $\lambda_{n_1}\lambda_{n_2} \in (k, 2k)$. By direct computations, one can easily see that

$$\Psi < \Phi < -2,$$

implying

$$Z < Y < -1 < X < W < 0.$$

Analogously to the previous case, we infer that system (6.11) admits real solutions (r, s) with $rs \neq 0$ if and only if

$$\frac{G}{X^2} < G + \nu < \frac{G}{W^2}.$$

Being $W^2 < 1$ and $X^2 < 1$, in the light of Lemma 6.5 we get

$$\mathfrak{m} = -g - \frac{kW^2(X - Y)}{\lambda_{n_1}(W^2 - 1)} < -\beta < -g - \frac{kX^2(X - Y)}{\lambda_{n_1}(X^2 - 1)} = \mathfrak{M}.$$

Moreover, system (6.12) admits real solutions (r, s) with $rs \neq 0$ if and only if

$$\frac{G+\nu}{Z^2} < G < \frac{G+\nu}{Y^2}.$$

Being $Z^2 > 1$ and $Y^2 > 1$, invoking Lemma 6.5 we conclude that

$$\mathfrak{m} = -g - \frac{k(X-Y)}{\lambda_{n_1}(1-Z^2)} < -\beta < -g - \frac{k(X-Y)}{\lambda_{n_1}(1-Y^2)} = \mathfrak{M}.$$

◇ *Case 3:* $\lambda_{n_1}(\lambda_{n_2} - \lambda_{n_1}) \in (2k, \infty)$. By direct computations, one can easily see that

$$\Phi < -2 \qquad \text{and} \qquad \Psi > 2,$$

implying

$$Y < -1 < X < 0 < Z < 1 < W.$$

Arguing as in the previous cases, system (6.11) admits real solutions (r, s) with $rs \neq 0$ if and only if

$$\frac{G}{W^2} < G + \nu < \frac{G}{X^2}.$$

Since $W^2 > 1$, the relation above reduces to

$$G + \nu < \frac{G}{X^2}.$$

Being $X^2 < 1$, making use of Lemma 6.5 we end up with

$$\mathfrak{M} = -g - \frac{kX^2(X-Y)}{\lambda_{n_1}(X^2-1)} < -\beta.$$

On the other hand, system (6.12) admits real solutions (r, s) with $rs \neq 0$ if and only if

$$\frac{G+\nu}{Y^2} < G < \frac{G+\nu}{Z^2}.$$

Again, since $0 < Z^2 < 1$, the relation above reduces to

$$\frac{G+\nu}{Y^2} < G.$$

Being $Y^2 > 1$, an exploitation of Lemma 6.5 leads to

$$\mathfrak{M} = -g - \frac{k(X-Y)}{\lambda_{n_1}(1-Y^2)} < -\beta.$$

The proof is finished. □

6.4. Classification of general bimodal solutions

In order to classify the general bimodal solutions, we introduce the (disjoint and possibly empty) subsets of $\mathbb{N} \times \mathbb{N}$, with $\mathfrak{m}$ and $\mathfrak{M}$ given by (6.5) and (6.6),

$$\mathbb{B}_1^\star = \{(n_1, n_2) : n_1 < n_2,\ \mathfrak{m} < -\beta < \mathfrak{M} \text{ and } \lambda_{n_1}\lambda_{n_2} \in (k, 2k)\},$$

and

$$\mathbb{B}_2^\star = \{(n_1, n_2) : n_1 < n_2,\ \mathfrak{M} < -\beta \text{ and } \lambda_{n_1}(\lambda_{n_2} - \lambda_{n_1}) \in (2k, \infty)\},$$

and we set

$$\mathbb{B}^\star = \mathbb{B}_1^\star \cup \mathbb{B}_2^\star.$$

Lemma 6.7. *We have the inclusion $\mathbb{B}^\star \subset \mathbb{E} \times \mathbb{E}$. In particular, $\mathbb{B}^\star$ has finite cardinality.*

Proof. By means of elementary computations, one can easily verify that the following implications hold:

$$\lambda_{n_1}\lambda_{n_2} \in (k, 2k) \quad \Rightarrow \quad \lambda_{n_2} < \mathfrak{m},$$
$$\lambda_{n_1}(\lambda_{n_2} - \lambda_{n_1}) \in (2k, \infty) \quad \Rightarrow \quad \lambda_{n_2} < \mathfrak{M}.$$

Therefore, by the very definitions of $\mathbb{B}^\star$ and $\mathbb{E}$,

$$(n_1, n_2) \in \mathbb{B}^\star \quad \Rightarrow \quad (n_1, n_2) \in \mathbb{E} \times \mathbb{E},$$

as claimed. □

We have now all the ingredients to state our main theorem.

Theorem 6.8. *System* (2.1) *admits bimodal solutions of not equidistributed energy if and only if the set $\mathbb{B}^\star$ is nonempty. More precisely, for every couple $(n_1, n_2) \in \mathbb{N} \times \mathbb{N}$ with $n_1 < n_2$, one of the following disjoint situations occurs.*

- *If $(n_1, n_2) \in \mathbb{B}^\star$, we have exactly 8 distinct bimodal solutions of not equidistributed energy: 4 of the form*

$$\begin{cases} u = re_{n_1} + te_{n_2}, \\ v = rXe_{n_1} + tWe_{n_2}, \end{cases}$$

where r, t solve system (6.9), *and 4 of the form*

$$\begin{cases} u = re_{n_1} + te_{n_2}, \\ v = rYe_{n_1} + tZe_{n_2}, \end{cases}$$

where r, t solve system (6.10).
- *If $(n_1, n_2) \notin \mathbb{B}^\star$, there are no bimodal solutions of not equidistributed energy involving the eigenvectors e_{n_1} and e_{n_2}.*

In summary, system (2.1) *admits $8|\mathbb{B}^\star|$ bimodal solutions of not equidistributed energy.*

Proof. Let us look for bimodal solutions of not equidistributed energy (u, v) of the form

$$\begin{cases} u = \alpha_{n_1} e_{n_1} + \alpha_{n_2} e_{n_2}, \\ v = \gamma_{n_1} e_{n_1} + \gamma_{n_2} e_{n_2}, \end{cases}$$

with $n_1 < n_2 \in \mathbb{N}$ and $\alpha_{n_i}, \gamma_{n_i} \in \mathbb{R} \setminus \{0\}$.

◇ *Step 1.* We preliminarily show that

$$\lambda_{n_1}\lambda_{n_2} \in (0, 2k) \setminus \{k\} \quad \text{or} \quad \lambda_{n_1}(\lambda_{n_2} - \lambda_{n_1}) \in (2k, \infty). \tag{6.14}$$

To this end, with reference to the weak formulation (2.3), choosing first $\phi = \psi = e_{n_1}$ and then $\phi = \psi = e_{n_2}$, we obtain the system

$$\begin{cases} \alpha_{n_1}(\lambda_{n_1}^2 + C_u\lambda_{n_1} + k) = k\gamma_{n_1}, \\ \gamma_{n_1}(\lambda_{n_1}^2 + C_v\lambda_{n_1} + k) = k\alpha_{n_1}, \\ \alpha_{n_2}(\lambda_{n_2}^2 + C_u\lambda_{n_2} + k) = k\gamma_{n_2}, \\ \gamma_{n_2}(\lambda_{n_2}^2 + C_v\lambda_{n_2} + k) = k\alpha_{n_2}. \end{cases} \tag{6.15}$$

Next, setting

$$\begin{cases} x_{n_1} = \dfrac{\lambda_{n_1}^2 + C_u\lambda_{n_1} + k}{k}, \\ y_{n_1} = \dfrac{\lambda_{n_1}^2 + C_v\lambda_{n_1} + k}{k}, \\ x_{n_2} = \dfrac{\lambda_{n_2}^2 + C_u\lambda_{n_2} + k}{k}, \\ y_{n_2} = \dfrac{\lambda_{n_2}^2 + C_v\lambda_{n_2} + k}{k}, \end{cases} \tag{6.16}$$

we get

$$\begin{cases} x_{n_1}y_{n_1} = 1, \\ x_{n_2}y_{n_2} = 1, \\ \zeta x_{n_1} - (\zeta - 1)\sigma = x_{n_2}, \\ \zeta y_{n_1} - (\zeta - 1)\sigma = y_{n_2}. \end{cases}$$

Observe that $\sigma \neq 0$, otherwise

$$\begin{cases} x_{n_1}y_{n_1} = 1, \\ x_{n_2}y_{n_2} = 1, \\ \zeta x_{n_1} = x_{n_2}, \\ \zeta y_{n_1} = y_{n_2}, \end{cases}$$

yielding $\zeta^2 = 1$ and contradicting the assumption $n_1 < n_2$. Therefore, we obtain

$$x_{n_1}y_{n_1} = 1, \tag{6.17}$$

$$x_{n_1} + y_{n_1} = \Phi, \tag{6.18}$$

$$x_{n_2}y_{n_2} = 1, \tag{6.19}$$
$$x_{n_2} + y_{n_2} = \Psi. \tag{6.20}$$

Clearly, the solutions are given by the four quadruplets

$$(X, Y, W, Z),$$
$$(X, Y, Z, W),$$
$$(Y, X, W, Z),$$
$$(Y, X, Z, W).$$

Since at least one (hence all) of the quadruplets has to have real components, making use of Lemma 6.2 we infer that

$$\lambda_{n_1}\lambda_{n_2} \in (0, 2k] \setminus \{k\} \quad \text{or} \quad \lambda_{n_1}(\lambda_{n_2} - \lambda_{n_1}) \in [2k, \infty).$$

In addition, due to the fact that (u, v) does not have equidistributed energy,

$$C_u \neq C_v \quad \Rightarrow \quad x_{n_1} \neq y_{n_1}.$$

Thus, an exploitation of Lemma 6.3 yields

$$\begin{cases} \lambda_{n_1}\lambda_{n_2} \neq 2k, \\ \lambda_{n_1}(\lambda_{n_2} - \lambda_{n_1}) \neq 2k, \end{cases}$$

and (6.14) follows.

⋄ *Step 2.* We now prove that, within (6.14), the coefficients α_{n_1} and α_{n_2} are solutions of system (6.9) or (6.10). Indeed, from (6.16) and recalling the definitions of f and g, four possibilities occur:

$$\begin{cases} C_u = f = \dfrac{kW - \lambda_{n_2}^2 - k}{\lambda_{n_2}}, \\ C_v = g = \dfrac{kZ - \lambda_{n_2}^2 - k}{\lambda_{n_2}}, \end{cases} \tag{6.21}$$

or

$$\begin{cases} C_u = f = \dfrac{kZ - \lambda_{n_2}^2 - k}{\lambda_{n_2}}, \\ C_v = g = \dfrac{kW - \lambda_{n_2}^2 - k}{\lambda_{n_2}}, \end{cases} \tag{6.22}$$

or

$$\begin{cases} C_u = g = \dfrac{kW - \lambda_{n_2}^2 - k}{\lambda_{n_2}}, \\ C_v = f = \dfrac{kZ - \lambda_{n_2}^2 - k}{\lambda_{n_2}}, \end{cases} \tag{6.23}$$

or

$$
\begin{cases}
C_u = g = \dfrac{kZ - \lambda_{n_2}^2 - k}{\lambda_{n_2}}, \\[2ex]
C_v = f = \dfrac{kW - \lambda_{n_2}^2 - k}{\lambda_{n_2}}.
\end{cases}
\tag{6.24}
$$

At this point, exploiting (6.14) and Lemma 6.3, we learn that $W \neq Z$. As a consequence, taking into account (6.8), we conclude that only systems (6.21) and (6.24) survive. Recalling the explicit forms of C_u and C_v given by (3.4), we remain with

$$
\begin{cases}
\varrho\alpha_{n_1}^2\lambda_{n_1} + \varrho\alpha_{n_2}^2\lambda_{n_2} + \beta = f, \\
\varrho\gamma_{n_1}^2\lambda_{n_1} + \varrho\gamma_{n_2}^2\lambda_{n_2} + \beta = g,
\end{cases}
$$

and

$$
\begin{cases}
\varrho\alpha_{n_1}^2\lambda_{n_1} + \varrho\alpha_{n_2}^2\lambda_{n_2} + \beta = g, \\
\varrho\gamma_{n_1}^2\lambda_{n_1} + \varrho\gamma_{n_2}^2\lambda_{n_2} + \beta = f.
\end{cases}
$$

Finally, due to (6.15), in the first case we infer that

$$
\begin{cases}
\gamma_{n_1} = X\alpha_{n_1}, \\
\gamma_{n_2} = W\alpha_{n_2},
\end{cases}
$$

while in the second one

$$
\begin{cases}
\gamma_{n_1} = Y\alpha_{n_1}, \\
\gamma_{n_2} = Z\alpha_{n_2}.
\end{cases}
$$

◇ *Step 3.* Collecting Steps 1-2 and Lemma 6.6, there exist bimodal solutions of not equidistributed energy (explicitly computed) if and only if the couple $(n_1, n_2) \in \mathbb{B}^\star$. □

6.5. *Two explicit examples*

We conclude by showing two explicit examples of bimodal solutions of not equidistributed energy. In what follows, in order to avoid the presence of unnecessary constants, we take for simplicity $\varrho = 1$, and we choose

$$
A = \frac{1}{\pi^2} L,
$$

being L the Laplace-Dirichlet operator of the concrete Example 2.2. Accordingly, the eigenvalues of A read

$$
\lambda_n = n^2,
$$

with corresponding eigenvectors

$$
e_n(x) = \sqrt{2}\,\sin(n\pi x).
$$

Example 6.9. Let

$$
k = 3 \qquad \text{and} \qquad (n_1, n_2) = (1, 2).
$$

In this situation, an easy computation shows that

$$
X = -2 + \sqrt{3},
$$

$$Y = -2 - \sqrt{3},$$
$$W = -7 + 4\sqrt{3},$$
$$Z = -7 - 4\sqrt{3},$$

and

$$\mathfrak{m} = \frac{61}{4} < 16 = \mathfrak{M}.$$

Accordingly, if β is such that

$$\frac{61}{4} < -\beta < 16,$$

the couple (n_1, n_2) belongs to $\mathbb{B}_1^\star$. Hence, there exist four solutions of the form

$$\begin{cases} u = \alpha_1 e_1 + \alpha_2 e_2, \\ v = (\sqrt{3} - 2)\alpha_1 e_1 + (4\sqrt{3} - 7)\alpha_2 e_2, \end{cases}$$

where $\alpha_1, \alpha_2 \in \mathbb{R}$ solve the system

$$\begin{cases} \alpha_1^2 + 4\alpha_2^2 = 3\sqrt{3} - 10 - \beta, \\ \alpha_1^2(\sqrt{3} - 2)^2 + 4\alpha_2^2(4\sqrt{3} - 7)^2 = -3\sqrt{3} - 10 - \beta, \end{cases} \tag{6.25}$$

and four solutions of the form

$$\begin{cases} u = \alpha_1 e_1 + \alpha_2 e_2, \\ v = -(\sqrt{3} + 2)\alpha_1 e_1 - (4\sqrt{3} + 7)\alpha_2 e_2, \end{cases}$$

where $\alpha_1, \alpha_2 \in \mathbb{R}$ solve the system

$$\begin{cases} \alpha_1^2 + 4\alpha_2^2 = -3\sqrt{3} - 10 - \beta, \\ \alpha_1^2(\sqrt{3} + 2)^2 + 4\alpha_2^2(4\sqrt{3} + 7)^2 = 3\sqrt{3} - 10 - \beta. \end{cases} \tag{6.26}$$

For instance, when $\beta = -31/2$, the solutions of system (6.25) are

$$(\pm\alpha_1, \pm\alpha_2) \qquad \text{and} \qquad (\pm\alpha_1, \mp\alpha_2),$$

with

$$\alpha_1 = -\sqrt{\frac{7\sqrt{3} - 12}{26\sqrt{3} - 45}} \approx -1.93185,$$

$$\alpha_2 = -\frac{1}{2}\sqrt{\frac{362\sqrt{3} - 627}{2(5042\sqrt{3} - 8733)}} \approx -1.31948,$$

while the solutions of system (6.26) are

$$(\pm\alpha_1, \pm\alpha_2) \qquad \text{and} \qquad (\pm\alpha_1, \mp\alpha_2),$$

with

$$\alpha_1 = -\sqrt{\frac{7\sqrt{3}+12}{26\sqrt{3}+45}} \approx -0.51763,$$

$$\alpha_2 = -\frac{1}{2}\sqrt{\frac{362\sqrt{3}+627}{2(5042\sqrt{3}+8733)}} \approx -0.09473.$$

Example 6.10. Let

$$k = 1 \qquad \text{and} \qquad (n_1, n_2) = (1, 2).$$

In this situation, an easy computation shows that

$$X = \frac{-4+\sqrt{7}}{3},$$
$$Y = \frac{-4-\sqrt{7}}{3},$$
$$W = \frac{11+4\sqrt{7}}{3},$$
$$Z = \frac{11-4\sqrt{7}}{3},$$

and

$$\mathfrak{M} = \frac{14}{3}.$$

Accordingly, if β is such that

$$\frac{14}{3} < -\beta,$$

the couple (n_1, n_2) belongs to $\mathbb{B}_2^\star$. Hence, there exist four solutions of the form

$$\begin{cases} u = \alpha_1 e_1 + \alpha_2 e_2, \\ v = \dfrac{\sqrt{7}-4}{3}\alpha_1 e_1 + \dfrac{4\sqrt{7}+11}{3}\alpha_2 e_2, \end{cases}$$

where $\alpha_1, \alpha_2 \in \mathbb{R}$ solve the system

$$\begin{cases} \alpha_1^2 + 4\alpha_2^2 = \dfrac{\sqrt{7}-4}{3} - 2 - \beta, \\ \alpha_1^2\Big(\dfrac{\sqrt{7}-4}{3}\Big)^2 + 4\alpha_2^2\Big(\dfrac{4\sqrt{7}+11}{3}\Big)^2 = -\dfrac{4+\sqrt{7}}{3} - 2 - \beta, \end{cases} \tag{6.27}$$

and four solutions of the form

$$\begin{cases} u = \alpha_1 e_1 + \alpha_2 e_2, \\ v = -\dfrac{4+\sqrt{7}}{3}\alpha_1 e_1 + \dfrac{11-4\sqrt{7}}{3}\alpha_2 e_2, \end{cases}$$

where $\alpha_1, \alpha_2 \in \mathbb{R}$ solve the system

$$\begin{cases} \alpha_1^2 + 4\alpha_2^2 = -\dfrac{4+\sqrt{7}}{3} - 2 - \beta, \\ \alpha_1^2\Big(\dfrac{4+\sqrt{7}}{3}\Big)^2 + 4\alpha_2^2\Big(\dfrac{11-4\sqrt{7}}{3}\Big)^2 = \dfrac{\sqrt{7}-4}{3} - 2 - \beta. \end{cases} \tag{6.28}$$

For instance, when $\beta = -5$, the solutions of system (6.27) are

$$(\pm\alpha_1, \pm\alpha_2) \qquad \text{and} \qquad (\pm\alpha_1, \mp\alpha_2),$$

with

$$\alpha_1 = -\frac{1}{3}\sqrt{\frac{31(28+11\sqrt{7})}{35+16\sqrt{7}}} \approx -1.59482,$$

$$\alpha_2 = -\frac{1}{6}\sqrt{\frac{883+316\sqrt{7}}{18011+6808\sqrt{7}}} \approx -0.03587,$$

while the solutions of system (6.28) are

$$(\pm\alpha_1, \pm\alpha_2) \qquad \text{and} \qquad (\pm\alpha_1, \mp\alpha_2),$$

with

$$\alpha_1 = -\frac{1}{3}\sqrt{\frac{31(11\sqrt{7}-28)}{16\sqrt{7}-35}} \approx -0.71992,$$

$$\alpha_2 = -\frac{1}{6}\sqrt{\frac{316\sqrt{7}-883}{6808\sqrt{7}-18011}} \approx -0.25809.$$

7. General Trimodal Solutions

Finally, we consider general trimodal solutions to system (2.1). As previously shown, trimodal EE-solutions exist. Then, one might ask if system (2.1) admits also trimodal solutions of not equidistributed energy. The answer to this question is negative.

Theorem 7.1. *Every trimodal solution is necessarily an* EE*-solution.*

Proof. Let (u, v) be a (general) trimodal solution. In particular, with reference to (3.3), $\alpha_{n_i} \neq 0$ and $\gamma_{n_i} \neq 0$ for every n_i. Assume by contradiction that (u, v) is not an EE-solution. Then, in the light of Lemma 5.3, the vectors

$$\begin{bmatrix}\alpha_{n_1}\\ \gamma_{n_1}\end{bmatrix}, \begin{bmatrix}\alpha_{n_2}\\ \gamma_{n_2}\end{bmatrix}, \begin{bmatrix}\alpha_{n_3}\\ \gamma_{n_3}\end{bmatrix}$$

are pairwise linearly independent. Accordingly, each of them can be written as a linear combination of the other two. In particular, there exist $a, b, c, d, e, f \neq 0$ such that

$$\begin{cases} \alpha_{n_3} = a\alpha_{n_1} + b\alpha_{n_2}, \\ \gamma_{n_3} = a\gamma_{n_1} + b\gamma_{n_2}, \end{cases} \tag{7.1}$$

$$\begin{cases} \alpha_{n_1} = c\alpha_{n_2} + d\alpha_{n_3}, \\ \gamma_{n_1} = c\gamma_{n_2} + d\gamma_{n_3}, \end{cases} \tag{7.2}$$

and

$$\begin{cases} \alpha_{n_2} = e\alpha_{n_1} + f\alpha_{n_3}, \\ \gamma_{n_2} = e\gamma_{n_1} + f\gamma_{n_3}. \end{cases} \tag{7.3}$$

Moreover, due to Lemma 5.2,

$$\begin{cases} \alpha_{n_1} + \gamma_{n_1} \neq 0, \\ \alpha_{n_2} + \gamma_{n_2} \neq 0, \\ \alpha_{n_3} + \gamma_{n_3} \neq 0. \end{cases} \tag{7.4}$$

Therefore, recalling (3.5),

$$\lambda_{n_1} = -\frac{C_u\alpha_{n_1} + C_v\gamma_{n_1}}{\alpha_{n_1} + \gamma_{n_1}}, \tag{7.5}$$

$$\lambda_{n_2} = -\frac{C_u\alpha_{n_2} + C_v\gamma_{n_2}}{\alpha_{n_2} + \gamma_{n_2}}, \tag{7.6}$$

$$\lambda_{n_3} = -\frac{C_u\alpha_{n_3} + C_v\gamma_{n_3}}{\alpha_{n_3} + \gamma_{n_3}}. \tag{7.7}$$

Substituting the expressions of α_{n_3} and γ_{n_3} given by (7.1) into (7.7), we obtain the identity

$$[a(\alpha_{n_1} + \gamma_{n_1}) + b(\alpha_{n_2} + \gamma_{n_2})]\lambda_{n_3} = -C_u[a\alpha_{n_1} + b\alpha_{n_2}] - C_v[a\gamma_{n_1} + b\gamma_{n_2}]$$

which, making use of (7.5)-(7.6), yields

$$\mathsf{A}\lambda_{n_1} + \mathsf{B}\lambda_{n_2} = (\mathsf{A} + \mathsf{B})\lambda_{n_3} \tag{7.8}$$

where

$$\mathsf{A} = a(\alpha_{n_1} + \gamma_{n_1}) \qquad \text{and} \qquad \mathsf{B} = b(\alpha_{n_2} + \gamma_{n_2}).$$

An analogous reasoning, exploiting now (7.2) and (7.3), provides the further equalities

$$\mathsf{C}\lambda_{n_2} + \mathsf{D}\lambda_{n_3} = (\mathsf{C} + \mathsf{D})\lambda_{n_1}, \tag{7.9}$$

$$\mathsf{E}\lambda_{n_1} + \mathsf{F}\lambda_{n_3} = (\mathsf{E} + \mathsf{F})\lambda_{n_2}, \tag{7.10}$$

having set

$$\mathsf{C} = c(\alpha_{n_2} + \gamma_{n_2}),$$
$$\mathsf{D} = d(\alpha_{n_3} + \gamma_{n_3}),$$
$$\mathsf{E} = e(\alpha_{n_1} + \gamma_{n_1}),$$
$$\mathsf{F} = f(\alpha_{n_3} + \gamma_{n_3}).$$

Since $a, b, c, d, e, f \neq 0$, from (7.4) we learn that $\mathsf{A}, \mathsf{B}, \mathsf{C}, \mathsf{D}, \mathsf{E}, \mathsf{F} \neq 0$. Then, introducing the matrix

$$\mathbf{M} = \begin{bmatrix} \mathsf{A} & \mathsf{B} & -(\mathsf{A} + \mathsf{B}) \\ -(\mathsf{C} + \mathsf{D}) & \mathsf{C} & \mathsf{D} \\ \mathsf{E} & -(\mathsf{E} + \mathsf{F}) & \mathsf{F} \end{bmatrix}$$

and the vector

$$\lambda = \begin{bmatrix} \lambda_{n_1} \\ \lambda_{n_2} \\ \lambda_{n_3} \end{bmatrix},$$

we rewrite (7.8)-(7.10) as

$$\mathbf{M}\lambda = \mathbf{0}.$$

Direct calculations show that Det($\mathbf{M}$) = 0, thus Rank($\mathbf{M}$) < 3.

◇ If Rank($\mathbf{M}$) = 2, in the light of the Rank-Nullity Theorem the solution set is a one-dimensional linear subspace of $\mathbb{R}^3$, explicitly given by

$$\text{Ker}(\mathbf{M}) = \left\{ \lambda = \begin{bmatrix} \lambda \\ \lambda \\ \lambda \end{bmatrix} : \lambda \in \mathbb{R} \right\}.$$

In particular, this forces $\lambda_{n_1} = \lambda_{n_2} = \lambda_{n_3}$, implying the desired contradiction.

◇ If Rank($\mathbf{M}$) = 1, there exists $\omega \neq 0$ such that

$$\begin{cases} \mathsf{A} = \omega \mathsf{B}, \\ (1+\omega)\mathsf{C} = \mathsf{D}. \end{cases}$$

Substituting the explicit expressions of A, B, C, D into the system above

$$a(\alpha_{n_1} + \gamma_{n_1}) = \omega b(\alpha_{n_2} + \gamma_{n_2}), \tag{7.11}$$

$$c(1+\omega)(\alpha_{n_2} + \gamma_{n_2}) = d(\alpha_{n_3} + \gamma_{n_3}). \tag{7.12}$$

Then, plugging (7.1) into (7.12) and exploiting (7.11) and (7.4),

$$c(1+\omega) = db(1+\omega).$$

Since $1 + \omega \neq 0$ (due to the fact that $\mathsf{D} \neq 0$), we end up with

$$c = db.$$

Appealing now to (7.1) and (7.2),

$$(1 + da)\begin{bmatrix} \alpha_{n_1} \\ \gamma_{n_1} \end{bmatrix} = 2d \begin{bmatrix} \alpha_{n_3} \\ \gamma_{n_3} \end{bmatrix},$$

meaning that the two vectors

$$\begin{bmatrix} \alpha_{n_1} \\ \gamma_{n_1} \end{bmatrix} \quad \text{and} \quad \begin{bmatrix} \alpha_{n_3} \\ \gamma_{n_3} \end{bmatrix}$$

are linearly dependent. □

Example 7.2. As a particular case, let us consider

$$A = L^{\frac{p+1}{2}}, \quad p \in \mathbb{N},$$

with L as in Example 2.2. In this situation, the eigenvalues read

$$\lambda_n = n^{p+1}\pi^{p+1}.$$

Accordingly, given a trimodal solution (which, as we know, is necessarily an EE-solution) and exploiting Corollary 5.6, we deduce the relation

$$n_1^{p+1} + n_2^{p+1} = n_3^{p+1}.$$

Therefore, when $p = 1$, they form a Pythagorean triplet. Otherwise the identity is impossible, due to the celebrated *Fermat's Last Theorem* proved by A. Wiles in recent years [25, 28]. Hence, for $p = 2, 3, 4, \ldots$, trimodal solutions do not exist.

8. Comparison with Single-Beam Equations

We conclude by comparing our results on the double-beam system (2.1) with some previous achievements on extensible single-beam equations. As customary, along the section, we will set

$$C_u = \beta + \varrho\|u\|_1^2. \tag{8.1}$$

The following theorem has been proved in [8].

Theorem 8.1. *The nontrivial solutions of the single-beam equation*

$$Au + C_u u = 0$$

are exactly $2|\mathbb{E}|$, *where, in the usual notation,*

$$\mathbb{E} = \{n : \lambda_n < -\beta\}$$

denotes the (finite) set of effective modes. Such solutions are unimodal, explicitly given by

$$u_n^{\pm} = \pm\sqrt{\frac{-\beta - \lambda_n}{\varrho\lambda_n}}\, e_n,$$

for every $n \in \mathbb{E}$.

Concerning the case of single beams which rely on an elastic foundation, the result reads as follows.

Theorem 8.2. *The nontrivial solutions of the single-beam equation*

$$A^2u + C_u Au + ku = 0 \tag{8.2}$$

can be either unimodal or bimodal (but not trimodal). In addition, the following hold.

- *Equation* (8.2) *admits nontrivial unimodal solutions if and only if the set*

$$\mathbb{F} = \left\{ n : \frac{k}{\lambda_n} + \lambda_n < -\beta \right\}$$

is nonempty. More precisely, for every $n \in \mathbb{N}$, one of the following disjoint situations occurs.

  - *If $n \in \mathbb{F}$, we have exactly 2 nontrivial unimodal solutions of the form*

$$u_n^{\pm} = \pm \sqrt{\frac{1}{\varrho \lambda_n} \left(-\beta - \frac{k}{\lambda_n} - \lambda_n \right)} e_n.$$

  - *If $n \notin \mathbb{F}$ all the unimodal solutions involving the eigenvector e_n are trivial.*

- *Equation* (8.2) *admits nontrivial bimodal solutions if and only if the set*

$$\mathbb{G} = \{(n_1, n_2) : n_1 < n_2,\ \lambda_{n_1} + \lambda_{n_2} < -\beta \text{ and } \lambda_{n_1}\lambda_{n_2} = k\}$$

is nonempty. More precisely, for every couple $(n_1, n_2) \in \mathbb{N}$ with $n_1 < n_2$, one of the following disjoint situations occurs.

  - *If $(n_1, n_2) \in \mathbb{G}$, we have exactly the (infinitely many) solutions of the form*

$$u = xe_{n_1} + ye_{n_2},$$

for all $(x, y) \in \mathbb{R}^2$ satisfying the equality

$$\varrho x^2 \lambda_{n_1} + \varrho y^2 \lambda_{n_2} + \lambda_{n_1} + \lambda_{n_2} + \beta = 0 \qquad \text{with} \qquad xy \neq 0.$$

  - *If $(n_1, n_2) \notin \mathbb{G}$, there are no nontrivial bimodal solutions involving the eigenvectors e_{n_1} and e_{n_2}.*

Theorem 8.2 has been proved in [3], in the concrete situation when $A = L$ (the Laplace-Dirichlet operator). We present here a short proof, which is valid even in our abstract setting.

Proof of Theorem 8.2. Let u be a weak solution[¶] to (8.2). Arguing as in the proof of Lemma 3.3, that is, writing

$$u = \sum_n \alpha_n e_n$$

for some $\alpha_n \in \mathbb{R}$, we obtain, for every $n \in \mathbb{N}$, the identity

$$\lambda_n^2 \alpha_n + C_u \lambda_n \alpha_n + k\alpha_n = 0.$$

Hence, if $\alpha_n \neq 0$, we infer that

$$\lambda_n^2 + C_u \lambda_n + k = 0.$$

[¶]Analogously to (2.3), $u \in \mathrm{H}^2$ is called a *weak solution* to (8.2) if, for every test $\phi \in \mathrm{H}^2$,

$$\langle u, \phi \rangle_2 + C_u \langle u, \phi \rangle_1 + k\langle u, \phi \rangle = 0.$$

Since the equation above admits at most two distinct solutions λ_{n_i}, we conclude that the nontrivial solutions to equation (8.2) can be either unimodal or bimodal (but not trimodal).

First, let us look for unimodal solutions u of the form

$$u = \alpha_n e_n$$

for a fixed $n \in \mathbb{N}$ and some coefficient $\alpha_n \neq 0$. Analogously to the proof of Theorem 4.2, from (8.2) we obtain

$$\lambda_n^2 + (\beta + \varrho\lambda_n\alpha_n^2)\lambda_n + k = 0,$$

which implies

$$\alpha_n^2 = \frac{1}{\varrho\lambda_n}\Big(-\beta - \frac{k}{\lambda_n} - \lambda_n\Big).$$

Therefore, there exist nontrivial unimodal solutions (explicitly computed) if and only if $n \in \mathbb{F}$.

Next, let us look for bimodal solutions u of the form

$$u = \alpha_{n_1} e_{n_1} + \alpha_{n_2} e_{n_2}$$

with $n_1 < n_2 \in \mathbb{N}$ and $\alpha_{n_i} \in \mathbb{R} \setminus \{0\}$. Similarly to the previous situation, from (8.2) we obtain the system

$$\begin{cases} \lambda_{n_1}^2 + C_u\lambda_{n_1} + k = 0, \\ \lambda_{n_2}^2 + C_u\lambda_{n_2} + k = 0. \end{cases}$$

Hence

$$\lambda_{n_1}\lambda_{n_2} = k$$

and the value C_u is determined by (8.1), which yields the relation

$$\varrho\alpha_{n_1}^2\lambda_{n_1} + \varrho\alpha_{n_2}^2\lambda_{n_2} + \lambda_{n_1} + \lambda_{n_2} + \beta = 0.$$

Therefore, there exist nontrivial bimodal solutions (explicitly computed) if and only if $(n_1, n_2) \in \mathbb{G}$. □

A closer look to Theorems 8.1 and 8.2 reveals that the set of steady states of the double-beam system (2.1) is very rich, and by no means represents a "double-copy" of the set of stationary solutions of a single-beam equation:

- According to §4, nonsymmetric unimodal solutions pop up, as well as unimodal solutions for which the elastic energy is not evenly distributed. This feature is illustrated in the forthcoming pictures[‖]. Moreover, not only a double series of bifurcations of the trivial solution occurs, but even buckled unimodal solutions suffer from a further bifurcation (see Lemma 4.1 and Figure 2 of §4).
- According to §5 and §6, system (2.1) admits infinitely many bimodal and trimodal EE-solutions, and also finitely many nonsymmetric bimodal solutions of not equidistributed energy.

[‖]The notation in the captions is the same as in §4.

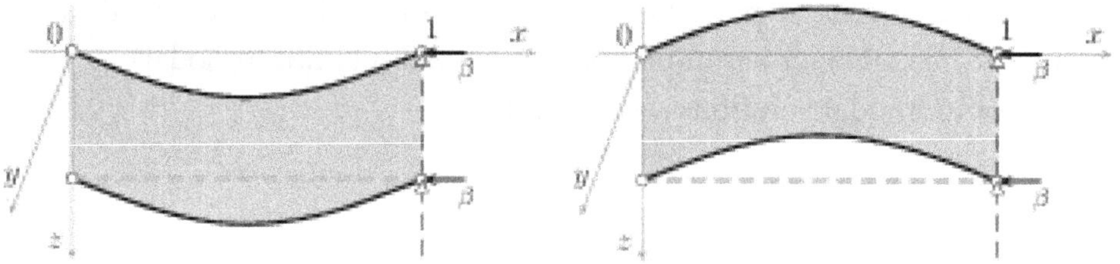

Figure 3. Symmetric in-phase unimodal solutions $(\alpha^{\pm}_{1,1}, \alpha^{\pm}_{1,1})$.

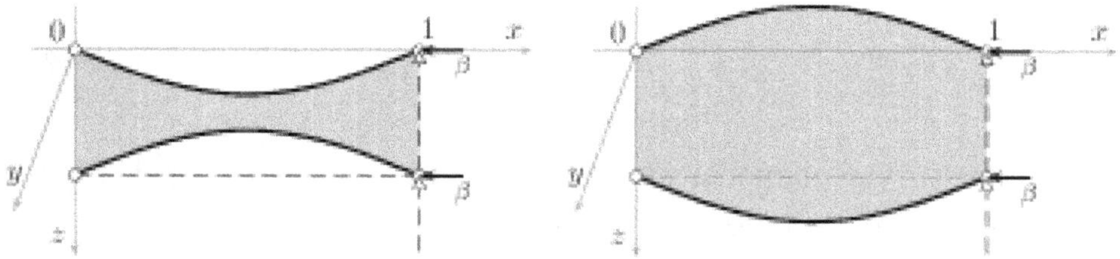

Figure 4. Symmetric out-of-phase unimodal solutions $(\alpha^{\pm}_{1,2}, \alpha^{\mp}_{1,2})$.

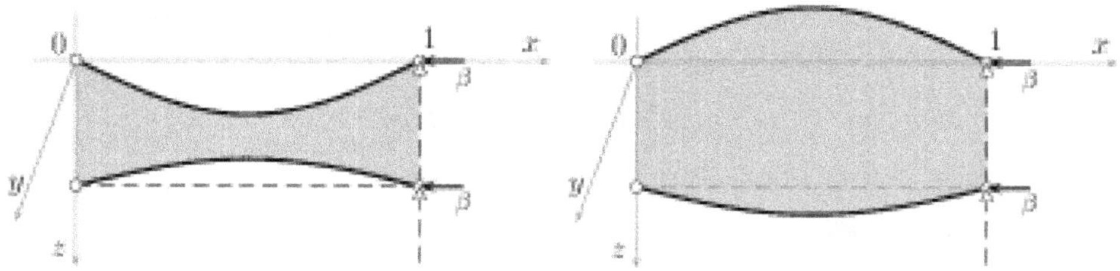

Figure 5. Nonsymmetric out-of-phase unimodal solutions $(\alpha^{\pm}_{1,3}, \alpha^{\mp}_{1,4})$.

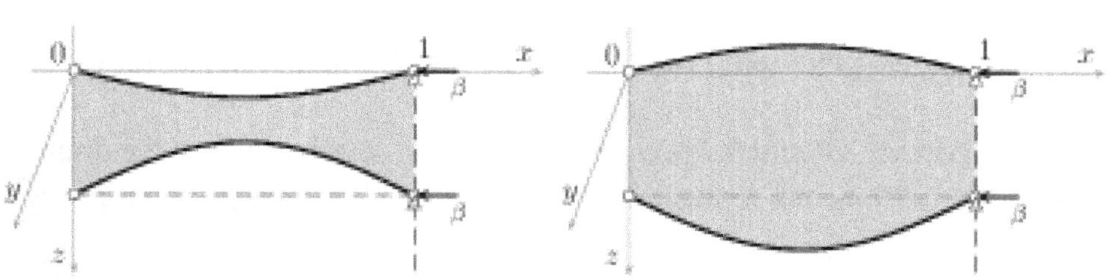

Figure 6. Nonsymmetric out-of-phase unimodal solutions $(\alpha^{\pm}_{1,4}, \alpha^{\mp}_{1,3})$.

9. Appendix: Dimensionless Models of Double-Beam Systems

Let us consider a thin and elastic Woinowsky-Krieger beam of natural length $\ell > 0$, uniform cross section Ω, and thickness $0 < h \ll \ell$. The beam is supposed to be homogeneous, of constant mass density $\rho > 0$ per unit volume, and symmetric with respect to the vertical plane $(\xi\text{-}z)$. Hence, we can restrict our attention to its rectangular section lying in the plane $y = 0$. Identifying the beam with such a section, we assume that its middle line at rest occupies the interval $[0,\ell]$ of the ξ-axis. According to the physical analysis carried out in [8, 13], in the isothermal case the motion equation for the vertical deflection of the midline of the beam

$$U : (\xi,\tau) \in [0,\ell] \times \mathbb{R}^+ \mapsto \mathbb{R}$$

reads

$$\mathfrak{L}U - \frac{Eh}{2\ell^2(1-\nu^2)}\Big(2D + \int_0^\ell |\partial_\xi U(s)|^2\,\mathrm{d}s\Big)\partial_{\xi\xi}U = \frac{G}{\ell|\Omega|}.$$

Here,

$$\mathfrak{L} = \rho\partial_{\tau\tau} - \frac{\rho h^2}{12}\partial_{\tau\tau\xi\xi} + \frac{Eh^3}{12\ell(1-\nu^2)}\partial_{\xi\xi\xi\xi}$$

denotes the evolution operator, while

- $|\Omega| > 0$ is the area of the cross section,
- $E > 0$ is the Young modulus (force per unit area),
- $\nu \in (-1, \frac{1}{2})$ is the Poisson ratio, which is negative for auxetic materials,
- $D \in \mathbb{R}$ is the axial displacement at the right end of the beam,
- $G : [0,\ell] \times \mathbb{R}^+ \to \mathbb{R}$ is the vertical body force applied on the section Ω.

We point out that the model is obtained by supposing the beam slender (i.e. $h \ll \ell$), and the modulus of the axial displacement D small when compared to the length of the beam (i.e. $|D| \ll \ell$ as well). See also [4, 5, 19] for more details.

Assuming that G is due to the distributed and mutual elastic action exerted between two equal Woinowsky-Krieger beams with vertical deflections $U = U(\xi,\tau)$ and $V = V(\xi,\tau)$, respectively, we let

$$G(\xi,\tau) = -\varkappa[U(\xi,\tau) - V(\xi,\tau)],$$

being $\varkappa > 0$ the uniform stiffness (force per unit length) of the elastic core. In this situation, the model describing the motion of the resulting elastically-coupled extensible double-beam nonlinear system becomes

$$\begin{cases} \mathfrak{L}U - \dfrac{Eh}{2\ell^2(1-\nu^2)}\Big(2D + \displaystyle\int_0^\ell |\partial_\xi U(s)|^2\,\mathrm{d}s\Big)\,\partial_{\xi\xi}U + \dfrac{\varkappa}{\ell|\Omega|}(U-V) = 0, \\ \mathfrak{L}V - \dfrac{Eh}{2\ell^2(1-\nu^2)}\Big(2D + \displaystyle\int_0^\ell |\partial_\xi V(s)|^2\,\mathrm{d}s\Big)\,\partial_{\xi\xi}V - \dfrac{\varkappa}{\ell|\Omega|}(U-V) = 0. \end{cases}$$

In order to rewrite the system in dimensionless form, we exploit the fact that the two beams have the same structural parameters. In particular, ℓ is viewed as the common *characteristic length* of the beams, while the *characteristic time* τ_0 is obtained by means of the well-known shear wave velocity c_0 in bulk elasticity, given by

$$c_0 = \sqrt{\frac{E}{2\rho(1+\nu)}}.$$

Then, the characteristic time τ_0 is equal to the ratio ℓ/c_0. Explicitly,

$$\tau_0 = \sqrt{\frac{2\ell^2\rho(1+\nu)}{E}}.$$

Consequently, introducing the dimensionless space and time variables

$$x = \frac{\xi}{\ell} \in [0, 1] \qquad \text{and} \qquad t = \frac{\tau}{\tau_0} \in \mathbb{R}^+,$$

along with the rescaled unknowns $u, v : [0, 1] \times \mathbb{R}^+ \to \mathbb{R}$ defined as

$$u(x, t) = \frac{U(\ell x, \tau_0 t)}{\ell} \qquad \text{and} \qquad v(x, t) = \frac{V(\ell x, \tau_0 t)}{\ell},$$

we end up with the dimensionless model

$$\begin{cases} \dfrac{\ell(1-\nu)}{h}\Big(\partial_{tt} - \dfrac{h^2}{12\ell^2}\partial_{ttxx}\Big)u + \delta\partial_{xxxx}u - (\chi + \|\partial_x u\|^2)\partial_{xx}u + \kappa(u - v) = 0, \\ \dfrac{\ell(1-\nu)}{h}\Big(\partial_{tt} - \dfrac{h^2}{12\ell^2}\partial_{ttxx}\Big)v + \delta\partial_{xxxx}v - (\chi + \|\partial_x v\|^2)\partial_{xx}v - \kappa(u - v) = 0, \end{cases}$$

where $\|\cdot\|$ denotes the L^2-norm on the unit interval $[0, 1]$, and

$$\delta = \frac{h^2}{6\ell^2} > 0, \qquad \chi = \frac{2D}{\ell} \in \mathbb{R}, \qquad \kappa = \frac{2\varkappa\ell^2(1-\nu^2)}{E|\Omega|h} > 0.$$

Under reasonably physical assumptions on the stiffness $\varkappa$ of the elastic core, and since D and h are comparable, we may conclude that $|\chi|$ and κ share the same order of magnitude h/ℓ, whereas δ is much smaller. Accordingly, $|\chi/\delta|$ and κ/δ may assume large values, for their order of magnitude is $\ell/h \gg 1$. Hence, all the stationary solutions exhibited in this paper are physically consistent.

Conflict of Interest

All authors declare no conflicts of interest in this paper.

References

1. I.V. Andrianov, *On the theory of berger plates*. J. Appl. Math. Mech., **47** (1983), 142-144.
2. J.M. Ball, *Stability theory for an extensible beam*. J. Differential Equations, **14** (1973), 399-418.
3. I. Bochicchio and E. Vuk, *Buckling and longterm dynamics of a nonlinear model for the extensible beam*. Math. Comput. Modelling, **51** (2010), 833-846.
4. P.G. Ciarlet, *A justification of the von Kármán equations*. Arch. Rational Mech. Anal., **73** (1980), 349-389.
5. P.G. Ciarlet and L. Gratie, *From the classical to the generalized von Kármán and Marguerre-von Kármán equations*. J. Comput. Appl. Math., **190** (2006), 470-486.
6. A. Ciekot and S. Kukla, *Frequency analysis of a double-nanobeam-system*. J. Appl. Math. Comput. Mech., **13** (2014), 23-31.
7. M. Coti Zelati, *Global and exponential attractors for the singularly perturbed extensible beam*. Discrete Contin. Dyn. Syst., **25** (2009), 1041-1060.

8. M. Coti Zelati, C. Giorgi and V. Pata, *Steady states of the hinged extensible beam with external load*. Math. Models Methods Appl. Sci., **20** (2010), 43-58.

9. J.M. Davies, Lightweight sandwich construction, Wiley-Blackwell, Oxford, 2001.

10. R.W. Dickey, *Free vibrations and dynamic buckling of the extensible beam*. J. Math. Anal. Appl., **29** (1970), 443-454.

11. R.W. Dickey, *Dynamic stability of equilibrium states of the extensible beam*. Proc. Amer. Math. Soc., **41** (1973), 94-102.

12. A. Eden and A.J. Milani, *Exponential attractors for extensible beam equations*. Nonlinearity, **6** (1993), 457-479.

13. C. Giorgi and M.G. Naso, *Modeling and steady state analysis of the extensible thermoelastic beam*. Math. Comp. Modelling, **53** (2011), 896-908.

14. C. Giorgi, V. Pata and E. Vuk, *On the extensible viscoelastic beam*. Nonlinearity, **21** (2008), 713-733.

15. P. Holmes and J. Marsden, *A partial differential equation with infinitely many periodic orbits: chaotic oscillations of a forced beam*. Arch. Rational Mech. Anal., **76** (1981), 135-165.

16. N. Kamiya, *Governing equations for large deflections of sandwich plates*. AIAA Journal, **14** (1976), 250-253.

17. S.G. Kelly and S. Srinivas, *Free vibrations of elastically connected stretched beams*. J. Sound Vibration, **326** (2009), 883-893.

18. W. Lacarbonara, Nonlinear structural mechanics. Theory, dynamical phenomena and modeling, Springer, New York, 2013.

19. J.E. Lagnese and J.L. Lions, Modelling analysis and control of thin plates, Masson, Paris, 1988.

20. P.J. McKenna, *Oscillations in suspension bridges, vertical and torsional*. Discrete Contin. Dyn. Syst. Ser. S, **7** (2014), 785-791.

21. T. Murmu and S. Adhikari, *Axial instability of a double-nanobeam-systems*. Phys. Lett. A, **375** (2011), 601-608.

22. Z. Oniszczuk, *Forced transverse vibrations of an elastically connected complex simply supported double-beam system*. J. Sound Vibration, **264** (2003), 273-286.

23. F.J. Plantema, Sandwich construction: the bending and buckling of sandwich beams, plates, and shells, John Wiley and Sons, New York, 1966.

24. E.L. Reiss and B.J. Matkowsky, *Nonlinear dynamic buckling of a compressed elastic column*. Quart. Appl. Math., **29** (1971), 245-260.

25. R. Taylor and A. Wiles, *Ring-theoretic properties of certain Hecke algebras*. Ann. of Math., **141** (1995), 553-572.

26. H.V. Vu, A.M. Ordóñez and B.K. Karnopp, *Vibration of a double-beam system*. J. Sound Vibration, **229** (2000), 807-822.

27. D.H. Wang and G.F. Wang, *Surface effects on the vibration and buckling of double-nanobeam-systems*. Journal of Nanomaterials, vol. 2011 (2011), Article ID 518706, 7 pages.

28. A. Wiles, *Modular elliptic curves and Fermat's last theorem*. Ann. of Math., **141** (1995), 443-551.

29. S. Woinowsky-Krieger, *The effect of an axial force on the vibration of hinged bars*. J. Appl. Mech., **17** (1950), 35-36.

30. D. Zenkert, An introduction to sandwich construction, EMAS Publications, West Midlands, United Kingdom, 1995.

31. Y.Q. Zhang, Y. Lu and G.W. Ma, *Effect of compressive axial load on forced transverse vibrations of a double-beam system*. Int. J. Mech. Sci., **50** (2008), 299-305.

32. Y.Q. Zhang, Y. Lu, S.L. Wang and X. Liu, *Vibration and buckling of a double-beam system under compressive axial loading*. J. Sound Vibration, **318** (2008), 341-352.

5

Solvability for the non-isothermal Kobayashi–Warren–Carter system

Ken Shirakawa [1,*] **and Hiroshi Watanabe** [2]

[1] Department of Mathematics, Faculty of Education, Chiba University, 1-33, Yayoi-cho, Inage-ku, Chiba, 263-8522, Japan

[2] Department of Computer Science and Intelligent Systems, Faculty of Engineering, Oita University, 700 Dannoharu, Oita, 870-1192, Japan

[*] **Correspondence:** Email: sirakawa@faculty.chiba-u.jp

Abstract: In this paper, a system of parabolic type initial-boundary value problems are considered. The system $(S)_\nu$ is based on the non-isothermal model of grain boundary motion by [38], which was derived as an extending version of the "Kobayashi–Warren–Carter model" of grain boundary motion by [23]. Under suitable assumptions, the existence theorem of L^2-based solutions is concluded, as a versatile mathematical theory to analyze various Kobayashi–Warren–Carter type models.

Keywords: Non-isothermal grain boundary motion; Kobayashi–Warren–Carter type model; existence of L^2-based solution; weighted total variation; time-discretization

1. Introduction

Let $0 < T < \infty$ be a constant of time, and let $N \in \mathbb{N}$ be a constant of spatial dimension such that $1 \leq N \leq 3$. Let $\Omega \subset \mathbb{R}^N$ be a bounded domain such that $\Gamma := \partial\Omega$ is smooth when $N > 1$. Besides, let us denote by $Q := (0, T) \times \Omega$ the product space of the time-interval $(0, T)$ and the spatial domain Ω, and similarly, let us set $\Sigma := (0, T) \times \Gamma$.

In this paper, we fix a constant $\nu \geq 0$, and consider the following system of initial-boundary value problems of parabolic types, denoted by $(S)_\nu$.

$(S)_\nu$:

$$\begin{cases} [u - \lambda(w)]_t - \Delta u = f \quad \text{in } Q, \\ Du \cdot \boldsymbol{n}_\Gamma + n_0(u - f_\Gamma) = 0 \quad \text{on } \Sigma, \\ u(0, x) = u_0(x), \quad x \in \Omega; \end{cases} \tag{1.1}$$

$$\begin{cases} w_t - \Delta w + \partial\gamma(w) + g_w(w,\eta) + \lambda'(w)u \\ \qquad\qquad +\alpha_w(w,\eta)|D\theta| + \nu^2\beta_w(w,\eta)|D\theta|^2 \ni 0 \text{ in } Q, \\ Dw \cdot \boldsymbol{n}_\Gamma = 0 \quad \text{on } \Sigma, \\ w(0,x) = w_0(x), \quad x \in \Omega; \end{cases} \tag{1.2}$$

$$\begin{cases} \eta_t - \Delta\eta + g_\eta(w,\eta) + \alpha_\eta(w,\eta)|D\theta| + \nu^2\beta_\eta(w,\eta)|D\theta|^2 = 0 \quad \text{in } Q, \\ D\eta \cdot \boldsymbol{n}_\Gamma = 0 \quad \text{on } \Sigma, \\ \eta(0,x) = \eta_0(x), \quad x \in \Omega; \end{cases} \tag{1.3}$$

$$\begin{cases} \alpha_0(w,\eta)\theta_t - \operatorname{div}\left(\alpha(w,\eta)\dfrac{D\theta}{|D\theta|} + 2\nu^2\beta(w,\eta)D\theta\right) = 0 \quad \text{in } Q, \\ (\alpha(w,\eta)\frac{D\theta}{|D\theta|} + 2\nu^2\beta(w,\eta)D\theta) \cdot \boldsymbol{n}_\Gamma = 0 \quad \text{on } \Sigma, \\ \theta(0,x) = \theta_0(x), \quad x \in \Omega. \end{cases} \tag{1.4}$$

Here, Du, Dw, $D\eta$ and $D\theta$ denote, respectively, the (distributional) gradients of the unknowns u, w, η and θ on Ω. $f = f(t,x)$ is the source term on Q, $f_\Gamma = f_\Gamma(t,x)$ is the boundary source on Σ. $u_0 = u_0(x)$, $w_0 = w_0(x)$, $\eta_0 = \eta_0(x)$ and $\theta_0 = \theta_0(x)$ are given initial data on Ω. $\partial\gamma$ is the subdifferential of a proper lower semi-continuous (l.s.c.) and convex function $\gamma = \gamma(w)$ on $\mathbb{R}$. $\lambda = \lambda(w)$, $g = g(w,\eta)$, $\alpha_0 = \alpha_0(w,\eta)$, $\alpha = \alpha(w,\eta)$ and $\beta = \beta(w,\eta)$ are given real-valued functions, and the scripts "$'$", "$_w$" and "$_\eta$" denote differentials with respect to the corresponding variables. n_0 is a given positive constant, and $\boldsymbol{n}_\Gamma$ is the unit outer normal on Γ.

The system $(\mathrm{S})_\nu$ is based on the non-isothermal model of grain boundary motion by Warren et al. [36], which was derived as an extending version of the "Kobayashi–Warren–Carter model" of grain boundary motion by Kobayashi et al. [22, 23]. Hence, the study of this paper is based on the previous works related to the Kobayashi–Warren–Carter model (e.g., [13, 15, 16, 17, 20, 21, 22, 23, 25, 26, 28, 29, 30, 31, 32, 36, 37, 39]).

According to the modeling method of [36], the system $(\mathrm{S})_\nu$ is roughly configured as a coupled system of the heat equation in (1.1), and a gradient system {(1.2)–(1.4)} of the following governing energy, called *free-energy*:

$$\begin{aligned} \mathscr{E}_\nu(u,w,\eta,\theta) &:= \frac{1}{2}\int_\Omega |Dw|^2\,dx + \int_\Omega \gamma(w)\,dx + \int_\Omega u\lambda(w)\,dx \\ &+\frac{1}{2}\int_\Omega |D\eta|^2\,dx + \int_\Omega g(w,\eta)\,dx + \int_\Omega \alpha(w,\eta)\,d|D\theta| + \int_\Omega \beta(w,\eta)|D(\nu\theta)|^2\,dx, \end{aligned} \tag{1.5}$$

$$\text{for } [u,w,\eta,\theta] \in L^2(\Omega) \times H^1(\Omega) \times H^1(\Omega) \times BV(\Omega) \text{ with } \nu\theta \in H^1(\Omega).$$

In this context, the unknown $u = u(t,x)$ is the relative temperature with the critical degree 0, and the unknown $w = w(t,x)$ is an order parameter to indicate the solidification order of the polycrystal. The term $u - \lambda(w)$ in (1.1) is the so-called *enthalpy*, and then the term $\lambda(w)$ corresponds to the effect of the *latent heat*. The unknowns $\eta = \eta(t,x)$ and $\theta = \theta(t,x)$ are components of the vector field

$$(t,x) \in Q \mapsto \eta(t,x)\left[\cos\theta(t,x), \sin\theta(t,x)\right] \in \mathbb{R}^2,$$

which was adopted in [22, 23] as a vectorial phase field to reproduce the crystalline orientation in Q. Here, the components η and θ are order parameters to indicate, respectively, the orientation order and angle of the grain. In particular, w and η are taken to satisfy the constraints $0 \leq w, \eta \leq 1$ in Q, and the cases $[w, \eta] \approx [1, 1]$ and $[w, \eta] \approx [0, 0]$ are respectively assigned to "the solidified-oriented phase" and "the liquefied-disoriented phase" which correspond to two stable phases in physical.

In view of these, we suppose that

(g0) the function $w \in [0, 1] \mapsto \lambda(w) \in \mathbb{R}$ is increasing, and if the temperature u is closed to the critical value, i.e. $u \approx 0$, then the function

$$[u, w, \eta] \in \mathbb{R}^2 \mapsto \gamma(w) + g(w, \eta) - \lambda(w)u \in (-\infty, \infty]$$

has two minimums, around $[1, 1]$ and $[0, 0]$.

Besides, referring to the previous works on phase transitions (e.g., [7, 8, 14, 18, 19, 34, 35]), we can exemplify the following settings as possible expressions of the functions λ, γ and g in the above (g0):

(g1) (constrained setting by logarithmic function; cf. [14, 34, 35])

$$\begin{cases} \lambda(w) = Lw, \quad \gamma(w) := \dfrac{1}{2}\big(w \log w + (1-w)\log(1-w)\big) \\ \quad \text{with } \gamma(0) = \gamma(1) := 1, \\ g(w, \eta) := -\dfrac{L}{2}\left(w - \dfrac{1}{2}\right)^2 + \dfrac{c}{2}(w - \eta)^2, \end{cases} \qquad \text{for } w, \eta \in \mathbb{R},$$

(g2) (setting with non-smooth constraint; cf. [7, 8, 18, 19, 35])

$$\begin{cases} \lambda(w) = Lw, \quad \gamma(w) := I_{[0,1]}(w), \\ g(w, \eta) := -\dfrac{L}{2}\left(w - \dfrac{1}{2}\right)^2 + \dfrac{c}{2}(w - \eta)^2, \end{cases} \qquad \text{for } w, \eta \in \mathbb{R}.$$

Here, L and c are positive constants, and $I_{[0,1]} : \mathbb{R} \to \{0, \infty\}$ is the indicator function on the compact interval $[0, 1]$.

Now, the objective of this study is to generalize the line of recent results [25, 26, 28, 29, 30, 31, 32, 37, 39], and to obtain an enhanced theory which enables the versatile analysis for Kobayashi–Warren–Carter type systems, under various situations. To this end, we set the goal of this paper to specify the assumptions, which can cover the settings as in (g1)–(g2), and can guarantee the validity of the following Main Theorem.

Main Theorem: the existence theorem of the solution $[u, w, \eta, \theta]$ to the systems $(\mathrm{S})_\nu$, for any $\nu \geq 0$, which behaves in the range of $C([0, T]; L^2(\Omega)^4)$, with the L^2-based sources $f \in L^2(0, T; L^2(\Omega))$ and $f_\Gamma \in L^2(0, T; L^2(\Gamma))$.

The main theorem is somehow to enhance the results [25, 31, 32] concerned with qualitative properties of isothermal/non-isothermal Kobayashi–Warren–Carter type systems.

2. Preliminaries

First we elaborate the notations which is used throughout this paper.

Notation 1 (Real analysis). *For arbitrary* $a_0, b_0 \in [-\infty, \infty]$, *we define*

$$a_0 \vee b_0 := \max\{a_0, b_0\} \text{ and } a_0 \wedge b_0 := \min\{a_0, b_0\}.$$

Fix $d \in \mathbb{N}$ *as a constant of dimension. Then, we denote by* $|\boldsymbol{x}|$ *and* $\boldsymbol{x} \cdot \boldsymbol{y}$ *the Euclidean norm of* $\boldsymbol{x} \in \mathbb{R}^d$ *and the standard scalar product of* $\boldsymbol{x}, \boldsymbol{y} \in \mathbb{R}^d$, *respectively, as usual, i.e.:*

$$|\boldsymbol{x}| := \sqrt{x_1^2 + \cdots + x_d^2} \text{ and } \boldsymbol{x} \cdot \boldsymbol{y} := x_1 y_1 + \cdots + x_d y_d$$
$$\text{for all } \boldsymbol{x} = [x_1, \ldots, x_d],\ \boldsymbol{y} = [y_1, \ldots, y_d] \in \mathbb{R}^d.$$

The d-dimensional Lebesgue measure is denoted by $\mathscr{L}^d$, *and unless otherwise specified, the measure theoretical phrases, such as "a.e.", "dt", "dx", and so on, are with respect to the Lebesgue measure in each corresponding dimension. Also, in the observations on a smooth surface* $S \subset \mathbb{R}^d$, *the phrase "a.e." is with respect to the Hausdorff measure in each corresponding Hausdorff dimension, and the area element on* S *is denoted by* dS.

For a (Lebesgue) measurable function $f : B \to [-\infty, \infty]$ *on a Borel subset* $B \subset \mathbb{R}^d$, *we denote by* $[f]^+$ *and* $[f]^-$, *respectively, the positive and negative parts of* f, *i.e.,*

$$[f]^+(x) := f(x) \vee 0 \text{ and } [f]^-(x) := -(f(x) \wedge 0), \text{ a.e. } x \in B.$$

Notation 2 (Abstract functional analysis). *For an abstract Banach space X, we denote by* $|\cdot|_X$ *the norm of X, and when X is a Hilbert space, we denote by* $(\cdot, \cdot)_X$ *its inner product. For a subset A of a Banach space X, we denote by* $\mathrm{int}(A)$ *and* $\overline{A}$ *the interior and the closure of A, respectively.*

Fix $1 < d \in \mathbb{N}$. *Then, for a Banach space X, the topology of the product Banach space* X^d *is endowed with the norm:*

$$|z|_{X^d} := \sum_{k=1}^{d} |z_k|_X, \text{ for } z = [z_1, \ldots, z_d] \in X^d.$$

However, if X is a Hilbert space, then the topology of the product Hilbert space X^d *is endowed with the inner product:*

$$(z, \tilde{z})_{X^d} := \sum_{k=1}^{d} (z_k, \tilde{z}_k)_X, \text{ for } z = [z_1, \ldots, z_d] \in X^d \text{ and } \tilde{z} = [\tilde{z}_1, \ldots, \tilde{z}_d] \in X^d,$$

and hence, the norm in this case is provided by

$$|z|_{X^d} := \sqrt{(z, z)_{X^d}} = \left(\sum_{k=1}^{d} |z_k|_X^2 \right)^{1/2}, \text{ for } z = [z_1, \ldots, z_d] \in X^d.$$

For a Banach space X, we denote the dual space by X^*. *For a single-valued operator* $\mathscr{A} : X \to X^*$, *we write*

$$\mathscr{A} z = [\mathscr{A} z_1, \ldots, \mathscr{A} z_d] \in [X^*]^d \text{ for any } z = [z_1, \ldots, z_d] \in X^d.$$

For any proper lower semi-continuous (l.s.c. hereafter) and convex function Ψ *defined on a Hilbert space* X*, we denote by* $D(\Psi)$ *its effective domain, and denote by* $\partial\Psi$ *its subdifferential. The subdifferential* $\partial\Psi$ *is a set-valued map corresponding to a weak differential of* Ψ*, and it has a maximal monotone graph in the product Hilbert space* X^2*. More precisely, for each* $z_0 \in X$*, the value* $\partial\Psi(z_0)$ *is defined as the set of all elements* $z_0^* \in X$ *that satisfy the variational inequality*

$$(z_0^*, z - z_0)_X \leq \Psi(z) - \Psi(z_0) \text{ for any } z \in D(\Psi),$$

and the set $D(\partial\Psi) := \{z \in X \mid \partial\Psi(z) \neq \emptyset\}$ *is called the domain of* $\partial\Psi$*. We often use the notation "*$[z_0, z_0^*] \in \partial\Psi$ *in* X^2*" to mean "*$z_0^* \in \partial\Psi(z_0)$ *in* X *with* $z_0 \in D(\partial\Psi)$*" by identifying the operator* $\partial\Psi$ *with its graph in* X^2*.*

Notation 3 (Basic elliptic operators)**.** *Let* $V = H^1(\Omega)$ *be a Hilbert space endowed with the inner product:*

$$(w, z)_V := \int_\Omega \nabla w \cdot \nabla z dx + n_0 \int_\Gamma wz\, d\Gamma, \text{ for } [w, z] \in V^2,$$

and let $C_V > 0$ *be the embedding constant of* $V \subset L^2(\Omega)$*.*

Let $\langle \cdot, \cdot \rangle$ *be the duality pairing between* V *and the dual space* V^**, and let* $F\colon V \to V^*$ *be the duality mapping defined by*

$$\langle Fw, z\rangle := (w, z)_V, \quad \text{for } [w, z] \in V^2.$$

Note that V^* *forms a Hilbert space endowed with the inner product:*

$$(w^*, z^*)_{V^*} := \langle w^*, F^{-1}z^*\rangle, \text{ for } [w^*, z^*] \in (V^*)^2.$$

For any $\varrho \in L^2(\Omega)$ *and any* $\varrho_\Gamma \in L^2(\Gamma)$*, we can regard the vectorial function* $\boldsymbol{\varrho}^* := [\varrho, \varrho_\Gamma] \in L^2(\Omega) \times L^2(\Gamma)$ *as an element of* V^**, via the following variational form:*

$$\langle \boldsymbol{\varrho}^*, z\rangle := (\varrho, z)_{L^2(\Omega)} + n_0(\varrho_\Gamma, z)_{L^2(\Gamma)} \text{ for } z \in V. \tag{2.1}$$

Note that for any $\boldsymbol{\varrho}^* = [\varrho, \varrho_\Gamma] \in L^2(\Omega) \times L^2(\Gamma)$*, the variational form* (2.1) *enables the following identification:*

$$F\omega = \boldsymbol{\varrho}^* \text{ in } V^*, \text{ iff. } \omega \in H^2(\Omega) \text{ and } \begin{cases} -\varDelta\omega = \varrho \text{ in } L^2(\Omega), \\ D\omega \cdot \boldsymbol{n}_\Gamma + n_0(\omega - \varrho_\Gamma) = 0 \text{ in } L^2(\Gamma). \end{cases}$$

On this basis, the product space $L^2(\Omega) \times L^2(\Gamma)$ *can be regarded as a subspace of* V^**, and the restriction* $F|_{H^2(\Omega)} : H^2(\Omega) \to L^2(\Omega) \times L^2(\Gamma)$ *can be regarded as a bijective linear operator associated with the Laplacian, subject to Robin type boundary condition (cf. [24]).*

In the meantime, we denote by $\varDelta_N$ *the Laplacian operator subject to the zero-Neumann boundary condition, i.e.,*

$$\varDelta_N : z \in W_N := \left\{ z \in H^2(\Omega) \;\middle|\; Dz \cdot \boldsymbol{n}_\Gamma = 0 \text{ in } L^2(\Gamma) \right\} \subset L^2(\Omega) \mapsto \varDelta z \in L^2(\Omega).$$

Remark 1. *We here show some representative examples of the subdifferentials, which is intimately related to our study.*

(Ex.1) *The quadratic functional* $u \in L^2(\Omega) \mapsto \frac{1}{2}|u|^2_{L^2(\Omega)}$ *can be regarded as a proper l.s.c. and convex function on* V^*, *via the standard* ∞*-extension, and then, the* V^**-subdifferential of this function coincides with the duality map* $F : V \to V^*$, *i.e.:*

$$[u, u^*] \in \partial[\tfrac{1}{2}|\cdot|^2_{L^2(\Omega)}] \text{ in } [V^*]^2, \text{ iff. } u \in V \text{ and } u^* = Fu \text{ in } V^*.$$

(Ex.2) *Let* $d \in \mathbb{N}$, *and let* $\gamma_0 : \mathbb{R}^d \to \mathbb{R}$ *be a convex function defined as*

$$y = [y_1, \ldots, y_d] \in \mathbb{R}^d \mapsto \gamma_0(y) := \gamma_1(y_1) + \gamma_2(y_2) + \cdots + \gamma_d(y_d),$$

by using proper l.s.c. and convex functions $\gamma_k : \mathbb{R} \to (-\infty, \infty]$, *for* $k = 1, \ldots, d$. *Let* $\Psi^d_{\gamma_0} : L^2(\Omega)^d \to (-\infty, \infty]$ *be a proper l.s.c. and convex function defined as:*

$$z \in L^2(\Omega)^d \mapsto \Psi^d_{\gamma_0}(z) := \begin{cases} \dfrac{1}{2}\displaystyle\int_\Omega |Dz|^2_{\mathbb{R}^{N\times d}}\,dx + \int_\Omega \gamma_0(z)\,dx, \\ \qquad \text{if } z \in H^1(\Omega)^d, \\ \infty, \quad \text{otherwise.} \end{cases}$$

Then, with regard to the subdifferential $\partial\Psi^d_{\gamma_0} \subset [L^2(\Omega)^d]^2$, *it is known (see, e.g., [4, 6]) that*

$$z \in L^2(\Omega)^d \mapsto \partial\Psi^d_{\gamma_0}(z) = \begin{cases} \left\{ z^* \in L^2(\Omega)^d \;\middle|\; \begin{array}{l} z^* + \Delta_N z \in \partial\gamma_0(z) \text{ in} \\ \mathbb{R}^d, \text{ a.e. in } \Omega \end{array} \right\}, \\ \qquad \text{if } z \in W^d_N, \\ \emptyset, \quad \text{otherwise.} \end{cases}$$

This fact is often summarized as $\partial\Psi^d_{\gamma_0} = -\Delta_N + \partial\gamma_0$ *in* $[L^2(\Omega)^d]^2$.

Notation 4 (BV theory; cf. [2, 3, 11, 12]). *Let* $d \in \mathbb{N}$, *and let* $U \subset \mathbb{R}^d$ *be an open set. We denote by* $\mathcal{M}(U)$ *the space of all finite Radon measures on* U. *The space* $\mathcal{M}(U)$ *is known as the dual space of the Banach space* $C_0(U)$, *i.e.,* $\mathcal{M}(U) = C_0(U)^*$, *where* $C_0(U)$ *is the closure of the class of test functions* $C^\infty_c(U)$ *in the topology of* $C(\overline{U})$.

A function $z \in L^1(U)$ *is called a function of bounded variation on* U, *iff. its distributional gradient* Dz *is a finite Radon measure on* U, *namely,* $Dz \in \mathcal{M}(U)^d$. *Here, for any* $z \in BV(U)$, *the Radon measure* Dz *is called the variation measure of* z, *and its total variation* $|Dz|$ *is called the total variation measure of* z. *Additionally, for any* $z \in BV(U)$, *it holds that*

$$|Dz|(U) = \sup\left\{ \int_U z \operatorname{div}\varphi\,dx \;\middle|\; \varphi \in C^1_c(U)^d \text{ and } |\varphi| \le 1 \text{ on } U \right\}.$$

The space $BV(U)$ *is a Banach space, endowed with the norm*

$$|z|_{BV(U)} := |z|_{L^1(U)} + |Dz|(U) \text{ for any } z \in BV(U),$$

and we say that $z_n \to z$ *weakly-$*$ in* $BV(U)$, *iff.* $z \in BV(U)$, $\{z_n\}_{n=1}^\infty \subset BV(U)$, $z_n \to z$ *in* $L^1(U)$ *and* $Dz_n \to Dz$ *weakly-$*$ in* $\mathcal{M}(U)^d$, *as* $n \to \infty$.

The space $BV(U)$ has another topology, called "strict topology", which is provided by the following distance (cf. [2, Definition 3.14]):

$$[\varphi, \psi] \in BV(U)^2 \mapsto |\varphi - \psi|_{L^1(U)} + \big||D\varphi|(U) - |D\psi|(U)\big|.$$

In this regard, we say that $z_n \to z$ strictly in $BV(U)$ iff. $z \in BV(U)$, $\{z_n\}_{n=1}^{\infty} \subset BV(U)$, $z_n \to z$ in $L^1(U)$ and $|Dz_n|(U) \to |Dz|(U)$, as $n \to \infty$.

Specifically, when the boundary ∂U is Lipschitz, the Banach space $BV(U)$ is continuously embedded into $L^{d/(d-1)}(U)$ and compactly embedded into $L^p(U)$ for any $1 \le p < d/(d-1)$ (see, e.g., [2, Corollary 3.49] or [3, Theorems 10.1.3–10.1.4]). Furthermore, if $1 \le q < \infty$, then the space $C^\infty(\overline{U})$ is dense in $BV(U) \cap L^q(U)$ for the intermediate convergence, i.e., for any $z \in BV(U) \cap L^q(U)$, there exists a sequence $\{z_n\}_{n=1}^{\infty} \subset C^\infty(\overline{U})$ such that $z_n \to z$ in $L^q(U)$ and strictly in $BV(U)$, as $n \to \infty$ (see, e.g., [3, Definition 10.1.3 and Theorem 10.1.2]).

Notation 5 (Weighted total variation; cf. [1, 2])**.** *For any nonnegative $\varrho \in H^1(\Omega) \cap L^\infty(\Omega)$ (i.e. any $0 \le \varrho \in H^1(\Omega) \cap L^\infty(\Omega)$) and any $z \in L^2(\Omega)$, we call the value* $\mathrm{Var}_\varrho(z) \in [0, \infty]$, *defined as,*

$$\mathrm{Var}_\varrho(v) := \sup\left\{ \int_\Omega v \,\mathrm{div}\, \varpi \, dx \;\middle|\; \begin{array}{l} \varpi \in L^\infty(\Omega)^N \text{ with a compact sup-} \\ \text{port, and } |\varpi| \le \varrho \text{ a.e. in } \Omega \end{array} \right\} \in [0, \infty],$$

"the total variation of v weighted by ϱ", or the "weighted total variation" in short.

Remark 2. *Referring to the general theories (e.g., [1, 2, 5]), we can confirm the following facts associated with the weighted total variations.*

(Fact 1) *(Cf. [5, Theorem 5]) For any $0 \le \varrho \in H^1(\Omega) \cap L^\infty(\Omega)$, the functional $z \in L^2(\Omega) \mapsto \mathrm{Var}_\varrho(z) \in [0, \infty]$ is a proper l.s.c. and convex function that coincides with the lower semi-continuous envelope of*

$$z \in W^{1,1}(\Omega) \cap L^2(\Omega) \mapsto \int_\Omega \varrho |Dz| \, dx \in [0, \infty).$$

(Fact 2) *(Cf. [1, Theorem 4.3] and [2, Proposition 5.48]) If $0 \le \varrho \in H^1(\Omega) \cap L^\infty(\Omega)$ and $z \in BV(\Omega) \cap L^2(\Omega)$, then there exists a Radon measure $|Dz|_\varrho \in \mathcal{M}(\Omega)$ such that*

$$|Dz|_\varrho(\Omega) = \int_\Omega d|Dz|_\varrho = \mathrm{Var}_\varrho(z),$$

and

$$\begin{cases} |Dz|_\varrho(A) \le |\varrho|_{L^\infty(\Omega)} |Dz|(A), \\[1ex] |Dz|_\varrho(A) = \inf\left\{ \liminf\limits_{n \to \infty} \displaystyle\int_A \varrho |D\tilde{z}_n| \, dx \;\middle|\; \begin{array}{l} \{\tilde{z}_n\}_{n=1}^{\infty} \subset W^{1,1}(A) \cap L^2(A) \text{ such} \\ \text{that } \tilde{z}_n \to z \text{ in } L^2(A) \text{ as } n \to \\ \infty \end{array} \right\}, \end{cases} \tag{2.2}$$

for any open set $A \subset \Omega$.

(Fact 3) *If $\varrho \in H^1(\Omega) \cap L^\infty(\Omega)$, $c_\varrho := \operatorname{ess\,inf}_{x\in\Omega} \varrho > 0$, and $z \in BV(\Omega) \cap L^2(\Omega)$, then for any open set $A \subset \Omega$, it follows that*

$$\begin{cases} |Dz|_\varrho(A) \geq c_\varrho |Dz|(A) \ \text{for any open set } A \subset \Omega, \\ D(\mathrm{Var}_\varrho) = BV(\Omega) \cap L^2(\Omega), \ \text{and} \\ \mathrm{Var}_\varrho(z) = \sup\left\{ \int_\Omega z \,\mathrm{div}\,(\varrho\varpi)\,dx \ \middle|\ \begin{array}{l} \varpi \in L^\infty(\Omega)^N \text{ with a} \\ \text{compact support, and} \\ |\varpi| \leq 1 \text{ a.e. in } \Omega \end{array} \right\}. \end{cases} \tag{2.3}$$

Moreover, the following properties can be inferred from (2.2)–(2.3):

- $|Dz|_c = c|Dz|$ *in $\mathcal{M}(\Omega)$ for any constant $c \geq 0$ and $z \in BV(\Omega) \cap L^2(\Omega)$;*
- $|Dz|_\varrho = \varrho|Dz|\mathcal{L}^N$ *in $\mathcal{M}(\Omega)$, if $0 \leq \varrho \in H^1(\Omega) \cap L^\infty(\Omega)$ and $z \in W^{1,1}(\Omega) \cap L^2(\Omega)$.*

Notation 6 (Generalized weighted total variation; cf. [25, Section 2]). *For any $\varrho \in H^1(\Omega) \cap L^\infty(\Omega)$ and any $z \in BV(\Omega) \cap L^2(\Omega)$, we define a real-valued Radon measure $[\varrho|Dz|] \in \mathcal{M}(\Omega)$, as follows:*

$$[\varrho|Dz|](B) := |Dz|_{[\varrho]^+}(B) - |Dz|_{[\varrho]^-}(B) \ \text{for any Borel set } B \subset \Omega.$$

Note that $[\varrho|Dz|](\Omega)$ can be configured as a generalized total variation *of $z \in BV(\Omega) \cap L^2(\Omega)$ by the possibly sign-changing weight $\varrho \in H^1(\Omega) \cap L^\infty(\Omega)$.*

Remark 3. *With regard to the generalized weighted total variations, the following facts are verified in [25, Section 2].*

(Fact 4) *(Strict approximation) Let $\varrho \in H^1(\Omega) \cap L^\infty(\Omega)$ and $z \in BV(\Omega) \cap L^2(\Omega)$ be arbitrary fixed functions, and let $\{z_n\}_{n=1}^\infty \subset C^\infty(\overline{\Omega})$ be a sequence such that*

$$z_n \to z \text{ in } L^2(\Omega) \ \text{ and } \ \text{strictly in } BV(\Omega) \text{ as } n \to \infty.$$

Then

$$\int_\Omega \varrho|Dz_n|\,dx \to \int_\Omega d[\varrho|Dz|] \ \text{ as } n \to \infty.$$

(Fact 5) *For any $z \in BV(\Omega) \cap L^2(\Omega)$, the mapping*

$$\varrho \in H^1(\Omega) \cap L^\infty(\Omega) \mapsto \int_\Omega d[\varrho|Dz|] \in \mathbb{R}$$

is a linear functional, and moreover, if $\varphi \in H^1(\Omega) \cap C(\overline{\Omega})$ and $\varrho \in H^1(\Omega) \cap L^\infty(\Omega)$, then

$$\int_\Omega d[\varphi\varrho|Dz|] = \int_\Omega \varphi\, d[\varrho|Dz|].$$

Finally, we mention the notion of functional convergences.

Definition 1 (Mosco convergence; cf. [27]). *Let X be an abstract Hilbert space. Let $\Psi : X \to (-\infty, \infty]$ be a proper l.s.c. and convex function, and let $\{\Psi_n\}_{n=1}^\infty$ be a sequence of proper l.s.c. and convex functions $\Psi_n : X \to (-\infty, \infty]$, $n = 1, 2, 3, \ldots$. We say that $\Psi_n \to \Psi$ on X, in the sense of Mosco, as $n \to \infty$, iff. the following two conditions are fulfilled.*

The condition of lower bound: $\liminf_{n\to\infty} \Psi_n(z_n^\circ) \geq \Psi(z^\circ)$, *if* $z^\circ \in X$, $\{z_n^\circ\}_{n=1}^\infty \subset X$, *and* $z_n^\circ \to z^\circ$ *weakly in* X *as* $n \to \infty$.

The condition of optimality: *for any* $z^\bullet \in D(\Psi)$, *there exists a sequence* $\{z_n^\bullet\}_{n=1}^\infty \subset X$ *such that* $z_n^\bullet \to z^\bullet$ *in* X *and* $\Psi_n(z_n^\bullet) \to \Psi(z^\bullet)$ *as* $n \to \infty$.

Definition 2 (Γ-convergence; cf. [9]). *Let* X *be an abstract Hilbert space,* $\Psi : X \to (-\infty, \infty]$ *be a proper functional, and* $\{\Psi_n\}_{n=1}^\infty$ *be a sequence of proper functionals* $\Psi_n : X \to (-\infty, \infty]$, $n = 1, 2, 3, \ldots$. *We say that* $\Psi_n \to \Psi$ *on* X*, in the sense of* Γ*-convergence, as* $n \to \infty$*, iff. the following two conditions are fulfilled.*

The condition of lower bound: $\liminf_{n\to\infty} \Psi_n(z_n^\circ) \geq \Psi(z^\circ)$, *if* $z^\circ \in X$, $\{z_n^\circ\}_{n=1}^\infty \subset X$, *and* $z_n^\circ \to z^\circ$ *(strongly) in* X *as* $n \to \infty$.

The condition of optimality: *for any* $z^\bullet \in D(\Psi)$, *there exists a sequence* $\{z_n^\bullet\}_{n=1}^\infty \subset X$ *such that* $z_n^\bullet \to z^\bullet$ *in* X *and* $\Psi_n(z_n^\bullet) \to \Psi(z^\bullet)$ *as* $n \to \infty$.

Remark 4. *Note that if the functionals are convex, then Mosco convergence implies* Γ*-convergence, i.e., the* Γ*-convergence of convex functions can be regarded as a weak version of Mosco convergence. Additionally, in the* Γ*-convergence of convex functions, we can see the following:*

(Fact 6) *Let* $\Psi : X \to (-\infty, \infty]$ *and* $\Psi_n : X \to (-\infty, \infty]$ *be proper l.s.c. and convex functions on a Hilbert space* X *such that* $\Psi_n \to \Psi$ *on* X*, in the sense of* Γ*-convergence, as* $n \to \infty$*. If it holds that:*

$$\begin{cases} [z, z^*] \in X^2, \quad [z_n, z_n^*] \in \partial\Psi_n \text{ in } X^2, \ n = 1, 2, 3, \ldots, \\ z_n \to z \text{ in } X \text{ and } z_n^* \to z^* \text{ weakly in } X, \text{ as } n \to \infty, \end{cases}$$

then $[z, z^*] \in \partial\Psi$ *in* X^2 *and* $\Psi_n(z_n) \to \Psi(z)$ *as* $n \to \infty$.

3. Main Theorem and the demonstration scenario

Throughout the paper, we set the following assumptions.

(A1) Let $f \in L^2(0, T; L^2(\Omega))$ and $f_\Gamma \in L^2(0, T; L^2(\Gamma))$ be given functions, and let $\boldsymbol{f}^* := [f, f_\Gamma] \in L^2(0, T; L^2(\Omega) \times L^2(\Gamma))$ be a time-dependent vectorial function which is regarded as $\boldsymbol{f}^* \in L^2(0, T; V^*)$, via (2.1) applied to $\boldsymbol{\varrho}^* = \boldsymbol{f}^*(t)$ for a.e. $t > 0$.

(A2) Let $\lambda \in W^{2,\infty}_{\mathrm{loc}}(\mathbb{R})$ be a function, and let $A_* > 0$ be a constant which is defined as:

$$A_* := \frac{1}{4(1 + C_V^2 |\lambda|^2_{W^{2,\infty}(0,1)})},$$

by using the embedding constant $C_V > 0$ of $V \subset L^2(\Omega)$.

(A3) Let $\alpha_0 \in W^{1,\infty}_{\mathrm{loc}}(\mathbb{R}^2)$ and $\alpha, \beta \in C^2(\mathbb{R}^2)$ be functions, such that:

- α and β are convex on $\mathbb{R}^2$;
- $\delta_* := \inf[\alpha_0(\mathbb{R}^2) \cup \alpha(\mathbb{R}^2) \cup \beta(\mathbb{R}^2)] > 0$;

- $\alpha_\eta(w, 0) \le 0$, $\beta_\eta(w, 0) \le 0$, $\alpha_\eta(w, 1) \ge 0$, and $\beta_\eta(w, 1) \ge 0$, for any $w \in [0, 1]$.

(A4) Let $\gamma : \mathbb{R} \to (-\infty, \infty]$ be a proper l.s.c. and convex function, such that $D(\gamma) = [0, 1]$.

(A5) Let $g \in C^2(\mathbb{R}^2)$ be a function such that

$$g_\eta(w, 0) \le 0 \text{ and } g_\eta(w, 1) \ge 0, \text{ for any } w \in [0, 1].$$

(A6) There exists a constant c_* such that $\gamma(w) + g(\boldsymbol{v}) \ge c_*$, for any $\boldsymbol{v} = [w, \eta] \in \mathbb{R}^2$.

(A7) Let $[u_0, \boldsymbol{v}_0, \theta_0] = [u_0, w_0, \eta_0, \theta_0]$ is a quartet of initial data, such that:

$$[u_0, w_0, \eta_0, \theta_0] \in \begin{cases} D_0 := \left\{ [\tilde{u}, \tilde{w}, \tilde{\eta}, \tilde{\theta}] \;\middle|\; \begin{array}{l} \tilde{u} \in L^2(\Omega),\ \tilde{w}, \tilde{\eta} \in H^1(\Omega), \\ \tilde{\theta} \in BV(\Omega) \cap L^\infty(\Omega), \text{ and} \\ 0 \le \tilde{w}, \tilde{\eta} \le 1 \text{ a.e. in } \Omega \end{array} \right\}, \\ \quad \text{if } \nu = 0, \\ D_1 := D_0 \cap [L^2(\Omega) \times H^1(\Omega) \times H^1(\Omega) \times H^1(\Omega)], \\ \quad \text{if } \nu > 0. \end{cases}$$

Now, for simplicity of description, we prepare the following notations:

$$\begin{cases} G(u; \boldsymbol{v}) = G(u; w, \eta) := g(w, \eta) + u\lambda(w), \\ [\nabla g](\boldsymbol{v}) = [\nabla g](w, \eta) := [g_w(w, \eta), g_\eta(w, \eta)], \\ [\nabla G](u; \boldsymbol{v}) = [\nabla G](u; w, \eta) := [g_w(w, \eta) + u\lambda'(w), g_\eta(w, \eta)], \end{cases}$$

and

$$\begin{cases} [\nabla \alpha](\boldsymbol{v}) = [\nabla \alpha](w, \eta) := [\alpha_w(w, \eta), \alpha_\eta(w, \eta)], \\ [\nabla \beta](\boldsymbol{v}) = [\nabla \beta](w, \eta) := [\beta_w(w, \eta), \beta_\eta(w, \eta)], \end{cases}$$
$$\text{for } u \in \mathbb{R} \text{ and } \boldsymbol{v} = [w, \eta] \in \mathbb{R}^2.$$

For any $\nu \ge 0$ and any $\boldsymbol{v} = [w, \eta] \in [H^1(\Omega) \cap L^\infty(\Omega)]^2$, we define a proper l.s.c. and convex function $\Phi_\nu(\boldsymbol{v}; \cdot)$ on $L^2(\Omega)$ by letting:

$$\theta \in L^2(\Omega) \mapsto \Phi_\nu(\boldsymbol{v}; \theta) = \Phi_\nu(w, \eta; \theta) := \begin{cases} \displaystyle\int_\Omega d[\alpha(\boldsymbol{v})|D\theta|] + \int_\Omega \beta(\boldsymbol{v})|D(\nu\theta)|^2\,dx, \\ \quad \text{if } \theta \in BV(\Omega) \text{ and } \nu\theta \in H^1(\Omega), \\ \infty, \quad \text{otherwise.} \end{cases}$$

Additionally, we set:

$$B_* := \frac{1 + A_*}{2}, \text{ by using the constant } A_* \text{ as in (A2)}, \tag{3.1}$$

and define a functional $\mathscr{F}_\nu$ on $L^2(\Omega)^4$ by letting:

$$\begin{aligned} [u, \boldsymbol{v}, \theta] = [u, w, \eta, \theta] \in L^2(\Omega)^4 \mapsto \mathscr{F}_\nu(u, \boldsymbol{v}, \theta) &= \mathscr{F}_\nu(u, w, \eta, \theta) \\ := B_*|u|^2_{L^2(\Omega)} + \Psi^2_\gamma(\boldsymbol{v}) + \int_\Omega (g(\boldsymbol{v}) - c_*)\,dx &+ \Phi_\nu(\boldsymbol{v}; \theta), \end{aligned} \tag{3.2}$$

where Ψ^2_γ is the convex function $\Psi^d_{\gamma_0}$ in Remark 1 in the case when $d = 2$ and $\gamma_0 = \gamma$. The above functional $\mathscr{F}_\nu$ is a modified version of the free-energy as in (1.5), and the assumptions (A3)–(A6) guarantee the non-negativity of this functional, i.e. $\mathscr{F}_\nu \geq 0$ on $L^2(\Omega)^4$.

Based on these, we define the solutions to the systems $(\mathrm{S})_\nu$, for $\nu \geq 0$, as follows.

Definition 3. *For any $\nu \geq 0$, a quartet $[u, \boldsymbol{v}, \theta] = [u, w, \eta, \theta] \in L^2(0, T; L^2(\Omega)^4)$ with $\boldsymbol{v} = [w, \eta]$ is called a solution to $(S)_\nu$, iff. $[u, \boldsymbol{v}, \theta]$ fulfills the following (S1)–(S6).*

(S1) $u \in W^{1,2}(0, T; V^*) \cap L^\infty(0, T; L^2(\Omega)) \cap L^2(0, T; V) \subset C([0, T]; L^2(\Omega))$.

(S2) $\boldsymbol{v} = [w, \eta] \in W^{1,2}(0, T; L^2(\Omega)^2) \cap L^\infty(0, T; H^1(\Omega)^2)$, *and*

$$0 \leq w(t, x) \leq 1 \text{ and } 0 \leq \eta(t, x) \leq 1, \text{ a.e. } (t, x) \in Q.$$

(S3) $\theta \in W^{1,2}(0, T; L^2(\Omega)) \cap L^\infty(Q)$, $|D\theta(\cdot)|(\Omega) \in L^\infty(0, T)$, $\nu\theta \in L^\infty(0, T; H^1(\Omega))$, *and* $|\theta| \leq |\theta_0|_{L^\infty(\Omega)}$ *a.e. in* Q.

(S4) u satisfies the following variational form:

$$\langle [u - \lambda(w)]_t(t), z\rangle + (Du(t), Dz)_{L^2(\Omega)^N} + n_0(u(t), z)_{L^2(\Gamma)}$$
$$= (f(t), z)_{L^2(\Omega)} + n_0(f_\Gamma(t), z)_{L^2(\Gamma)}, \text{ for any } z \in V, \text{ and a.e. } t \in (0, T),$$

with the initial condition $u(0) = u_0$ in $L^2(\Omega)$.

(S5) $\boldsymbol{v} = [w, \eta]$ satisfies the following two variational forms:

$$(w_t(t) + g_w(\boldsymbol{v})(t) + u(t)\lambda'(w(t)), w(t) - \varphi)_{L^2(\Omega)} + (Dw(t), D(w(t) - \varphi))_{L^2(\Omega)^N}$$
$$+ \int_\Omega d[(w(t) - \varphi)\alpha_w(\boldsymbol{v}(t))|D\theta(t)|] + \int_\Omega (w(t) - \varphi)\beta_w(\boldsymbol{v}(t))|D(\nu\theta)(t)|^2\, dx$$
$$+ \int_\Omega \gamma(w(t))\, dx \leq \int_\Omega \gamma(\varphi) dx, \text{ for any } \varphi \in H^1(\Omega) \cap L^\infty(\Omega) \text{ and a.e. } t \in (0, T),$$

and

$$\left(\eta_t(t) + g_\eta(\boldsymbol{v})(t), \psi\right)_{L^2(\Omega)} + (D\eta(t), D\psi)_{L^2(\Omega)^N}$$
$$+ \int_\Omega d[\psi\alpha_\eta(\boldsymbol{v}(t))|D\theta(t)|] + \int_\Omega \psi\beta_\eta(\boldsymbol{v}(t))|D(\nu\theta)(t)|^2\, dx = 0,$$
$$\text{for any } \psi \in H^1(\Omega) \cap L^\infty(\Omega) \text{ and a.e. } t \in (0, T),$$

with the initial condition $\boldsymbol{v}(0) = [w(0), \eta(0)] = \boldsymbol{v}_0 = [w_0, \eta_0]$ in $L^2(\Omega)^2$.

(S6) θ satisfies the following variational form:

$$(\alpha_0(\boldsymbol{v}(t))\theta_t(t), \theta(t) - \omega)_{L^2(\Omega)} + \Phi_\nu(\boldsymbol{v}(t); \theta(t)) \leq \Phi_\nu(\boldsymbol{v}(t); \omega),$$
$$\text{for any } \omega \in D(\Phi_\nu(\boldsymbol{v}(t); \cdot)) \text{ and a.e. } t \in (0, T),$$

with the initial condition $\theta(0) = \theta_0$ in $L^2(\Omega)$.

Remark 5. *The variational identity in the above (S4) can be reformulated as:*

$$[u - \lambda(w)]_t(t) + Fu(t) = \boldsymbol{f}^*(t) \quad in\ V^*, \quad for\ a.e.\ t \in (0, T). \tag{3.3}$$

Also, two variational forms in (S5) can be reduced to:

$$\begin{aligned}
&(\boldsymbol{v}_t(t) + [\nabla G](u; \boldsymbol{v}(t)), \boldsymbol{v}(t) - \boldsymbol{\varpi})_{L^2(\Omega)^2} \\
&+(D\boldsymbol{v}(t), D(\boldsymbol{v}(t) - \boldsymbol{\varpi}))_{L^2(\Omega)^{N\times 2}} \\
&+ \int_\Omega d[|D\theta(t)|(\boldsymbol{v}(t) - \boldsymbol{\varpi}) \cdot [\nabla\alpha](\boldsymbol{v}(t))] \\
&+ \int_\Omega |D(\nu\theta)(t)|^2(\boldsymbol{v}(t) - \boldsymbol{\varpi}) \cdot [\nabla\beta](\boldsymbol{v}(t))\, dx \\
&+ \int_\Omega \gamma(\boldsymbol{v}(t))\, dx \le \int_\Omega \gamma(\boldsymbol{\varpi})\, dx, \\
&for\ any\ \boldsymbol{\varpi} = [\varphi, \psi] \in [H^1(\Omega) \cap L^\infty(\Omega)]^2\ and\ a.e.\ t \in (0, T),
\end{aligned} \tag{3.4}$$

by using the identification

$$\gamma(\tilde{\boldsymbol{v}}) := \gamma(\tilde{w}), \quad for\ all\ \tilde{\boldsymbol{v}} = [\tilde{w}, \tilde{\eta}] \in \mathbb{R}^2,$$

and by using the abbreviation:

$$\begin{aligned}
&\int_\Omega d[|D\tilde{\theta}|\boldsymbol{\varpi} \cdot \tilde{\boldsymbol{v}}] := \int_\Omega d[\varphi\tilde{w}|D\tilde{\theta}|] + \int_\Omega d[\psi\tilde{\eta}|D\tilde{\theta}|], \\
&for\ \tilde{\boldsymbol{v}} = [\tilde{w}, \tilde{\eta}],\ \boldsymbol{\varpi} = [\varphi, \psi] \in [H^1(\Omega) \cap L^\infty(\Omega)]^2\ and\ \tilde{\theta} \in BV(\Omega) \cap L^2(\Omega).
\end{aligned} \tag{3.5}$$

Furthermore, the variational form in (S6) is equivalent to the following evolution equation:

$$\alpha_0(\boldsymbol{v}(t))\theta_t(t) + \partial\Phi_\nu(\boldsymbol{v}(t); \theta(t)) \ni 0\ in\ L^2(\Omega),\ a.e.\ t \in (0, T), \tag{3.6}$$

governed by the subdifferential $\partial\Phi_\nu(\boldsymbol{v}(t); \cdot) \subset L^2(\Omega)^2$ *of the time-dependent convex function* $\Phi_\nu(\boldsymbol{v}(t); \cdot)$, *for* $t \in (0, T)$.

Now, our Main Theorem is stated as follows.

Main Theorem *Let* $\nu \ge 0$ *be a fixed constant. Then, under (A1)-(A7), the system* $(S)_\nu$ *admits at least one solution* $[u, \boldsymbol{v}, \theta] = [u, w, \eta, \theta] \in L^2(0, T; L^2(\Omega)^4)$ *with* $\boldsymbol{v} = [w, \eta]$.

Remark 6. *Note that the presence of mobilities* $\alpha_0 = \alpha_0(w, \eta)$, $\alpha = \alpha(w, \eta)$ *and* $\beta = \beta(w, \eta)$ *makes the uniqueness problems for the systems* $(S)_\nu$, $\nu \ge 0$, *be quite tough. In fact, even if we overview the kindred works to this study, we can find only two cases [15, Theorem 2.2] and [40, Theorem 2.2] that obtained the uniqueness results under some restricted situations.*

Finally, we devote the remaining part of this Section to show the sketch of the demonstration scenario, since the proof of the Main Theorem is going to be extended.

In this paper, the Main Theorem will be obtained as a consequence of some approximating approaches, and then, the approximating problems will be associated with the time-discretization versions of (3.3)–(3.6), under positive setting of the constant ν. Hence, when we consider the approximating problems, we suppose $\nu > 0$, and fix the constant of time-step $h \in (0,1]$. Also, we denote by $[f]_0^{\mathrm{ex}} \in L^2(\mathbb{R}; L^2(\Omega))$, $[f_\Gamma]_0^{\mathrm{ex}} \in L^2(\mathbb{R}; L^2(\Gamma))$ and $[f^*]_0^{\mathrm{ex}} \in L^2(\mathbb{R}; V^*)$ the zero-extensions of f, f_Γ and f^* $(= [f, f_\Gamma])$, respectively.

On this basis, the approximating problem for our system $(\mathrm{S})_\nu$ is denoted by $(\mathrm{AP})_h^\nu$, and stated as follows.

$(\mathrm{AP})_h^\nu$: to find a sequence $\{[u_i^\nu, \boldsymbol{v}_i^\nu, \theta_i^\nu]\}_{i=1}^\infty \subset D_1$ with $\{\boldsymbol{v}_i^\nu\}_{i=1}^\infty = \{[w_i^\nu, \eta_i^\nu]\}_{i=1}^\infty$, which fulfills that

$$\frac{u_i^\nu - u_{i-1}^\nu}{h} - \lambda'(w_i^\nu)\frac{w_i^\nu - w_{i-1}^\nu}{h} + F u_i^\nu = [f_i^*]^h \text{ in } V^*, \tag{3.7}$$

$$\begin{aligned}
&\frac{1}{h}(\boldsymbol{v}_i^\nu - \boldsymbol{v}_{i-1}^\nu, \boldsymbol{v}_i^\nu - \boldsymbol{\varpi})_{L^2(\Omega)^2} + (D\boldsymbol{v}_i^\nu, D(\boldsymbol{v}_i^\nu - \boldsymbol{\varpi}))_{L^2(\Omega)^{N\times 2}} \\
&\quad + ([\nabla G](u_i^\nu; \boldsymbol{v}_i^\nu), \boldsymbol{v}_i^\nu - \boldsymbol{\varpi})_{L^2(\Omega)^2} + \int_\Omega \gamma(\boldsymbol{v}_i^\nu)\, dx \\
&\quad + \int_\Omega (\boldsymbol{v}_i^\nu - \boldsymbol{\varpi}) \cdot \left(|D\theta_{i-1}^\nu|[\nabla\alpha](\boldsymbol{v}_i^\nu) + \nu^2|D\theta_{i-1}^\nu|^2[\nabla\beta](\boldsymbol{v}_i^\nu)\right) dx \\
&\quad \le \int_\Omega \gamma(\boldsymbol{\varpi})\, dx, \text{ for any } \boldsymbol{\varpi} \in [H^1(\Omega) \cap L^\infty(\Omega)]^2,
\end{aligned} \tag{3.8}$$

$$\alpha_0(\boldsymbol{v}_i^\nu)\frac{\theta_i^\nu - \theta_{i-1}^\nu}{h} + \partial\Phi_\nu(\boldsymbol{v}_i^\nu; \theta_i^\nu) \ni 0 \text{ in } L^2(\Omega), \tag{3.9}$$

for $i = 1, 2, 3, \dots$, starting from the initial data:

$$[u_0^\nu, \boldsymbol{v}_0^\nu, \theta_0^\nu] \in D_1 \text{ with } \boldsymbol{v}_0^\nu = [w_0^\nu, \eta_0^\nu].$$

In the context, for any $i \in \mathbb{N}$, $[f_i^*]^h = [f_i^h, f_{\Gamma,i}^h] \in L^2(\Omega) \times L^2(\Gamma)\ (\subset V^*)$, consists of the components:

$$f_i^h := \frac{1}{h}\int_{(i-1)h}^{ih} [f]_0^{\mathrm{ex}}(\tau)\, d\tau \text{ in } L^2(\Omega) \text{ and } f_{\Gamma,i}^h := \frac{1}{h}\int_{(i-1)h}^{ih} [f_\Gamma]_0^{\mathrm{ex}}(\tau)\, d\tau \text{ in } L^2(\Gamma).$$

Hence, before the proof of Main Theorem, it will be needed to verify the following theorem.

Theorem 1 (Solvability of the approximating problem). *There exists a small constant $h_1^\circ \in (0,1]$ such that if $\nu > 0$ and $h \in (0, h_1^\circ]$, then the approximating problem $(AP)_h^\nu$ admits a unique solution $\{[u_i^\nu, \boldsymbol{v}_i^\nu, \theta_i^\nu]\}_{i=1}^\infty \subset D_1$, and moreover,*

$$\begin{aligned}
&\frac{A_*}{2h}|u_i^\nu - u_{i-1}^\nu|_{V^*}^2 + \frac{1}{2h}|\boldsymbol{v}_i^\nu - \boldsymbol{v}_{i-1}^\nu|_{L^2(\Omega)^2}^2 + \frac{1}{h}\left|\sqrt{\alpha_0(\boldsymbol{v}_i^\nu)}(\theta_i^\nu - \theta_{i-1}^\nu)\right|_{L^2(\Omega)}^2 + \frac{h}{2}|u_i^\nu|_V^2 \\
&\quad + \mathscr{F}_\nu(u_i^\nu, \boldsymbol{v}_i^\nu, \theta_i^\nu) \le \mathscr{F}_\nu(u_{i-1}^\nu, \boldsymbol{v}_{i-1}^\nu, \theta_{i-1}^\nu) + h|[f_i^*]^h|_{V^*}^2, \text{ for } i = 1, 2, 3, \dots,
\end{aligned} \tag{3.10}$$

where A_* is the constant as in (A2).

However, due to the presence of L^1-terms $\nu^2|D\theta_{i-1}|^2[\nabla\beta](\boldsymbol{v}_i^\nu) \in L^1(\Omega)^2$, $i = 1, 2, 3, \ldots$, in (3.8), the above Theorem 1 will not be a straightforward consequence of standard variational method, and in fact, this theorem will be obtained via further approximating approach by means of some relaxed systems for $(\mathrm{AP})_h^\nu$.

In the observation of the relaxed system, we first fix a large constant $M > (N+2)/2$, and fix a small constant $\varepsilon \in (0, 1]$ as the relaxation index. Besides, we define

$$D_M := D_1 \cap [L^2(\Omega) \times H^1(\Omega) \times H^1(\Omega) \times H^M(\Omega)],$$

and for any $\tilde{\boldsymbol{v}} \in L^2(\Omega)^2$, we define a relaxed functional $\Phi_\varepsilon^\nu(\tilde{\boldsymbol{v}};\cdot)$ for $\Phi_\nu(\tilde{\boldsymbol{v}};\cdot)$, by letting:

$$\theta \in L^2(\Omega) \mapsto \Phi_\varepsilon^\nu(\tilde{\boldsymbol{v}};\theta) := \begin{cases} \Phi_\nu(\tilde{\boldsymbol{v}};\theta) + \dfrac{\varepsilon^2}{2}|\theta|_{H^M(\Omega)}^2, & \text{if } \theta \in H^M(\Omega), \\ \infty, & \text{otherwise.} \end{cases}$$

Note that for any $\tilde{\boldsymbol{v}} \in L^2(\Omega)^2$, the functional $\Phi_\varepsilon^\nu(\tilde{\boldsymbol{v}};\cdot)$ is proper l.s.c. and convex on $L^2(\Omega)$, such that:

$$D(\Phi_\varepsilon^\nu(\tilde{\boldsymbol{v}};\cdot)) = H^M(\Omega) \subset W^{1,\infty}(\Omega),$$

and hence, the L^2-subdifferential $\partial\Phi_\varepsilon^\nu(\tilde{\boldsymbol{v}};\cdot)$ is a maximal monotone graph in $L^2(\Omega)^2$.

On this basis, we denote by $(\mathrm{RX})_\varepsilon$ the relaxed system for $(\mathrm{AP})_h^\nu$, and prescribe the system $(\mathrm{RX})_\varepsilon$ as follows.

$(\mathrm{RX})_\varepsilon$: to find a sequence $\{[u_{\varepsilon,i}^\nu, \boldsymbol{v}_{\varepsilon,i}^\nu, \theta_{\varepsilon,i}^\nu]\}_{i=1}^\infty \subset D_M$ with $\{\boldsymbol{v}_{\varepsilon,i}^\nu\}_{i=1}^\infty = \{[w_{\varepsilon,i}^\nu, \eta_{\varepsilon,i}^\nu]\}_{i=1}^\infty$, which fulfills that

$$\frac{u_{\varepsilon,i}^\nu - u_{\varepsilon,i-1}^\nu}{h} - \lambda'(w_{\varepsilon,i}^\nu)\frac{w_{\varepsilon,i}^\nu - w_{\varepsilon,i-1}^\nu}{h} + Fu_{\varepsilon,i}^\nu = [f_i^*]^h \text{ in } V^*, \tag{3.11}$$

$$\begin{aligned} &\frac{\boldsymbol{v}_{\varepsilon,i}^\nu - \boldsymbol{v}_{\varepsilon,i-1}^\nu}{h} - \Delta_N \boldsymbol{v}_{\varepsilon,i}^\nu + \partial\gamma(\boldsymbol{v}_{\varepsilon,i}^\nu) + [\nabla G](u_{\varepsilon,i}^\nu;\boldsymbol{v}_{\varepsilon,i}^\nu) \\ &\quad + |D\theta_{\varepsilon,i-1}^\nu|[\nabla\alpha](\boldsymbol{v}_{\varepsilon,i}^\nu) + \nu^2|D\theta_{\varepsilon,i-1}^\nu|^2[\nabla\beta](\boldsymbol{v}_{\varepsilon,i}^\nu) \ni 0 \text{ in } L^2(\Omega)^2, \end{aligned} \tag{3.12}$$

$$\alpha_0(\boldsymbol{v}_{\varepsilon,i}^\nu)\frac{\theta_{\varepsilon,i}^\nu - \theta_{\varepsilon,i-1}^\nu}{h} + \partial\Phi_\varepsilon^\nu(\boldsymbol{v}_{\varepsilon,i}^\nu;\theta_{\varepsilon,i}^\nu) \ni 0 \text{ in } L^2(\Omega), \tag{3.13}$$

for $i = 1, 2, 3, \ldots$, starting from the initial data:

$$[u_{\varepsilon,0}^\nu, \boldsymbol{v}_{\varepsilon,0}^\nu, \theta_{\varepsilon,0}^\nu] \in D_M \text{ with } \boldsymbol{v}_{\varepsilon,0}^\nu = [w_{\varepsilon,0}^\nu, \eta_{\varepsilon,0}^\nu].$$

Then, we can see that

$$|D\theta_{\varepsilon,i-1}^\nu| \in L^\infty(\Omega) \text{ and } \nu^2|D\theta_{\varepsilon,i-1}^\nu|^2[\nabla\beta](\boldsymbol{v}_{\varepsilon,i}^\nu) \in L^\infty(\Omega)^2, \ i = 1, 2, 3, \ldots.$$

It implies that the general theories of L^2-subdifferentials will be available for the relaxed system $(\mathrm{RX})_\varepsilon$.

Thus, it will be needed to verify the following proposition, as the first task to proving the Main Theorem.

Proposition 1. *There exists a small constant* $h_0^\circ \in (0, 1]$, *such that if* $h \in (0, h_0^\circ]$, *then the system* $(\mathrm{RX})_\varepsilon$ *admits a unique solution* $\{[u_{\varepsilon,i}^\nu, \boldsymbol{v}_{\varepsilon,i}^\nu, \theta_{\varepsilon,i}^\nu]\}_{i=1}^\infty \subset D_M$ *with* $\{\boldsymbol{v}_{\varepsilon,i}^\nu\}_{i=1}^\infty = \{[w_{\varepsilon,i}^\nu, \eta_{\varepsilon,i}]^\nu\}_{i=1}^\infty$.

In view of these, we set the demonstration scenario of the Main Theorem, by assigning the proofs of Proposition 1, Theorem 1 and Main Theorem to Sections 4, 5 and 6, respectively.

4. Proof of Proposition 1

Before we start the proof, we need to show some lemmas.

Lemma 1. *Let us put* $\Delta^\bullet := [0,1] \times [-1,2] \subset \mathbb{R}^2$, *and let us assume*

$$0 < h \leq h_2^\circ := \frac{1}{2(1 + |g|_{C^2(\Delta^\circ)} + 5|\lambda|^2_{W^{2,\infty}(0,1)})}. \tag{4.1}$$

Let us fix $f_0^* \in V^*$, $[u_0^\circ, \eta_0^\circ, w_0^\circ, \theta_0^\circ] \in L^2(\Omega) \times H^1(\Omega) \times H^1(\Omega) \times W^{1,\infty}(\Omega)$ *and* $w^\circ \in H^1(\Omega)$, *and let us assume that* $0 \leq w_0^\circ, w^\circ \leq 1$ *a.e. in* Ω. *Then, the following auxiliary system:*

$$\frac{u - u_0^\circ}{h} - \lambda'(w^\circ)\frac{w - w_0^\circ}{h} + Fu = f_0^* \text{ in } V^*, \tag{4.2}$$

$$\begin{aligned} &\frac{w - w_0^\circ}{h} - \Delta_N w + \partial\gamma(w) + g_w(w,\eta) \\ &\quad + \alpha_w(w,\eta)|D\theta_0^\circ| + \nu^2\beta_w(w,\eta)|D\theta_0^\circ|^2 \ni -\lambda'(w^\circ)u \text{ in } L^2(\Omega), \end{aligned} \tag{4.3}$$

$$\begin{aligned} &\frac{\eta - \eta_0^\circ}{h} - \Delta_N \eta + \partial I_{[-1,2]}(\eta) + g_\eta(w,\eta) \\ &\quad + \alpha_\eta(w,\eta)|D\theta_0^\circ| + \nu^2\beta_\eta(w,\eta)|D\theta_0^\circ|^2 = 0 \text{ in } L^2(\Omega), \end{aligned} \tag{4.4}$$

admits a unique solution $[u, w, \eta] \in V \times H^1(\Omega)^2$, *where* $\partial I_{[-1,2]}$ *is the subdifferential of the indicator function* $I_{[-1,2]} : \mathbb{R} \to \{0, \infty\}$ *on the compact interval* $[-1,2]$, *and this is an additional term to guarantee the boundedness of the range* $\eta(\Omega)$ *for the component* η.

Proof. First, we put:

$$e := u - \lambda'(w^\circ)w, \; e_0^\circ := u_0^\circ - \lambda'(w^\circ)w_0^\circ, \text{ and } \boldsymbol{v}_0^\circ = [w_0^\circ, \eta_0^\circ],$$

$$[\tilde{w}, \tilde{\eta}] \in \mathbb{R} \mapsto \gamma^\bullet(\tilde{w}, \tilde{\eta}) := \gamma(\tilde{w}) + I_{[-1,2]}(\tilde{\eta}),$$

and reformulate the system {(4.2)–(4.4)} as follows:

$$\frac{e - e_0^\circ}{h} + F(e + \lambda'(w^\circ)w) = f_0^* \text{ in } V^*, \tag{4.5}$$

$$\begin{aligned} &\frac{\boldsymbol{v} - \boldsymbol{v}_0^\circ}{h} + \partial\Psi^2_{\gamma^\bullet}(\boldsymbol{v}) + [\nabla g](w,\eta) \\ &\quad + |D\theta_0^\circ|[\nabla\alpha](\boldsymbol{v}) + \nu^2|D\theta_0^\circ|^2[\nabla\beta](\boldsymbol{v}) \ni \begin{bmatrix} -\lambda'(w^\circ)(e + \lambda'(w^\circ)w) \\ 0 \end{bmatrix} \text{ in } L^2(\Omega)^2, \end{aligned} \tag{4.6}$$

where $\Psi^2_{\gamma^\bullet}$ is the functional $\Psi^d_{\gamma_0}$ as in Remark 1 (Ex.2), in the case when $d = 2$ and $\gamma_0 = \gamma^\bullet$ on $\mathbb{R}^2$, and $\partial\Psi^2_{\gamma^\bullet}$ is the subdifferential of $\Psi^2_{\gamma^\bullet}$ in $L^2(\Omega)^2$. Then, in the light of Remark 1, we can associate the auxiliary system $\{(4.2)–(4.4)\}$ with a minimization problem for the following functional:

$$
\begin{aligned}
[e, \boldsymbol{v}] &= [e, w, \eta] \in V^* \times L^2(\Omega)^2 \mapsto \Psi_0^\bullet(w^\circ; e, \boldsymbol{v}) = \Psi_0^\bullet(w^\circ; e, w, \eta) \\
&:= \begin{cases}
\frac{1}{2h}|e - e_0^\circ|_{V^*}^2 + \frac{1}{2h}|\boldsymbol{v} - \boldsymbol{v}_0^\circ|_{L^2(\Omega)^2}^2 + \frac{1}{2}|e + \lambda'(w^\circ)w|_{L^2(\Omega)}^2 \\
\quad + \Psi_{\gamma^\bullet}^2(\boldsymbol{v}) + \int_\Omega \left(\alpha(\boldsymbol{v})|D\theta_0^\circ| + v^2\beta(\boldsymbol{v})|D\theta_0^\circ|^2\right) dx \\
\quad + \int_\Omega g(\boldsymbol{v})\, dx - (f_0^*, e)_{V^*}, \\
\quad \text{if } [e, \boldsymbol{v}] = [e, w, \eta] \in L^2(\Omega) \times D(\Psi_{\gamma^\bullet}^2), \\
\infty, \quad \text{otherwise,}
\end{cases}
\end{aligned}
\tag{4.7}
$$

via its stationary system $\{(4.5)–(4.6)\}$. Then, taking into account (A2)–(A6), (4.1) and (4.7), we find a positive constant C_0°, independent of the variables $[e, \boldsymbol{v}] = [e, w, \eta]$ and w°, such that:

$$
\Psi_0^\bullet(w^\circ; e, \boldsymbol{v}) \geq C_0^\circ(|e|_{L^2(\Omega)}^2 + |\boldsymbol{v}|_{H^1(\Omega)^2}^2 - 1), \text{ for all } [e, \boldsymbol{v}] \in D(\Psi_0^\bullet(w^\circ; \cdot)). \tag{4.8}
$$

Now, the above coercivity enables us to apply the standard minimization argument to $\Psi_0^\bullet$ (cf. [3, 10]), and to obtain the solution $[u, w, \eta] = [e + \lambda'(w^\circ)w, w, \eta]$ to $\{(4.2)–(4.4)\}$, via the minimizer $[e, \boldsymbol{v}] = [e, w, \eta] \in V \times H^1(\Omega)^2$ of $\Psi_0^\bullet(w^\circ; \cdot)$, with $\boldsymbol{v} = [w, \eta] \in D(\Psi_{\gamma^\bullet}^2)$.

In the meantime, the uniqueness can be seen by using the standard method, i.e. by taking the difference of two solutions $[e_k, \boldsymbol{v}_k] = [e_k, w_k, \eta_k] \in V^* \times L^2(\Omega)^2$ with $\boldsymbol{v}_k = [w_k, \eta_k] \in D(\Psi_\gamma^2)$, $k = 1, 2$, to the stationary system $\{(4.5)–(4.6)\}$. In fact, multiplying the both sides of the subtraction of (4.5) by $e_1 - e_2$ in V^*, multiplying the both sides of the subtraction of (4.6) by $\boldsymbol{v}_1 - \boldsymbol{v}_2$ in $L^2(\Omega)^2$, and using (A2)–(A5), (4.8) and Schwartz's inequality, we have:

$$
\begin{aligned}
&\frac{1}{h}|e_1 - e_2|_{V^*}^2 + \frac{1}{h}\left(1 - h|[\nabla g]|_{W^{1,\infty}(\Delta^\bullet)^2}\right)|\boldsymbol{v}_1 - \boldsymbol{v}_2|_{L^2(\Omega)^2}^2 \\
&+ |D(\boldsymbol{v}_1 - \boldsymbol{v}_2)|_{L^2(\Omega)^{N\times 2}}^2 + |(e_1 - e_2) + \lambda'(w^\circ)(w_1 - w_2)|_{L^2(\Omega)}^2 \leq 0.
\end{aligned}
\tag{4.9}
$$

Since the assumption (4.1) implies $(1 - h|[\nabla g]|_{W^{1,\infty}(\Delta^\bullet)^2}) \geq \frac{1}{2}$, we can deduce from (4.9) the uniqueness for the system $\{(4.2)–(4.4)\}$. □ □

Lemma 2. *Let $w^\circ \in H^1(\Omega)$ be the function as in Lemma 1, and let $\Psi_0^\bullet(w^\circ; \cdot)$ be the functional on $V^* \times L^2(\Omega)^2$ given in (4.7). Also, let us take a sequence $\{w_n^\circ\}_{n=1}^\infty \subset H^1(\Omega)$ such that $0 \leq w_n^\circ \leq 1$ a.e. in Ω, for $n = 1, 2, 3, \ldots$, and let us define a sequence $\{\Psi_0^\bullet(w_n^\circ; \cdot)\}_{n=1}^\infty$ of functionals on $V^* \times L^2(\Omega)^2$, by putting $w^\circ = w_n^\circ$ in (4.7), for $n = 1, 2, 3, \ldots$. Besides, let us assume that:*

$$
w_n^\circ \to w^\circ \text{ in the pointwise sense a.e. in } \Omega, \text{ as } n \to \infty. \tag{4.10}
$$

Then, $\Psi_0^\bullet(w_n^\circ; \cdot) \to \Psi_0^\bullet(w^\circ; \cdot)$ on $V^* \times L^2(\Omega)^2$, in the sense of Γ-convergence, as $n \to \infty$.

Proof. The condition of lower-bound will be seen, immediately, from the lower semi-continuity of the following functional (of 4-variables):

$$
[w^\circ, e, \boldsymbol{v}] \in L^2(\Omega) \times V^* \times L^2(\Omega)^2 \mapsto \Psi_0^\bullet(w^\circ; e, \boldsymbol{v}) \in (-\infty, \infty].
$$

The condition of optimality will be verified by taking the singleton $\{[e, \boldsymbol{v}]\}$ for any $[e, \boldsymbol{v}] \in D(\Psi_0^\bullet(w^\circ; \cdot)) = D(\Psi_0^\bullet(w_n^\circ; \cdot))$ for all $n \geq 1$ as the sequence corresponding to $\{z_n^\bullet\}_{n=1}^\infty$ in Definition 2. □ □

Lemma 3. *Under the assumptions as in the previous Lemmas 1–2, let us take the solution* $[e, \boldsymbol{v}] = [e, w, \eta] \in V \times H^1(\Omega)^2$ *to the stationary system* $\{(4.5)–(4.6)\}$ *with* $\boldsymbol{v} = [w, \eta]$*, and for any* $n \in \mathbb{N}$*, let us denote by* $[e_n, \boldsymbol{v}_n] = [e_n, w_n, \eta_n] \in V \times H^1(\Omega)^2$ *the solution to* $\{(4.5)–(4.6)\}$ *with* $\boldsymbol{v}_n = [w_n, \eta_n]$*, when* $w^\circ = w_n^\circ$*. Then, the assumption* (4.10) *implies that:*

$$\begin{aligned}[e_n, \boldsymbol{v}_n] = [e_n, w_n, \eta_n] \to [e, \boldsymbol{v}] = [e, w, \eta] \text{ in } V^* \times L^2(\Omega)^2, \\ \text{and weakly in } L^2(\Omega) \times H^1(\Omega)^2, \text{ as } n \to \infty.\end{aligned} \tag{4.11}$$

Proof. In the light of Lemma 1 (including the proof), we can see that:

$$\begin{aligned}&\Psi_0^\bullet(\tilde{w}^\circ; \tilde{e}, \tilde{\boldsymbol{v}}) = \Psi_0^\bullet(\tilde{w}^\circ; \tilde{e}, \tilde{w}, \tilde{\eta}) \le \Psi_0^\bullet(\tilde{w}^\circ; 0, 0, 0) \\ &\le C_1^\circ := \frac{1}{2h}(|e_0^\circ|_{V^*}^2 + |\boldsymbol{v}_0^\circ|_{L^2(\Omega)^2}^2) + \mathscr{L}^N(\Omega)(|\gamma(0)| + |g(0,0)|) \\ &\qquad +\alpha(0,0)|\theta_0^\circ|_{W^{1,1}(\Omega)} + \nu^2\beta(0,0)|\theta_0^\circ|_{H^1(\Omega)}^2,\end{aligned} \tag{4.12}$$

for any $\tilde{w}^\circ \in H^1(\Omega)$ with $0 \le \tilde{w}^\circ \le 1$ a.e. in Ω, and
any solution $[\tilde{e}, \tilde{\boldsymbol{v}}] = [\tilde{e}, \tilde{w}, \tilde{\eta}]$ to $\{(4.5)–(4.6)\}$ with $\tilde{\boldsymbol{v}} = [\tilde{w}, \tilde{\eta}]$ when $w^\circ = \tilde{w}^\circ$.

Since the constant C_1° is independent of the choice of $\tilde{w}^\circ$, the convergence (4.11) will be observed by taking into account (4.8), (4.12) and the uniqueness for $\{(4.5)–(4.6)\}$, and by applying Lemma 2, and the general theories of the compact embeddings (cf. [3, 11]) and the Γ convergence (cf. [9]). □ □

Lemma 4. *Let* h_2° *be the constant as in* (4.1)*. Let* $f_0^* \in V^*$*,* $u_0^\circ \in L^2(\Omega)$*,* $\boldsymbol{v}_0^\circ = [w_0^\circ, \eta_0^\circ] \in H^1(\Omega)^2$ *and* $\theta_0^\circ \in W^{1,\infty}(\Omega)$ *be the functions as in Lemma 1. Then, for any* $h \in (0, h_2^\circ]$*, the following system:*

$$\frac{u - u_0^\circ}{h} - \lambda'(w)\frac{w - w_0^\circ}{h} + Fu = f_0^* \text{ in } V^*, \tag{4.13}$$

$$\begin{aligned}&\frac{\boldsymbol{v} - \boldsymbol{v}_0^\circ}{h} - \varDelta_N \boldsymbol{v} + \begin{bmatrix} \partial\gamma(w) \\ 0 \end{bmatrix} + [\nabla g](\boldsymbol{v}) \\ &\quad + |D\theta_0^\circ|[\nabla\alpha](\boldsymbol{v}) + \nu^2|D\theta_0^\circ|^2[\nabla\beta](\boldsymbol{v}) \ni \begin{bmatrix} -\lambda'(w)u \\ 0 \end{bmatrix} \text{ in } L^2(\Omega)^2,\end{aligned} \tag{4.14}$$

admits at least one solution $[u, \boldsymbol{v}] = [u, w, \eta] \in V \times H^1(\Omega)^2$ *with* $\boldsymbol{v} = [w, \eta]$.

Proof. Let us set a compact set $K_1^\bullet$ in $L^2(\Omega)$, by letting:

$$K_1^\bullet := \left\{ \tilde{w} \in H^1(\Omega) \;\middle|\; \begin{array}{l} 0 \le \tilde{w} \le 1 \text{ a.e. in } \Omega, \text{ and} \\ \dfrac{1}{2h}|\tilde{w} - w_0^\circ|_{L^2(\Omega)}^2 + \dfrac{1}{2}|D\tilde{w}|_{L^2(\Omega)^N}^2 \\ \le C_1^\circ + |c_*|\mathscr{L}^N(\Omega) + \dfrac{1}{2h}|e_0^\circ|_{V^*}^2 + h|f_0^*|_{V^*}^2 \end{array} \right\},$$

and let us consider an operator $P_1^\bullet : K_1^\bullet \to L^2(\Omega)$, which maps any $w^\circ \in K_1^\bullet$ to the component w of the solution $[u, w, \eta] \in V \times H^1(\Omega) \times H^1(\Omega)$ to $\{(4.2)–(4.4)\}$. Then, on account of (A3), (A6), Lemma 3, (4.7) and (4.12), it will be seen that $P_1^\bullet K_1^\bullet \subset K_1^\bullet$ and $P_1^\bullet$ is a continuous operator in the topology of $L^2(\Omega)$.

So, applying Schauder's fixed point theorem, we find a fixed point $w^\bullet \in K_1^\bullet$ for $P_1^\bullet$, i.e. $w^\bullet = P_1^\bullet w^\bullet$ in $L^2(\Omega)$.

Now, let us denote by $[u^\bullet, w^\bullet, \eta^\bullet] \in V \times H^1(\Omega) \times H^1(\Omega)$ the solution to {(4.2)–(4.4)}, involved in the fixed point $w^\bullet$. Then, this triplet $[u^\bullet, w^\bullet, \eta^\bullet]$ must be a special solution to {(4.2)–(4.4)} such that $w^\bullet = w^\circ$. Hence, our remaining task will be to show that

$$0 \leq \eta^\bullet \leq 1 \text{ a.e. in } \Omega, \tag{4.15}$$

namely, the subdifferential $\partial I_{[-1,2]}$ in (4.4) will not affect for $\eta^\bullet$. To this end, we check two inequalities:

$$\frac{0 - \eta_0^\circ}{h} + g_\eta(w^\bullet, 0) + |D\theta_0^\circ|\alpha_\eta(w^\bullet, 0) + \nu^2|D\theta_0^\circ|^2\beta_\eta(w^\bullet, 0) \leq 0 \text{ in } L^2(\Omega), \tag{4.16}$$

$$\frac{1 - \eta_0^\circ}{h} + g_\eta(w^\bullet, 1) + |D\theta_0^\circ|\alpha_\eta(w^\bullet, 1) + \nu^2|D\theta_0^\circ|^2\beta_\eta(w^\bullet, 1) \geq 0 \text{ in } L^2(\Omega), \tag{4.17}$$

with use of the assumptions (A3), (A5) and $0 \leq \eta_0^\circ \leq 1$ a.e. in Ω.

On this basis, let us take the difference from (4.16) to (4.4) when $\eta = \eta^\bullet$ and $w = w^\circ = w^\bullet$ (resp. from (4.4) to (4.17) when $\eta = \eta^\bullet$ and $w = w^\circ = w^\bullet$), and multiply the both sides by $[-\eta^\bullet]^+$ (resp. by $[\eta^\bullet - 1]^+$). Then, with the assumptions (A3), (A5) and $\partial I_{[-1,2]}(0) = \{0\}$ (resp. $\partial I_{[-1,2]}(1) = \{0\}$) in mind, it is inferred that:

$$\begin{gathered}\frac{1}{h}\left(1 - h|g_{\eta\eta}|_{C(\Delta^\bullet)}\right)|[-\eta^\bullet]^+|^2_{L^2(\Omega)} + |D[-\eta^\bullet]^+|^2_{L^2(\Omega)^N} \leq 0 \\ \left(\text{resp. } \frac{1}{h}\left(1 - h|g_{\eta\eta}|_{C(\Delta^\bullet)}\right)|[\eta^\bullet - 1]^+|^2_{L^2(\Omega)} + |D[\eta^\bullet - 1]^+|^2_{L^2(\Omega)^N} \leq 0\right).\end{gathered} \tag{4.18}$$

Since the assumption (4.1) implies $1 - h|g_{\eta\eta}|_{C(\Delta^\bullet)} \geq \frac{1}{2}$, we can deduce (4.15) from (4.18), and conclude that the triplet $[u^\bullet, v^\bullet] = [u^\bullet, \eta^\bullet, w^\bullet]$ with $v^\bullet := [w^\bullet, \eta^\bullet]$ solves the system {(4.13)–(4.14)}. □ □

Lemma 5. *Let $f_0^* \in V^*$ and $\theta_0^\circ \in H^M(\Omega)$ be fixed functions, and let $[u, v] = [u, w, \eta] \in V \times H^1(\Omega)^2$ be a solution to the system {(4.13)–(4.14)} with $v = [w, \eta]$. Then, the inclusion*

$$\alpha_0(v)\frac{\theta - \theta_0^\circ}{h} + \partial\Phi_\varepsilon^\nu(v; \theta) \ni 0 \text{ in } L^2(\Omega) \tag{4.19}$$

admits a unique solution $\theta \in H^M(\Omega)$.

Proof. We omit the proof, because this lemma is obtained, immediately, just as a direct consequence of [31, Lemma 3.4]. □ □

Lemma 6. *Under the assumption (4.1), let us take a quartet $[u, v, \theta] = [u, w, \eta, \theta] \in D_M$ with $v = [w, \eta] \in H^1(\Omega)^2$, which solves the coupled system {(4.13)–(4.14),(4.19)}. Then, the following energy-inequality holds:*

$$\begin{aligned}&\frac{A_*}{2h}|u - u_0^\circ|^2_{V^*} + \frac{1}{2h}|v - v_0^\circ|^2_{L^2(\Omega)^2} + \frac{1}{h}|\sqrt{\alpha_0(v)}(\theta - \theta_0^\circ)|^2_{L^2(\Omega)} \\ &+\frac{h}{2}|u|_V^2 + \mathscr{F}_\varepsilon^\nu(u, v, \theta) \leq \mathscr{F}_\varepsilon^\nu(u_0^\circ, v_0^\circ, \theta_0^\circ) + h|f_0^*|^2_{V^*},\end{aligned} \tag{4.20}$$

where $A_* > 0$ is the constant as in (A2), and $\mathscr{F}_\varepsilon^\nu$ is the relaxed version of the functional $\mathscr{F}_\nu$, defined as:

$$\begin{aligned}[u, \boldsymbol{v}, \theta] = [u, w, \eta, \theta] \in L^2(\Omega)^4 \mapsto \mathscr{F}_\varepsilon^\nu(u, \boldsymbol{v}, \theta) = \mathscr{F}_\varepsilon^\nu(u, w, \eta, \theta) \\ = B_*|u|_{L^2(\Omega)}^2 + \Psi_\gamma^2(\boldsymbol{v}) + \int_\Omega (g(\boldsymbol{v}) - c_*)\, dx + \Phi_\varepsilon^\nu(\boldsymbol{v}; \theta),\end{aligned} \tag{4.21}$$

with the constant $B_* = (1 + A_*)/2$ as in (3.1).

Proof. First, let us rewrite the equation (4.13) as follows:

$$\begin{aligned}(u - u_0^\circ, z)_{L^2(\Omega)} + h\langle Fu, z\rangle = h\langle f_0^*, z\rangle \\ +(\lambda'(w)(w - w_0^\circ), z)_{L^2(\Omega)}, \text{ for any } z \in V,\end{aligned} \tag{4.22}$$

and let us put $z = u$. Then, by using Schwarz's inequality, we have:

$$\frac{1}{2}|u|_{L^2(\Omega)}^2 + \frac{h}{2}|u|_V^2 \le \frac{1}{2}|u_0^\circ|_{L^2(\Omega)}^2 + \frac{h}{2}|f_0^*|_{V^*}^2 + (\lambda'(w)(w - w_0^\circ), u)_{L^2(\Omega)}. \tag{4.23}$$

Alternatively, if we rewrite the equation (4.13) to:

$$\begin{aligned}\frac{1}{h}(u - u_0^\circ, z^*)_{V^*} + \langle z^*, u\rangle = (f_0^*, z^*)_{V^*} + \frac{1}{h}(\lambda'(w)(w - w_0^\circ), z^*)_{V^*}, \\ \text{for any } z^* \in V^*,\end{aligned}$$

and put $z^* = A_*(u - u_0^\circ) \in V$, then we also see that:

$$\frac{A_*}{2h}|u - u_0^\circ|_{V^*}^2 + \frac{A_*}{2}|u|_{L^2(\Omega)}^2 \le \frac{A_*}{2}|u_0^\circ|_{L^2(\Omega)}^2 + A_* h|f_0^*|_{V^*}^2 + \frac{1}{4h}|w - w_0^\circ|_{L^2(\Omega)}^2. \tag{4.24}$$

Next, let us multiply the both sides of the inclusion (4.14) by $\boldsymbol{v} - \boldsymbol{v}_0^\circ$. Then, in the light of (A2)–(A5) and Taylor's theorem, we infer that:

$$\begin{aligned}&\frac{1}{h}\left(1 - \frac{h}{2}|g|_{C^2([0,1]^2)}\right)|\boldsymbol{v} - \boldsymbol{v}_0^\circ|_{L^2(\Omega)^2}^2 + \frac{1}{2}|D\boldsymbol{v}|_{L^2(\Omega)^{N\times 2}}^2 + \int_\Omega \gamma(w)\, dx + \int_\Omega g(\boldsymbol{v})\, dx \\ &\quad + \int_\Omega \alpha(\boldsymbol{v})|D\theta_0^\circ|\, dx + \nu^2 \int_\Omega \beta(\boldsymbol{v})|D\theta_0^\circ|^2\, dx \\ &\le \frac{1}{2}|D\boldsymbol{v}_0^\circ|_{L^2(\Omega)^{N\times 2}}^2 + \int_\Omega \gamma(w_0^\circ)\, dx + \int_\Omega g(\boldsymbol{v}_0^\circ)\, dx \\ &\quad + \int_\Omega \alpha(\boldsymbol{v}_0^\circ)|D\theta_0^\circ|\, dx + \nu^2 \int_\Omega \beta(\boldsymbol{v}_0^\circ)|D\theta_0^\circ|^2\, dx - (\lambda'(w)(w - w_0^\circ), u)_{L^2(\Omega)}.\end{aligned} \tag{4.25}$$

Furthermore, applying the both sides of (4.19) by $\theta - \theta_0^\circ$, it follows that:

$$\frac{1}{h}|\sqrt{\alpha_0(\boldsymbol{v})}(\theta - \theta_0^\circ)|_{L^2(\Omega)}^2 + \Phi_\varepsilon^\nu(\boldsymbol{v}; \theta) \le \Phi_\varepsilon^\nu(\boldsymbol{v}; \theta_0^\circ). \tag{4.26}$$

Now, since (4.1) implies $1 - \frac{h}{2}|g|_{C^2([0,1]^2)^2} \ge \frac{3}{4}$, the energy-inequality (4.20) can be obtained by taking the sum of (4.23)–(4.26) with (A2) in mind. □ □

Lemma 7. *By the restriction $1 \leq N \leq 3$ of the spatial dimension, there exists a positive constant C_2°, such that under the notations and assumptions as in Lemma 6, the condition:*

$$C_2^\circ h^{\frac{1}{3}}(1 + 2(\mathscr{F}_\varepsilon^\gamma(u_0^\circ, \boldsymbol{v}_0^\circ, \theta_0^\circ) + h|f_0^*|_{V^*}^2)^{\frac{2}{3}}) \leq \frac{1}{2}, \text{ and } 0 < h \leq h_2^\circ, \tag{4.27}$$

implies the uniqueness of the solution $[u, \boldsymbol{v}, \theta] = [u, w, \eta, \theta] \in D_M$ to the system {(4.13)–(4.14), (4.19)} with $\boldsymbol{v} = [w, \eta]$.

Proof. In the light of the uniqueness of θ as in Lemma 5, it is enough to focus only on the uniqueness for the component $[u, \boldsymbol{v}] = [u, w, \eta] \in V \times H^1(\Omega)^2$ with $\boldsymbol{v} = [w, \eta]$. To this end, we take two triplets $[u_k, \boldsymbol{v}_k] = [u_k, w_k, \eta_k] \in D_M$ with $\boldsymbol{v}_k = [w_k, \eta_k]$, $k = 1, 2$, that fulfill (4.13)–(4.14).

First, with the equivalence of (4.13) and (4.22) in mind, we take the difference between two variational forms (4.22) for u_k, $k = 1, 2$, and put $z = u_1 - u_2$ in V. Then:

$$\begin{aligned}|u_1 - u_2|_{L^2(\Omega)}^2 + h|u_1 - u_2|_V^2 &= (\lambda'(w_1)w_1 - \lambda'(w_2)w_2, u_1 - u_2)_{L^2(\Omega)} \\ &\quad -((\lambda'(w_1) - \lambda'(w_2))w_0^\circ, u_1 - u_2)_{L^2(\Omega)},\end{aligned}$$

so that by using (A2) and Schwarz's inequality, we have:

$$\frac{1}{4}|u_1 - u_2|_{L^2(\Omega)}^2 + h|u_1 - u_2|_V^2 \leq 3|\lambda'|_{W^{1,\infty}(0,1)}^2 |w_1 - w_2|_{L^2(\Omega)}^2. \tag{4.28}$$

Secondly, let us take the difference between two inclusions (4.14) for $\boldsymbol{v}_k = [w_k, \eta_k]$, $k = 1, 2$, and multiply the both sides by $\boldsymbol{v}_1 - \boldsymbol{v}_2$ in $L^2(\Omega)^2$. Then, by using (A2)–(A5) and Schwarz's inequality, it is computed that:

$$\begin{aligned}&\frac{1}{h}\left(1 - h|[\nabla g]|_{C^1([0,1]^2)^2}\right)|\boldsymbol{v}_1 - \boldsymbol{v}_2|_{L^2(\Omega)^2}^2 + |D(\boldsymbol{v}_1 - \boldsymbol{v}_2)|_{L^2(\Omega)^{N\times 2}}^2 \\ &\leq -(\lambda'(w_1)u_1 - \lambda'(w_2)u_2, w_1 - w_2)_{L^2(\Omega)} \\ &\leq |\lambda'|_{L^\infty(0,1)}|u_1 - u_2|_{L^2(\Omega)}|w_1 - w_2|_{L^2(\Omega)} + \left(u_1(\lambda'(w_1) - \lambda'(w_2)), w_1 - w_2\right)_{L^2(\Omega)} \\ &\leq \frac{1}{8}|u_1 - u_2|_{L^2(\Omega)}^2 + 2|\lambda'|_{L^\infty(0,1)}^2|w_1 - w_2|_{L^2(\Omega)}^2 + |\lambda''|_{L^\infty(0,1)}\int_\Omega |u_1||w_1 - w_2|^2\,dx.\end{aligned} \tag{4.29}$$

Here, the dimensional restriction $1 \leq N \leq 3$ and the assumption (4.27) enable to apply the analytic technique as in [19, Lemma 3.1], and to find a constant $C_2^\circ > 0$, independent of h and triplets $[u_0^\circ, \boldsymbol{v}_0^\circ]$ and $[u_k, \boldsymbol{v}_k]$, $k = 1, 2$, such that:

$$|\lambda''|_{L^\infty(0,1)}\int_\Omega |u_1||w_1 - w_2|^2 dx \leq \frac{1}{2}|D(w_1 - w_2)|_{L^2(\Omega)}^2 + C_2^\circ(1 + |u_1|_V^{\frac{4}{3}})|w_1 - w_2|_{L^2(\Omega)}^2. \tag{4.30}$$

Furthermore, under (4.27), the inequality (4.20) enables to derive that:

$$\begin{aligned}C_2^\circ(1 + |u_1|_V^{\frac{4}{3}})|w_1 - w_2|_{L^2(\Omega)}^2 &= C_2^\circ h^{\frac{1}{3}}(h^{\frac{2}{3}} + (h|u_1|_V^2)^{\frac{2}{3}}) \cdot \frac{1}{h}|w_1 - w_2|_{L^2(\Omega)}^2 \\ &\leq C_2^\circ h^{\frac{1}{3}}(1 + 2(\mathscr{F}_\varepsilon^\gamma(u_0^\circ, \boldsymbol{v}_0^\circ, \theta_0^\circ) + h|f_0^*|_{V^*}^2)^{\frac{2}{3}}) \cdot \frac{1}{h}|w_1 - w_2|_{L^2(\Omega)}^2 \\ &\leq \frac{1}{2h}|w_1 - w_2|_{L^2(\Omega)}^2.\end{aligned} \tag{4.31}$$

Now, taking sum of (4.28)–(4.29) with (4.30)–(4.31) in mind, we obtain that:

$$\begin{aligned}
&\frac{1}{8}|u_1 - u_2|^2_{L^2(\Omega)} + h|u_1 - u_2|^2_V \\
&\quad + \frac{1}{h}\left(\frac{1}{2} - h\left(|g|_{C^2([0,1]^2)} + 5|\lambda|^2_{W^{2,\infty}(0,1)}\right)\right)|\boldsymbol{v}_1 - \boldsymbol{v}_2|^2_{L^2(\Omega)^2} \\
&\quad + \frac{1}{2}|D(\boldsymbol{v}_1 - \boldsymbol{v}_2)|^2_{L^2(\Omega)^{N\times 2}} \leq 0.
\end{aligned} \tag{4.32}$$

This implies the required uniqueness, because $\frac{1}{2} - h(|g|_{C^2([0,1]^2)^2} + 5|\lambda|^2_{W^{2,\infty}(0,1)}) > 0$ follows from the assumption (4.1) and (4.27). □ □

Proof of Proposition 1. Let us take a positive constant h_2° defined by (4.1). Let us set a positive constant h_0°, so small to satisfy that:

$$C_2^\circ(h_0^\circ)^{\frac{1}{3}}(1 + 2(\mathscr{F}_\varepsilon^\nu(u_0^\circ, \boldsymbol{v}_0^\circ, \theta_0^\circ) + h_0^\circ|[f^*]_0^{\mathrm{ex}}|^2_{L^2(0,T;V^*)})^{\frac{2}{3}}) \leq \frac{1}{2}, \text{ and } 0 < h_0^\circ \leq h_2^\circ.$$

Then, from (4.20), it will be observed that:

$$\begin{aligned}
&C_2^\circ h^{\frac{1}{3}}(1 + 2(\mathscr{F}_\varepsilon^\nu(u^\nu_{\varepsilon,i-1}, \boldsymbol{v}^\nu_{\varepsilon,i-1}, \theta^\nu_{\varepsilon,i-1}) + h|[f_i^*]^h|^2_{V^*})^{\frac{2}{3}}) \\
&\quad \leq C_2^\circ h^{\frac{1}{3}}(1 + 2(\mathscr{F}_\varepsilon^\nu(u^\nu_{\varepsilon,i-2}, \boldsymbol{v}^\nu_{\varepsilon,i-2}, \theta^\nu_{\varepsilon,i-2}) + h(|[f_i^*]^h|^2_{V^*} + |[f_{i-1}^*]^h|^2_{V^*}))^{\frac{2}{3}}) \\
&\quad \leq \cdots \leq C_2^\circ h^{\frac{1}{3}}(1 + 2(\mathscr{F}_\varepsilon^\nu(u^\nu_{\varepsilon,0}, \boldsymbol{v}^\nu_{\varepsilon,0}, \theta^\nu_{\varepsilon,0}) + |[f^*]_0^{\mathrm{ex}}|^2_{L^2(0,T;V^*)})^{\frac{2}{3}}) \\
&\quad \leq \frac{1}{2}, \text{ for all } 0 < h \leq h_0^\circ\ (\leq h_2^\circ) \text{ and } i = 1, 2, 3, \ldots.
\end{aligned} \tag{4.33}$$

In view of this, the Proposition 1 will be concluded by means of the following algorithm.

(Step 0) Let $h \in (0, h_0^\circ]$, let $i = 1$, and let $[u^\nu_{\varepsilon,0}, \boldsymbol{v}^\nu_{\varepsilon,0}, \theta^\nu_{\varepsilon,0}] \in D_M$.

(Step 1) Obtain the quartet $[u^\nu_{\varepsilon,i}, \boldsymbol{v}^\nu_{\varepsilon,i}, \theta^\nu_{\varepsilon,i}] \in D_M$ as the unique solution to the system {(4.13)–(4.14), (4.19)}, by applying Lemmas 4–7 to the case when:

$$f_0^* = [f_{i-1}^*]^h \text{ in } V^*,\ u_0^\circ = u^\nu_{\varepsilon,i-1} \text{ in } L^2(\Omega),$$
$$\boldsymbol{v}_0^\circ = \boldsymbol{v}^\nu_{\varepsilon,i-1} \text{ in } H^1(\Omega)^2 \text{ and } \theta_0^\circ = \theta^\nu_{\varepsilon,i-1} \text{ in } H^M(\Omega).$$

(Step 2) Iterate the value of i and return to (Step 1).

Note that (4.33) let the assumption $h \in (0, h_0^\circ]$ be a uniform condition to make sense (Step 1), for all $i = 1, 2, 3, \ldots$. □

5. Proof of Theorem 1

Let us set $h_1^\circ := h_0^\circ$ i.e. the constant as in Proposition 1, and let us fix $\nu > 0$, $h \in (0, h_1^\circ]$ and the initial value $[u_0^\nu, \boldsymbol{v}_0^\nu, \theta_0^\nu] = [u_0^\nu, w_0^\nu, \eta_0^\nu, \theta_0^\nu] \in D_1$ with $\boldsymbol{v}_0^\nu = [w_0^\nu, \eta_0^\nu]$. Besides, we recall the following lemmas obtained in [31].

Lemma 8. (cf. [31, Lemma 4.1]) *Assume* $v^\circ \in [H^1(\Omega) \cap L^\infty(\Omega)]^2$, $\{v^\circ_\varepsilon\}_{0<\varepsilon\leq 1} \subset [H^1(\Omega) \cap L^\infty(\Omega)]^2$, *and*

$$\begin{cases} v^\circ_\varepsilon \to v^\circ \text{ in the pointwise sense a.e. in } \Omega \text{ as } \varepsilon \downarrow 0, \\ \{v^\circ_\varepsilon\}_{0<\varepsilon\leq 1} \text{ is bounded in } L^\infty(\Omega)^2. \end{cases}$$

Then, for the sequence of convex functions $\{\Phi^\nu_\varepsilon(v^\circ_\varepsilon; \cdot)\}_{0<\varepsilon\leq 1}$, *it holds that* $\Phi^\nu_\varepsilon(v^\circ_\varepsilon; \cdot) \to \Phi_\nu(v^\circ; \cdot)$ *on* $L^2(\Omega)$, *in the sense of Mosco, as* $\varepsilon \downarrow 0$.

Lemma 9. (cf. [31, Lemma 4.2]) *Assume that*

$$\begin{cases} v^\circ \in [H^1(\Omega) \cap L^\infty(\Omega)]^2,\ \{v^\circ_\varepsilon\}_{0<\varepsilon\leq 1} \subset [H^1(\Omega) \cap L^\infty(\Omega)]^2, \\ \{v^\circ_\varepsilon\}_{0<\varepsilon\leq 1} \text{ is bounded in } L^\infty(\Omega)^2, \\ v^\circ_\varepsilon \to v^\circ \text{ in the pointwise sense, a.e. in } \Omega, \text{ as } \varepsilon \downarrow 0, \end{cases}$$

and

$$\begin{cases} \theta^\circ \in H^1(\Omega),\ \{\theta^\circ_\varepsilon\}_{0<\varepsilon\leq 1} \subset H^1(\Omega), \\ \theta^\circ_\varepsilon \to \theta^\circ \text{ in } L^2(\Omega) \text{ and } \Phi^\nu_\varepsilon(v^\circ_\varepsilon; \theta^\circ_\varepsilon) \to \Phi_\nu(v^\circ; \theta^\circ), \text{ as } \varepsilon \downarrow 0. \end{cases}$$

Then, $\theta^\circ_\varepsilon \to \theta^\circ$ *in* $H^1(\Omega)$ *and* $\varepsilon\theta^\circ_\varepsilon \to 0$ *in* $H^M(\Omega)$, *as* $\varepsilon \downarrow 0$.

Lemma 10. (cf. [31, Lemma 4.4]) *Let* $v^\circ \in [H^1(\Omega) \cap L^\infty(\Omega)]^2$ *and* $\check{\theta}^\circ_0, \hat{\theta}^\circ_0 \in H^1(\Omega)$ *be fixed functions, and let* $[\check{\theta}, \check{\theta}^*]$, $[\hat{\theta}, \hat{\theta}^*] \in L^2(\Omega)^2$ *be pairs of functions such that*

$$\begin{cases} [\check{\theta}, \check{\theta}^*] \in \partial\Phi_\nu(v^\circ; \cdot) \text{ in } L^2(\Omega)^2 \text{ and } \dfrac{1}{h}\alpha_0(v^\circ)(\check{\theta} - \check{\theta}^\circ_0) + \check{\theta}^* \leq 0 \text{ a.e. in } \Omega, \\ [\hat{\theta}, \hat{\theta}^*] \in \partial\Phi_\nu(v^\circ; \cdot) \text{ in } L^2(\Omega)^2 \text{ and } \dfrac{1}{h}\alpha_0(v^\circ)(\hat{\theta} - \hat{\theta}^\circ_0) + \hat{\theta}^* \geq 0 \text{ a.e. in } \Omega, \end{cases} \tag{5.1}$$

respectively. Then, it follows that

$$|\sqrt{\alpha_0(v^\circ)}[\check{\theta} - \hat{\theta}]^+|^2_{L^2(\Omega)} \leq |\sqrt{\alpha_0(v^\circ)}[\check{\theta}^\circ_0 - \hat{\theta}^\circ_0]^+|^2_{L^2(\Omega)},$$

and therefore, if $\check{\theta}^\circ_0 \leq \hat{\theta}^\circ_0$ *in* Ω, *then the inequality* $\check{\theta} \leq \hat{\theta}$ *a.e. in* Ω *also follows from (A3).*

Moreover, if the both inequalities in (5.1) *hold as equalities, then:*

$$|\sqrt{\alpha_0(v^\circ)}(\check{\theta} - \hat{\theta})|^2_{L^2(\Omega)} \leq |\sqrt{\alpha_0(v^\circ)}(\check{\theta}^\circ_0 - \hat{\theta}^\circ_0)|^2_{L^2(\Omega)},$$

$$\text{i.e. } \check{\theta}^\circ_0 = \hat{\theta}^\circ_0 \text{ implies } \check{\theta} = \hat{\theta} \text{ in } L^2(\Omega).$$

Based on these, we divide the proof of Theorem 1 in two parts: the part of existence; the part of uniqueness and energy inequality.

The part of existence. Let $\nu > 0$ be a fixed constant. By Lemma 8, there exists a sequence $\{\tilde{\theta}^\nu_{\varepsilon,0}\}_{0<\varepsilon\leq 1} \subset H^M(\Omega)$ such that

$$\tilde{\theta}^\nu_{\varepsilon,0} \to \theta^\nu_0 \text{ in } H^1(\Omega) \text{ and } \Phi^\nu_\varepsilon(v^\nu_0; \tilde{\theta}^\nu_{\varepsilon,0}) \to \Phi_\nu(v^\nu_0; \theta^\nu_0) \text{ as } \varepsilon \downarrow 0.$$

So, by virtue of Proposition 1 we can take a class $\{[\tilde{u}^\nu_{\varepsilon,i}, \tilde{v}^\nu_{\varepsilon,i}, \tilde{\theta}^\nu_{\varepsilon,i}] \,|\, i \in \mathbb{N},\ \varepsilon \in (0,1]\}$ consisting of solutions $\{[\tilde{u}^\nu_{\varepsilon,i}, \tilde{v}^\nu_{\varepsilon,i}, \tilde{\theta}^\nu_{\varepsilon,i}]\}_{i=1}^\infty = \{[\tilde{u}^\nu_{\varepsilon,i}, \tilde{w}^\nu_{\varepsilon,i}, \tilde{\eta}^\nu_{\varepsilon,i}, \tilde{\theta}^\nu_{\varepsilon,i}]\}_{i=1}^\infty \subset D_M$ to $(\mathrm{RX})_\varepsilon$ with $\{\tilde{v}^\nu_{\varepsilon,i}\}_{i=1}^\infty = \{[\tilde{w}^\nu_{\varepsilon,i}, \tilde{\eta}^\nu_{\varepsilon,i}]\}_{i=1}^\infty$, starting from the initial data $[u^\nu_{\varepsilon,0}, \boldsymbol{v}^\nu_{\varepsilon,0}, \theta^\nu_{\varepsilon,0}] = [u^\nu_0, \boldsymbol{v}^\nu_0, \tilde{\theta}^\nu_{\varepsilon,0}]$ for $0 < \varepsilon \leq 1$. Then, with Lemma 6 and the algorithm (Step 0)–(Step 2) in mind, we remark the following energy-inequality:

$$\begin{aligned}
&\frac{A_*}{2h}|\tilde{u}^\nu_{\varepsilon,i} - \tilde{u}^\nu_{\varepsilon,i-1}|^2_{V^*} + \frac{1}{2h}|\tilde{\boldsymbol{v}}^\nu_{\varepsilon,i} - \tilde{\boldsymbol{v}}^\nu_{\varepsilon,i-1}|^2_{L^2(\Omega)^2} + \frac{1}{h}\left|\sqrt{\alpha_0(\tilde{\boldsymbol{v}}^\nu_{\varepsilon,i})}(\tilde{\theta}^\nu_{\varepsilon,i} - \tilde{\theta}^\nu_{\varepsilon,i-1})\right|^2_{L^2(\Omega)} \\
&+\frac{h}{2}|\tilde{u}^\nu_{\varepsilon,i}|^2_V + \mathscr{F}^\nu_\varepsilon(\tilde{u}^\nu_{\varepsilon,i}, \tilde{\boldsymbol{v}}^\nu_{\varepsilon,i}, \tilde{\theta}^\nu_{\varepsilon,i}) \leq \mathscr{F}^\nu_\varepsilon(\tilde{u}^\nu_{\varepsilon,i-1}, \tilde{\boldsymbol{v}}^\nu_{\varepsilon,i-1}, \tilde{\theta}^\nu_{\varepsilon,i-1}) + h|[f^*_i]^h|^2_{V^*}, \\
&\qquad\qquad \text{for all } 0 < \varepsilon \leq 1 \text{ and } i = 1, 2, 3, \ldots.
\end{aligned} \tag{5.2}$$

In the light of (A3)–(A6), (4.21) and (5.2), the class $\{[\tilde{u}^\nu_{\varepsilon,i}, \tilde{\boldsymbol{v}}^\nu_{\varepsilon,i}, \tilde{\theta}^\nu_{\varepsilon,i}] \mid i \in \mathbb{N},\ \varepsilon \in (0,1]\}$ is bounded in $V \times H^1(\Omega)^3$. Therefore, applying a diagonal argument and the general theories of compactness (cf. [3, 11]), we find sequences $\{\varepsilon_n\}_{n=1}^\infty \subset (0,1]$, $\{[u^\nu_i, \boldsymbol{v}^\nu_i, \theta^\nu_i]\}_{i=1}^\infty = \{[u^\nu_i, w^\nu_i, \eta^\nu_i, \theta^\nu_i]\}_{i=1}^\infty \subset V \times H^1(\Omega)^2 \times H^1(\Omega)$, with $\{\boldsymbol{v}^\nu_i\}_{i=1}^\infty = \{[w^\nu_i, \eta^\nu_i]\}_{i=1}^\infty$, such that

$$\begin{cases}
1 \geq \varepsilon_1 > \cdots > \varepsilon_n \downarrow 0 \text{ as } n \to \infty, \\
\tilde{u}^\nu_{i,n} := \tilde{u}^\nu_{\varepsilon_n,i} \to u^\nu_i \text{ in } L^2(\Omega), \text{ weakly in } V \text{ as } n \to \infty, \\
\tilde{\boldsymbol{v}}^\nu_{i,n} := \tilde{\boldsymbol{v}}^\nu_{\varepsilon_n,i} \to \boldsymbol{v}^\nu_i \text{ in } L^2(\Omega)^2, \text{ weakly in } H^1(\Omega)^2, \text{ weakly-}* \text{ in } L^\infty(\Omega)^2, \\
\qquad \text{and in the pointwise sense a.e. in } \Omega, \text{ as } n \to \infty, \\
\tilde{\theta}^\nu_{i,n} \equiv \tilde{\theta}^\nu_{\varepsilon_n,i} \to \theta^\nu_i \text{ in } L^2(\Omega), \text{ weakly in } H^1(\Omega) \\
\qquad \text{and in the pointwise sense a.e. in } \Omega, \text{ as } n \to \infty, \\
0 \leq w^\nu_i \leq 1 \text{ and } 0 \leq \eta^\nu_i \leq 1 \text{ a.e. in } \Omega; \text{ for all } i = 0, 1, 2, \ldots.
\end{cases} \tag{5.3}$$

Moreover, by (3.13), (5.3), Lemmas 8–9 and Remark 4 (Fact 6), we infer that

$$\begin{cases}
[\theta^\nu_i, -\frac{1}{h}\alpha_0(\boldsymbol{v}^\nu_i)(\theta^\nu_i - \theta^\nu_{i-1})] \in \partial\Phi_\nu(\boldsymbol{v}^\nu_i; \cdot) \text{ in } L^2(\Omega)^2, \\
\Phi^\nu_{\varepsilon_n}(\tilde{\boldsymbol{v}}^\nu_{i,n}; \tilde{\theta}^\nu_{i,n}) \to \Phi_\nu(\boldsymbol{v}^\nu_i; \theta^\nu_i),\ \tilde{\theta}^\nu_{i,n} \to \theta^\nu_i \text{ in } H^1(\Omega) \\
\qquad \text{and } \varepsilon_n \tilde{\theta}^\nu_{i,n} \to 0 \text{ in } H^M(\Omega), \text{ as } n \to \infty,
\end{cases} \quad \text{for } i = 0, 1, 2, \ldots. \tag{5.4}$$

Also, since

$$[c, 0] \in \partial\Phi_\nu(\boldsymbol{v}^\nu_i; \cdot) \quad \text{in } L^2(\Omega)^2, \text{ for all } c \in \mathbb{R} \text{ and } i = 0, 1, 2, \ldots,$$

it is observed that

$$\theta^\nu_i \leq |\theta^\nu_{i-1}|_{L^\infty(\Omega)} \text{ (resp. } \theta^\nu_i \geq -|\theta^\nu_{i-1}|_{L^\infty(\Omega)}) \text{ a.e. in } \Omega, \text{ for any } i \in \mathbb{N},$$

by applying Lemma 10 as the case when

$$\begin{cases}
\boldsymbol{v}^\circ = \boldsymbol{v}^\nu_i, \\
\check{\theta}^\circ_0 = \theta^\nu_{i-1},\ \hat{\theta}^\circ_0 = |\theta^\nu_{i-1}|_{L^\infty(\Omega)} \text{ (resp. } \check{\theta}^\circ_0 = -|\theta^\nu_{i-1}|_{L^\infty(\Omega)},\ \hat{\theta}^\circ_0 = \theta^\nu_{i-1}), \\
[\check{\theta}, \check{\theta}^*] = [\theta^\nu_i, -\frac{1}{h}\alpha_0(\boldsymbol{v}^\nu_i)(\theta^\nu_i - \theta^\nu_{i-1})] \text{ (resp. } [\check{\theta}, \check{\theta}^*] = [-|\theta^\nu_{i-1}|_{L^\infty(\Omega)}, 0]), \\
[\hat{\theta}, \hat{\theta}^*] = [|\theta^\nu_{i-1}|_{L^\infty(\Omega)}, 0] \left(\text{resp. } [\hat{\theta}, \hat{\theta}^*] = [\theta^\nu_i, -\frac{1}{h}\alpha_0(\boldsymbol{v}^\nu_i)(\theta^\nu_i - \theta^\nu_{i-1})]\right).
\end{cases}$$

Having in mind (A2)–(A5), (3.11)–(3.12) and (5.3)–(5.4), we can see that

$$\begin{aligned}
&\frac{1}{h}(u_i^\nu - u_{i-1}^\nu, z)_{L^2(\Omega)} - \frac{1}{h}(\lambda'(w_i^\nu)(w_i^\nu - w_{i-1}^\nu), z)_{L^2(\Omega)} + (u_i^\nu, z)_V \\
&= \lim_{n\to\infty}\left[\frac{1}{h}(\tilde{u}_{i,n}^\nu - \tilde{u}_{i-1,n}^\nu, z)_{L^2(\Omega)} - \frac{1}{h}(\lambda'(\tilde{w}_{i,n}^\nu)(\tilde{w}_{i,n}^\nu - \tilde{w}_{i-1,n}^\nu), z)_{L^2(\Omega)} + (\tilde{u}_{i,n}^\nu, z)_V\right] \\
&= \langle [f_i^*]^h, z\rangle, \text{ for any } z \in V \text{ and } i = 1, 2, 3, \ldots,
\end{aligned}$$

and

$$\begin{aligned}
&(D\boldsymbol{v}_i^\nu, D(\boldsymbol{v}_i^\nu - \varpi))_{L^2(\Omega)^{N\times 2}} + \int_\Omega \gamma(w_i^\nu)\,dx - \int_\Omega \gamma(\varphi)\,dx \\
&\leq \liminf_{n\to\infty}(D\tilde{\boldsymbol{v}}_{i,n}^\nu, D(\tilde{\boldsymbol{v}}_{i,n}^\nu - \varpi))_{L^2(\Omega)^{N\times 2}} + \liminf_{n\to\infty}\int_\Omega \gamma(\tilde{w}_{i,n}^\nu)dx - \int_\Omega \gamma(\varphi)\,dx \\
&\leq \limsup_{n\to\infty}(D\tilde{\boldsymbol{v}}_{i,n}^\nu, D(\tilde{\boldsymbol{v}}_{i,n}^\nu - \varpi))_{L^2(\Omega)^{N\times 2}} + \liminf_{n\to\infty}\int_\Omega \gamma(\tilde{w}_{i,n}^\nu)dx - \int_\Omega \gamma(\varphi)\,dx \\
&\leq -\lim_{n\to\infty}\left(\frac{1}{h}(\tilde{\boldsymbol{v}}_{i,n}^\nu - \tilde{\boldsymbol{v}}_{i-1,n}^\nu, \tilde{\boldsymbol{v}}_{i,n}^\nu - \varpi)_{L^2(\Omega)^2} + \int_\Omega [\nabla G](\tilde{u}_{i,n}^\nu; \tilde{\boldsymbol{v}}_{i,n}^\nu)\cdot(\tilde{\boldsymbol{v}}_{i,n}^\nu - \varpi)dx\right) \\
&\quad - \lim_{n\to\infty}\int_\Omega ([\nabla\alpha](\tilde{\boldsymbol{v}}_{i,n}^\nu)|D\tilde{\theta}_{i-1,n}^\nu| + \nu^2[\nabla\beta](\tilde{\boldsymbol{v}}_{i,n}^\nu)|D\tilde{\theta}_{i-1,n}^\nu|^2)\cdot(\tilde{\boldsymbol{v}}_{i,n}^\nu - \varpi)dx \\
&\leq -\frac{1}{h}(\boldsymbol{v}_i^\nu - \boldsymbol{v}_{i-1}^\nu, \boldsymbol{v}_i^\nu - \varpi)_{L^2(\Omega)^2} - \int_\Omega [\nabla G](u_i^\nu; \boldsymbol{v}_i^\nu)\cdot(\boldsymbol{v}_i^\nu - \varpi)dx \\
&\quad - \int_\Omega ([\nabla\alpha](\boldsymbol{v}_i^\nu)|D\theta_{i-1}^\nu| + \nu^2[\nabla\beta](\boldsymbol{v}_i^\nu)|D\theta_{i-1}^\nu|^2)\cdot(\boldsymbol{v}_i^\nu - \varpi)dx, \\
&\text{for any } \varpi = [\varphi, \psi] \in [H^1(\Omega)\cap L^\infty(\Omega)]^2, \text{ and } i = 1, 2, 3, \ldots.
\end{aligned} \tag{5.5}$$

The above calculations imply that the limiting sequence $\{[u_i^\nu, \boldsymbol{v}_i^\nu, \theta_i^\nu]\}_{i=1}^\infty$ is a solution to the approximating system $(\mathrm{AP})_h^\nu$. □

The part of uniqueness and energy inequality. By putting $\varpi = \boldsymbol{v}_i^\nu$ in (5.5), for $i \in \mathbb{N}$, one can see from (5.3) that:

$$\begin{aligned}
&|D\boldsymbol{v}_i^\nu|_{L^2(\Omega)^{N\times 2}}^2 \leq \liminf_{n\to\infty}|D\tilde{\boldsymbol{v}}_{i,n}^\nu|_{L^2(\Omega)^{N\times 2}}^2 \leq \limsup_{n\to\infty}|D\tilde{\boldsymbol{v}}_{i,n}^\nu|_{L^2(\Omega)^{N\times 2}}^2 \\
\leq\; &\lim_{n\to\infty}(D\tilde{\boldsymbol{v}}_{i,n}^\nu, D\boldsymbol{v}_i^\nu)_{L^2(\Omega)^{N\times 2}} + \int_\Omega \gamma(w_i^\nu)\,dx - \liminf_{n\to\infty}\int_\Omega \gamma(\tilde{w}_{i,n}^\nu)\,dx \\
\leq\; &|D\boldsymbol{v}_i^\nu|_{L^2(\Omega)^{N\times 2}}^2, \text{ for } i = 1, 2, 3, \ldots.
\end{aligned} \tag{5.6}$$

By the convergences (5.3) and (5.6), the uniform convexity of L^2-based topologies enable to say:

$$\tilde{\boldsymbol{v}}_{i,n}^\nu \to \boldsymbol{v}_i^\nu \text{ in } H^1(\Omega)^2 \text{ as } n \to \infty, \text{ for } i = 1, 2, 3, \ldots. \tag{5.7}$$

Hence, the energy-inequality (3.10) will be obtained, immediately, by putting $\varepsilon = \varepsilon_n$ in (5.2), for $n \in \mathbb{N}$, and letting $n \to \infty$ with (5.3) and (5.7) in mind.

In the meantime, we note that the condition (4.33) is still available in the proof of Theorem 1. Also, the regularity $\theta_0^\circ \in H^M(\Omega)$ will not necessary in the calculations (4.29)–(4.32), and the line of these calculations will work even if $\theta_0^\circ \in H^1(\Omega)$.

In view of these, the verification part of the uniqueness for $(\mathrm{AP})_h^\nu$ will be a slight modification of that as in (Step 1) in the previous section. Then, the principal modifications will be to replace the application parts of Lemma 5 and the energy-inequality (4.20), by those of Lemma 10 and (3.10), respectively. □

6. Proof of Main Theorem

Let $\nu \geq 0$ be a fixed constant, and let $h_1^\circ \in (0, 1]$ be the constant as in Theorem 1. Also, we refer to [31] to recall the following lemma.

Lemma 11. (Γ-convergence; [31, Lemma 6.2]) *Assume $\boldsymbol{v}^\bullet \in [H^1(\Omega) \cap L^\infty(\Omega)]^2$, $\{\boldsymbol{v}_{\tilde{\nu}}^\bullet\}_{\tilde{\nu}>0} \subset [H^1(\Omega) \cap L^\infty(\Omega)]^2$, and*

$$\begin{cases} \boldsymbol{v}_{\tilde{\nu}}^\bullet \to \boldsymbol{v}^\bullet \text{ in the pointwise sense, a.e. in } \Omega, \text{ as } \tilde{\nu} \downarrow 0, \\ \{\boldsymbol{v}_{\tilde{\nu}}^\bullet\}_{\tilde{\nu}>0} \text{ is bounded in } L^\infty(\Omega)^2. \end{cases}$$

Then, for the sequence of convex functions $\{\Phi_{\tilde{\nu}}(\boldsymbol{v}_{\tilde{\nu}}^\bullet; \cdot)\}_{\tilde{\nu}>0}$, it holds that $\Phi_{\tilde{\nu}}(\boldsymbol{v}_{\tilde{\nu}}^\bullet; \cdot) \to \Phi_0(\boldsymbol{v}^\bullet; \cdot)$ on $L^2(\Omega)$, in the sense of Γ-convergence, as $\tilde{\nu} \downarrow 0$.

Based on Lemma 11 and [31, Remark 6.1], we take a sequence $\{\vartheta_0^{\tilde{\nu}}\}_{\tilde{\nu}>0} \subset H^1(\Omega)$, such that:

$$|\theta_0^{\tilde{\nu}}| \leq |\theta_0|_{L^\infty(\Omega)} \text{ a.e. in } \Omega, \text{ for any } \tilde{\nu} > 0,$$

and

$$\begin{cases} \vartheta_0^{\tilde{\nu}} \to \theta_0 \text{ in } L^2(\Omega) \text{ and } \Phi_{\tilde{\nu}}(\boldsymbol{v}_0; \vartheta_0^{\tilde{\nu}}) \to \Phi_0(\boldsymbol{v}_0; \theta_0), \text{ as } \tilde{\nu} \downarrow 0, \text{ if } \nu = 0, \\ \vartheta_0^{\tilde{\nu}} = \theta_0 \text{ in } H^1(\Omega) \text{ for } \tilde{\nu} > 0, \text{ if } \nu > 0, \end{cases}$$

and for any $h \in (0, h_1^\circ]$ and any $\tilde{\nu} \in (0, \nu + 1]$, let us take the solution $\{[u_i^{\tilde{\nu}}, \boldsymbol{v}_i^{\tilde{\nu}}, \theta_i^{\tilde{\nu}}]\}_{i=0}^\infty$ to $(\mathrm{AP})_h^{\tilde{\nu}}$ with $\{\boldsymbol{v}_i^{\tilde{\nu}}\}_{i=1}^\infty = \{[w_i^{\tilde{\nu}}, \eta_i^{\tilde{\nu}}]\}_{i=1}^\infty$, under the initial condition $[u_0^{\tilde{\nu}}, \boldsymbol{v}_0^{\tilde{\nu}}, \theta_0^{\tilde{\nu}}] = [u_0, \boldsymbol{v}_0, \vartheta_0^{\tilde{\nu}}] \in D_1$ with $\boldsymbol{v}_0^{\tilde{\nu}} = [w_0^{\tilde{\nu}}, \eta_0^{\tilde{\nu}}] = [w_0, \eta_0]$. As is easily seen,

$$F_0^\nu := \sup_{0<\tilde{\nu}\leq\nu+1} \mathscr{F}_{\tilde{\nu}}(u_0, \boldsymbol{v}_0, \vartheta_0^{\tilde{\nu}}) < \infty.$$

For any $h \in (0, h_1^\circ]$ and any $\tilde{\nu} \in (0, \nu + 1]$, we define the following time-interpolations:

$$f_h^*(t) = [f_h(t), f_{\Gamma,h}(t)] := [f_i^*]^h = [f_i^h, f_{\Gamma,i}^h] \text{ in } V^* \text{ (in } L^2(\Omega) \times L^2(\Gamma)),$$

$$\text{for all } t \geq 0 \text{ and } 0 \leq i \in \mathbb{Z} \text{ satisfying } t \in ((i-1)h, ih],$$

and

$$\begin{cases} [\overline{u}_h^{\tilde{\nu}}(t), \overline{\boldsymbol{v}}_h^{\tilde{\nu}}(t), \overline{\theta}_h^{\tilde{\nu}}(t)] = [\overline{u}_h^{\tilde{\nu}}(t), \overline{w}_h^{\tilde{\nu}}(t), \overline{\eta}_h^{\tilde{\nu}}(t), \overline{\theta}_h^{\tilde{\nu}}(t)] := [u_i^{\tilde{\nu}}, \boldsymbol{v}_i^{\tilde{\nu}}, \theta_i^{\tilde{\nu}}] \text{ in } L^2(\Omega)^4, \\ \qquad \text{for all } t \geq 0 \text{ and } 0 \leq i \in \mathbb{Z} \text{ satisfying } t \in ((i-1)h, ih], \\ [\underline{u}_h^{\tilde{\nu}}(t), \underline{\boldsymbol{v}}_h^{\tilde{\nu}}(t), \underline{\theta}_h^{\tilde{\nu}}(t)] = [\underline{u}_h^{\tilde{\nu}}(t), \underline{w}_h^{\tilde{\nu}}(t), \underline{\eta}_h^{\tilde{\nu}}(t), \underline{\theta}_h^{\tilde{\nu}}(t)] := [u_{i-1}^{\tilde{\nu}}, \boldsymbol{v}_{i-1}^{\tilde{\nu}}, \theta_{i-1}^{\tilde{\nu}}] \text{ in } L^2(\Omega)^4, \\ \qquad \text{for all } t \geq 0 \text{ and } 0 \leq i \in \mathbb{Z} \text{ satisfying } t \in [(i-1)h, ih), \\ [\widehat{u}_h^{\tilde{\nu}}(t), \widehat{\boldsymbol{v}}_h^{\tilde{\nu}}(t), \widehat{\theta}_h^{\tilde{\nu}}(t)] = [\widehat{u}_h^{\tilde{\nu}}(t), \widehat{w}_h^{\tilde{\nu}}(t), \widehat{\eta}_h^{\tilde{\nu}}(t), \widehat{\theta}_h^{\tilde{\nu}}(t)] \\ \qquad := \dfrac{ih-t}{h}[u_{i-1}^{\tilde{\nu}}(t), \boldsymbol{v}_{i-1}^{\tilde{\nu}}(t), \theta_{i-1}^{\tilde{\nu}}(t)] + \dfrac{t-(i-1)h}{h}[u_i^{\tilde{\nu}}, \boldsymbol{v}_i^{\tilde{\nu}}, \theta_i^{\tilde{\nu}}] \text{ in } L^2(\Omega)^4, \\ \qquad \text{for all } t \geq 0 \text{ and } 0 \leq i \in \mathbb{Z} \text{ satisfying } t \in [(i-1)h, ih). \end{cases} \tag{6.1}$$

Besides, we define:

$$D_\nu(\theta_0) := \begin{cases} \left\{ [\tilde{u}, \tilde{\boldsymbol{v}}, \tilde{\theta}] \in D_0 \;\middle|\; |\tilde{\theta}|_{L^\infty(\Omega)} \leq |\theta_0|_{L^\infty(\Omega)} \right\}, \text{ if } \nu = 0, \\ \left\{ [\tilde{u}, \tilde{\boldsymbol{v}}, \tilde{\theta}] \in D_1 \;\middle|\; |\tilde{\theta}|_{L^\infty(\Omega)} \leq |\theta_0|_{L^\infty(\Omega)} \right\}, \text{ if } \nu > 0, \end{cases}$$

and we note that:

$$\begin{aligned} &\{[\overline{u}_h^{\tilde{\nu}}(t), \overline{\boldsymbol{v}}_h^{\tilde{\nu}}(t), \overline{\theta}_h^{\tilde{\nu}}(t)], [\underline{u}_h^{\tilde{\nu}}(t), \underline{\boldsymbol{v}}_h^{\tilde{\nu}}(t), \underline{\theta}_h^{\tilde{\nu}}(t)], [\widehat{u}_h^{\tilde{\nu}}(t), \widehat{\boldsymbol{v}}_h^{\tilde{\nu}}(t), \widehat{\theta}_h^{\tilde{\nu}}(t)]\} \\ &\subset D_\nu(\theta_0), \quad \text{for all } t \geq 0, 0 < h \leq h_1^\circ \text{ and } 0 < \tilde{\nu} \leq \nu + 1. \end{aligned}$$

Then, from the energy-inequality (3.10) in Theorem 1, it is checked that

$$\begin{aligned} &\frac{A_*}{2}\int_s^t |(\widehat{u}_h^{\tilde{\nu}})_t|_{V^*}\, d\tau + \frac{1}{2}\int_s^t |(\widehat{\boldsymbol{v}}_h^{\tilde{\nu}})_t(\tau)|_{L^2(\Omega)^2}^2 d\tau + \int_s^t |\sqrt{\alpha_0(\overline{\boldsymbol{v}}_h^{\tilde{\nu}})}(\widehat{\theta}_h^{\tilde{\nu}})_t(\tau)|_{L^2(\Omega)}^2 d\tau \\ &+\frac{1}{2}\int_s^t |\overline{u}_h^{\tilde{\nu}}(\tau)|_V^2 d\tau + \mathscr{F}_{\tilde{\nu}}(\overline{u}_h^{\tilde{\nu}}, \overline{\boldsymbol{v}}_h^{\tilde{\nu}}, \overline{\theta}_h^{\tilde{\nu}})(t) \leq \mathscr{F}_{\tilde{\nu}}(\underline{u}_h^{\tilde{\nu}}, \underline{\boldsymbol{v}}_h^{\tilde{\nu}}, \underline{\theta}_h^{\tilde{\nu}})(s) + \int_s^t |f_h^*(\tau)|_{V^*}^2 d\tau \\ &\qquad \text{for all } 0 \leq s \leq t \leq T, 0 < h \leq h_1^\circ \text{ and } 0 < \tilde{\nu} \leq \nu + 1, \end{aligned}$$

and additionally, from (A1)–(A6) and (3.2), it follows that

$$\begin{aligned} &B_*|\overline{u}_h^{\tilde{\nu}}(t)|_{L^2(\Omega)}^2 + \frac{1}{2}|D\overline{\boldsymbol{v}}_h^{\tilde{\nu}}(t)|_{L^2(\Omega)^{N\times 2}}^2 + \delta_*\left(|D\overline{\theta}_h^{\tilde{\nu}}(t)|(\Omega) + |D(\tilde{\nu}\overline{\theta}_h^{\tilde{\nu}})(t)|_{L^2(\Omega)^{N\times 2}}^2\right) \\ &\leq \mathscr{F}_{\tilde{\nu}}(\overline{u}_h^{\tilde{\nu}}, \overline{\boldsymbol{v}}_h^{\tilde{\nu}}, \overline{\theta}_h^{\tilde{\nu}})(t) \vee \mathscr{F}_{\tilde{\nu}}(\underline{u}_h^{\tilde{\nu}}, \underline{\boldsymbol{v}}_h^{\tilde{\nu}}, \underline{\theta}_h^{\tilde{\nu}})(t) \\ &\leq F_*^\nu := F_0^\nu + |f^*|_{L^2(0,T;V^*)}^2, \quad \text{for all } 0 \leq t \leq T \text{ and } 0 < \tilde{\nu} \leq \nu + 1. \end{aligned} \tag{6.2}$$

Based on these, one can see that:

(♯1) the class $\{\widehat{u}_h^{\tilde{\nu}} \,|\, h \in (0, h_1^\circ],\ \tilde{\nu} \in (0, \nu+1]\}$ is bounded in the space $W^{1,2}(0,T;V^*) \cap C([0,T];L^2(\Omega)) \cap L^2(0,T;V)$.

(♯2) the class $\{\widehat{\boldsymbol{v}}_h^{\tilde{\nu}} \,|\, h \in (0, h_1^\circ],\ \tilde{\nu} \in (0, \nu+1]\}$ is bounded in the space $W^{1,2}(0,T;L^2(\Omega)^2) \cap L^\infty(0,T; H^1(\Omega)^2) \cap L^\infty(Q)^2$.

(♯3) the class $\{\widehat{\theta}_h^{\tilde{\nu}} \,|\, h \in (0, h_1^\circ],\ \tilde{\nu} \in (0, \nu+1]\}$ is bounded in the space $W^{1,2}(0,T;L^2(\Omega)) \cap L^\infty(Q)$, and $\{\Phi_{\tilde{\nu}}(\boldsymbol{v}_h^{\tilde{\nu}}; \theta_h^{\tilde{\nu}}) \,|\, h \in (0, h_1^\circ],\ \tilde{\nu} \in (0, \nu+1]\}$ is bounded in $L^\infty(0,T)$, i.e. $\{|D\overline{\theta}_h^{\tilde{\nu}}(\,\cdot\,)|(\Omega) \,|\, h \in (0, h_1^\circ],\ \tilde{\nu} \in (0, \nu+1]\}$ is bounded in $L^\infty(0,T)$, and $\{D(\tilde{\nu}\overline{\theta}_h^{\tilde{\nu}}) \,|\, h \in (0, h_1^\circ],\ \tilde{\nu} \in (0, \nu+1]\}$ is bounded in $L^\infty(0,T;L^2(\Omega)^N)$.

Hence, by applying the general theories of compactness, as in [2, 3, 11, 33], we find a quartet of functions $[u, \boldsymbol{v}, \theta] = [u, w, \eta, \theta] \in L^2(0,T;L^2(\Omega)^4)$ with $\boldsymbol{v} = [w, \eta]$ and sequences $\{h_n\}_{n=1}^\infty \subset (0, h_1^\circ]$ and $\{\nu_n\}_{n=1}^\infty \subset (0, \nu+1]$, with the subsequences:

$$\begin{cases} \{[\overline{u}_n, \overline{\boldsymbol{v}}_n, \overline{\theta}_n]\}_{n=1}^\infty = \{[\overline{u}_n, \overline{w}_n, \overline{\eta}_n, \overline{\theta}_n]\}_{n=1}^\infty := \{[\overline{u}_{h_n}^{\nu_n}, \overline{\boldsymbol{v}}_{h_n}^{\nu_n}, \overline{\theta}_{h_n}^{\nu_n}]\}_{n=1}^\infty, \\ \{[\underline{u}_n, \underline{\boldsymbol{v}}_n, \underline{\theta}_n]\}_{n=1}^\infty = \{[\underline{u}_n, \underline{w}_n, \underline{\eta}_n, \underline{\theta}_n]\}_{n=1}^\infty := \{[\underline{u}_{h_n}^{\nu_n}, \underline{\boldsymbol{v}}_{h_n}^{\nu_n}, \underline{\theta}_{h_n}^{\nu_n}]\}_{n=1}^\infty, \\ \{[\widehat{u}_n, \widehat{\boldsymbol{v}}_n, \widehat{\theta}_n]\}_{n=1}^\infty = \{[\widehat{u}_n, \widehat{w}_n, \widehat{\eta}_n, \widehat{\theta}_n]\}_{n=1}^\infty := \{[\widehat{u}_{h_n}^{\nu_n}, \widehat{\boldsymbol{v}}_{h_n}^{\nu_n}, \widehat{\theta}_{h_n}^{\nu_n}]\}_{n=1}^\infty, \end{cases}$$

such that:

$$h_1^\circ \geq h_1 > h_2 > \cdots > h_n \downarrow 0 \text{ and } \nu_n \to \nu, \text{ as } n \to \infty, \tag{6.3}$$

$$\begin{cases} u \in W^{1,2}(0,T;V^*) \cap L^\infty(0,T;L^2(\Omega)) \cap L^2(0,T;V) \subset C([0,T];L^2(\Omega)), \\ \boldsymbol{v} \in W^{1,2}(0,T;L^2(\Omega)^2) \cap L^\infty(0,T;H^1(\Omega)^2) \cap L^\infty(Q)^2, \\ \theta \in W^{1,2}(0,T;L^2(\Omega)) \cap L^\infty(Q),\ \Phi_\nu(\boldsymbol{v};\theta) \in L^\infty(0,T), \\ [u(t), \boldsymbol{v}(t), \theta(t)] \in D_\nu(\theta_0) \text{ for all } t \geq 0, \\ [u(0), \boldsymbol{v}(0), \theta(0)] = [u_0, \boldsymbol{v}_0, \theta_0] \text{ in } L^2(\Omega)^4, \end{cases} \tag{6.4}$$

$$\begin{cases} \widehat{u}_n \to u \text{ in } L^2(I;L^2(\Omega)), \text{ weakly in } W^{1,2}(I;V^*), \\ \qquad \text{weakly-}* \text{ in } L^\infty(I;V), \\ \widehat{\boldsymbol{v}}_n \to \boldsymbol{v} \text{ in } C(\overline{I};L^2(\Omega)^2), \text{ weakly in } W^{1,2}(I;L^2(\Omega)^2), \\ \qquad \text{weakly-}* \text{ in } L^\infty(I;H^1(\Omega)^2) \text{ and weakly-}* \text{ in } L^\infty(Q)^2, \\ \widehat{\theta}_n \to \theta \text{ in } C(\overline{I};L^2(\Omega)), \text{ weakly in } W^{1,2}(I;L^2(\Omega)), \\ \qquad \text{weakly-}* \text{ in } L^\infty(Q), \\ \nu_n\widehat{\theta}_n \to \nu\theta \text{ weakly in } L^2(I;H^1(\Omega)), \end{cases} \tag{6.5}$$

$$\boldsymbol{f}_{h_n}^* \to \boldsymbol{f}^* \text{ in } L^2(I;V^*) \quad ([f_{h_n}, f_{\Gamma,h_n}] \to [f, f_\Gamma] \text{ in } L^2(I;L^2(\Omega) \times L^2(\Gamma))), \tag{6.6}$$

as $n \to \infty$, for any open interval $I \subset (0,T)$, and

$$\begin{cases} \overline{u}_n(t) \to u(t) \text{ and } \underline{u}_n(t) \to u(t) \text{ in } L^2(\Omega), \text{ weakly in } V, \\ \overline{\boldsymbol{v}}_n(t) \to \boldsymbol{v}(t) \text{ and } \underline{\boldsymbol{v}}_n(t) \to \boldsymbol{v}(t) \text{ in } L^2(\Omega)^2, \text{ weakly in } H^1(\Omega)^2 \\ \qquad \text{and weakly-}* \text{ in } L^\infty(\Omega)^2, \\ \overline{\theta}_n(t) \to \theta(t) \text{ in } L^2(\Omega), \text{ weakly-}* \text{ in } BV(\Omega), \\ \nu_n\overline{\theta}_n(t) \to \nu\theta(t) \text{ weakly in } H^1(\Omega), \end{cases} \tag{6.7}$$

as $n \to \infty$ for a.e. $t \in (0,T)$.

Now, we recall some lemmas which will act key-roles in the proof of Main Theorem.

Lemma 12. *Let $I \subset (0,T)$ be an open interval, and let $\nu \geq 0$ and $\{\nu_n\}_{n=1}^{\infty}$ be the sequence as in (6.3). Let $\zeta \in L^2(I; L^2(\Omega))$ be a function such that*

$$|D\zeta(\,\cdot\,)|(\Omega) \in L^1(I) \ \text{ and } \ \nu\zeta \in L^2(I; H^1(\Omega)).$$

Then, there exists a sequence $\{\tilde{\zeta}_n\}_{n=1}^{\infty} \subset C^{\infty}(\overline{Q})$, such that:

$$\tilde{\zeta}_n \to \zeta \text{ in } L^2(I; L^2(\Omega)), \ \int_I \left| \int_\Omega |D\tilde{\zeta}_n(t)|\, dx - \int_\Omega d|D\zeta(t)| \right| dt \to 0,$$
$$\text{and } \nu_n\tilde{\zeta}_n \to \nu\zeta \text{ in } L^2(I; H^1(\Omega)), \ \text{ as } n \to \infty.$$

Proof. When $\nu > 0$, the standard C^{∞}-approximation in $L^2(I; H^1(\Omega))$ will correspond to the required sequence. Meanwhile, when $\nu = 0$, this lemma is verified by taking the C^{∞}-approximation as in [25, Lemma 5] and [29, Remark 2], and by applying the diagonal argument as in [25, Lemma 8]. □ □

Lemma 13. *Let $I \subset (0,T)$ be any open interval. Assume that*

$$\begin{cases} \varrho \in C(\overline{I}; L^2(\Omega)) \cap L^{\infty}(I; H^1(\Omega)), \ \log\varrho \in L^{\infty}(I \times \Omega), \\ \varrho_n \in C(\overline{I}; L^2(\Omega)) \cap L^{\infty}(I; H^1(\Omega)), \ \log\varrho_n \in L^{\infty}(I \times \Omega), \ \textit{for } n = 1, 2, 3, \ldots, \\ \varrho_n(t) \to \varrho(t) \textit{ in } L^2(\Omega) \textit{ and weakly in } H^1(\Omega) \textit{ as } n \to \infty, \textit{ for a.e. } t \in I, \end{cases}$$

and

$$\begin{cases} \zeta \in L^2(I; L^2(\Omega)) \textit{ with } |D\zeta(\,\cdot\,)|(\Omega) \in L^1(I), \\ \{\zeta_n\}_{n=1}^{\infty} \subset L^2(I; L^2(\Omega)) \textit{ with } \{|D\zeta_n(\,\cdot\,)|(\Omega)\}_{n=1}^{\infty} \subset L^1(I), \\ \zeta_n(t) \to \zeta(t) \textit{ in } L^2(\Omega) \textit{ as } n \to \infty, \ \textit{a.e. } t \in I. \end{cases}$$

Then the following items hold.

(I) The functions:

$$t \in I \mapsto \int_\Omega d[\varrho(t)|D\zeta(t)|]\, dt \textit{ and } t \in I \mapsto \int_\Omega d[\varrho_n(t)|D\zeta_n(t)|]\, dt, \textit{ for } n = 1, 2, 3, \ldots,$$

are integrable, and

$$\liminf_{n\to\infty} \int_I \int_\Omega d[\varrho_n(t)|D\zeta_n(t)|]\, dt \geq \int_I \int_\Omega d[\varrho(t)|D\zeta(t)|]\, dt.$$

(II) If:

$$\int_I \int_\Omega d[\varrho_n(t)|D\zeta_n(t)|]\, dt \to \int_I \int_\Omega d[\varrho(t)|D\zeta(t)|]\, dt \textit{ as } n \to \infty$$

and

$$\begin{cases} \omega \in L^{\infty}(I; H^1(\Omega)) \cap L^{\infty}(I \times \Omega), \ \{\omega_n\}_{n=1}^{\infty} \subset L^{\infty}(I; H^1(\Omega)) \cap L^{\infty}(I \times \Omega), \\ \{\omega_n\}_{n=1}^{\infty} \textit{ is a bounded sequence in } L^{\infty}(I \times \Omega), \\ \omega_n(t) \to \omega(t) \textit{ in } L^2(\Omega) \textit{ and weakly in } H^1(\Omega) \textit{ as } n \to \infty, \textit{ a.e. } t \in I, \end{cases}$$

then

$$\int_I \int_\Omega \omega_n(t)|D\zeta_n(t)|\, dx\, dt \to \int_I \int_\Omega d[\omega(t)|D\zeta(t)|] \textit{ as } n \to \infty.$$

Proof. This lemma is verified, immediately, as a consequence of [26, Lemmas 4.2–4.4] (see also [25, Section 2]). □ □

Proof of Main Theorem. We show that the quartet $[u, v, \theta] = [u, w, \eta, \theta] \in L^2(0,T; L^2(\Omega)^4)$ as in (6.4) fulfills the conditions (S1)–(S6) in Definition 3. Then, since (6.4) directly guarantees the conditions (S1)–(S3), we focus on the verifications of remaining (S4)–(S6).

To this end, let us fix arbitrary open interval $I \subset (0,T)$, and let us review (3.7)–(3.9) and (6.1), to check that:

$$\begin{aligned}\int_I \langle (\widehat{u}_n)_t(t), z\rangle\,dt + \int_I (\overline{u}_n(t), z)_V\,dt = \int_I (\lambda'(\overline{w}_n(t))(\widehat{w}_n)_t(t), z)_{L^2(\Omega)}\,dt \\ + \int_I \langle f^*_{h_n}(t), z\rangle\,dt, \quad \text{for any } z \in V \text{ and } n = 1,2,3,\ldots,\end{aligned} \tag{6.8}$$

$$\begin{aligned}&\int_I ((\widehat{v}_n)_t(t), \overline{v}_n(t) - \varpi)_{L^2(\Omega)^2} dt \\ &\quad + \int_I (D\overline{v}_n(t), D(\overline{v}_n(t) - \varpi))_{L^2(\Omega)^{N\times 2}}\,dt \\ &\quad + \int_I ([\nabla G](\overline{u}_n; \overline{v}_n)(t), \overline{v}_n(t) - \varpi)_{L^2(\Omega)^2} dt \\ &\quad + \int_I\int_\Omega [\nabla\alpha](\overline{v}_n(t)) \cdot (\overline{v}_n(t) - \varpi)|D\underline{\theta}_n(t)|dxdt \\ &\quad + \nu_n^2 \int_I\int_\Omega [\nabla\beta](\overline{v}_n(t)) \cdot (\overline{v}_n(t) - \varpi)|D\underline{\theta}_n(t)|^2 dxdt \\ &\quad + \int_I\int_\Omega \gamma(\overline{w}_n(t))dxdt \le \int_I\int_\Omega \gamma(\varphi)dxdt,\end{aligned} \tag{6.9}$$

for any $\varpi = [\varphi, \psi] \in [H^1(\Omega) \cap L^\infty(\Omega)]^2$ and $n = 1,2,3,\ldots,$

and

$$\begin{aligned}&\int_I (\alpha_0(\overline{v}_n(t))(\widehat{\theta}_n)_t(t), \overline{\theta}_n(t) - \zeta(t))_{L^2(\Omega)} dt \\ &\quad + \int_I\int_\Omega \alpha(\overline{v}_n(t))|D\overline{\theta}_n(t)|\,dxdt + \nu_n^2 \int_I\int_\Omega \beta(\overline{v}_n(t))|D\overline{\theta}_n(t)|^2\,dxdt \\ &\le \int_I\int_\Omega \alpha(\overline{v}_n(t))|D\zeta(t)|\,dxdt + \nu_n^2 \int_I\int_\Omega \beta(\overline{v}_n(t))|D\zeta(t)|^2\,dxdt\end{aligned} \tag{6.10}$$

for any $\zeta \in L^2(I; H^1(\Omega))$ and $n = 1,2,3,\ldots.$

Now, let us first take the limit of (6.10) as $n \to \infty$. Then, from (A3), (♯2)–(♯3), (6.4)–(6.5), (6.7) and Lemma 13 (I), it is seen that

$$\int_I (\alpha_0(\boldsymbol{v}(t))\theta_t(t), \theta(t) - \zeta(t))_{L^2(\Omega)} dt + \int_I \Phi_\nu(\boldsymbol{v}(t); \theta(t))\, dt$$

$$\leq \lim_{n\to\infty} \int_I (\alpha_0(\overline{\boldsymbol{v}}_n)(\widehat{\theta}_n)_t(t), \overline{\theta}_n(t) - \zeta(t))_{L^2(\Omega)}\, dt$$

$$+ \liminf_{n\to\infty} \left[\int_I \int_\Omega \alpha(\overline{\boldsymbol{v}}_n(t))|D\overline{\theta}_n(t)|\, dxdt + \int_I \int_\Omega \beta(\overline{\boldsymbol{v}}_n(t))|D(\nu_n \overline{\theta}_n)(t)|^2\, dxdt \right]$$

$$\leq \lim_{n\to\infty} \left[\int_I \int_\Omega \alpha(\overline{\boldsymbol{v}}_n(t))|D\zeta(t)|\, dxdt + \int_I \int_\Omega \beta(\overline{\boldsymbol{v}}_n(t))|D(\nu_n \zeta)(t)|^2\, dxdt \right]$$

$$= \int_I \Phi_\nu(\boldsymbol{v}(t); \zeta(t))\, dt, \text{ for any } \zeta \in L^2(I; H^1(\Omega)).$$

Since the open interval $I \subset (0, T)$ is arbitrary, the above inequality implies that

$$(\alpha_0(\boldsymbol{v}(t))\theta_t(t), \theta(t) - \omega)_{L^2(\Omega)} + \Phi_\nu(\boldsymbol{v}(t); \theta(t)) \leq \Phi_\nu(\boldsymbol{v}(t); \omega)$$
$$\text{for any } \omega \in H^1(\Omega) \text{ and a.e. } t \in (0, T).$$

Additionally, in the light of Remark 3 (Fact 4), we can say the above inequality holds for $\omega \in BV(\Omega) \cap L^2(\Omega)$. Thus, (S6) is verified.

Next, with (6.4) and Lemma 12 in mind, let us take a sequence $\{\tilde{\theta}_n\}_{n=1}^{\infty} \subset C^\infty(\overline{I \times \Omega})$ such that

$$\tilde{\theta}_n \to \theta \text{ in } L^2(I; L^2(\Omega)), \quad \int_I |D\tilde{\theta}_n|\, dxdt \to \int_I d|D\theta(t)|\, dt,$$
$$\nu_n \tilde{\theta}_n \to \nu\theta \text{ in } L^2(I; H^1(\Omega)), \text{ as } n \to \infty.$$

Then, putting $\zeta = \tilde{\theta}_n$ in (6.10) and letting $n \to \infty$, it is observed from (♯2)–(♯3), (6.4)–(6.5), (6.7) and Lemma 13 that:

$$\int_I \int_\Omega d[\alpha(\boldsymbol{v}(t))|D\theta(t)|]\, dt + \int_I \int_\Omega \beta(\boldsymbol{v}(t))|D(\nu\theta)(t)|^2\, dxdt$$

$$\leq \liminf_{n\to\infty} \int_I \int_\Omega \alpha(\overline{\boldsymbol{v}}_n(t))|D\overline{\theta}_n(t)|\, dxdt + \liminf_{n\to\infty} \int_I \int_\Omega \beta(\overline{\boldsymbol{v}}_n(t))|D(\nu_n \overline{\theta}_n)(t)|^2\, dxdt$$

$$\leq \limsup_{n\to\infty} \left[\int_I \int_\Omega \alpha(\overline{\boldsymbol{v}}_n(t))|D\overline{\theta}_n(t)| dx\, dt + \int_I \int_\Omega \beta(\overline{\boldsymbol{v}}_n(t))|D(\nu_n \overline{\theta}_n)(t)|^2\, dxdt \right]$$

$$\leq \lim_{n\to\infty} \left[\int_I \int_\Omega \alpha(\overline{\boldsymbol{v}}_n(t))|D\tilde{\theta}_n(t)| dx\, dt + \int_I \int_\Omega \beta(\overline{\boldsymbol{v}}_n(t))|D(\nu_n \tilde{\theta}_n)(t)|^2\, dxdt \right]$$

$$- \lim_{n\to\infty} \int_I (\alpha_0(\overline{\boldsymbol{v}}_n)(\widehat{\theta}_n)_t(t), \overline{\theta}_n(t) - \tilde{\theta}_n(t))_{L^2(\Omega)}\, dt$$

$$= \int_I \int_\Omega d[\alpha(\boldsymbol{v}(t))|D\theta(t)|]\, dt + \int_I \int_\Omega \beta(\boldsymbol{v}(t))|D(\nu\theta)(t)|^2\, dxdt.$$

The above inequality implies that:

$$\lim_{n\to\infty} \int_I \int_\Omega \alpha(\overline{\boldsymbol{v}}_n(t))|D\overline{\theta}_n(t)| dxdt = \int_I \int_\Omega d[\alpha(\boldsymbol{v}(t))|D\theta(t)|] dt, \tag{6.11}$$

and

$$\lim_{n\to\infty}\int_I\int_\Omega \beta(\overline{v}_n(t))|D(v_n\overline{\theta}_n)(t)|^2dxdt = \int_I\int_\Omega \beta(v(t))|D(v\theta)(t)|^2\,dt. \tag{6.12}$$

By virtue of (♯2)–(♯3), (6.4)–(6.5), (6.7) and (6.11), we can apply Lemma 13 to see that:

$$\int_I\int_\Omega |D\overline{\theta}_n(t)|dxdt \to \int_I\int_\Omega d|D\theta(t)|\,dt, \text{ as } n\to\infty.$$

Besides, (6.1)–(6.2) and (6.5) enable to check:

$$\left|\int_I\int_\Omega |D\overline{\theta}_n|dxdt - \int_I\int_\Omega |D\underline{\theta}_n|dxdt\right| \le \frac{2F_*^\nu}{\delta_*}h_n \to 0, \text{ as } n\to\infty,$$

and (6.5), (6.7) and the above convergence further enable to show that:

$$\begin{aligned}&\lim_{n\to\infty}\int_I\int_\Omega (\overline{v}_n(t)-\varpi)\cdot[\nabla\alpha](\overline{v}_n(t))|D\underline{\theta}_n(t)|dxdt \\ &= \int_I\int_\Omega d[(\overline{v}_n(t)-\varpi)\cdot[\nabla\alpha](v(t))|D\theta(t)|]dt \text{ for any } \varpi\in[H^1(\Omega)\cap L^\infty(\Omega)]^2,\end{aligned} \tag{6.13}$$

by applying Lemma 13 (II).

Similarly, from (6.12) and the uniform convexity of L^2-based topology, one can see that

$$\begin{cases} \sqrt{\beta(\overline{v}_n)}D(v_n\overline{\theta}_n) \to \sqrt{\beta(v)}D(v\theta) \text{ in } L^2(I;L^2(\Omega)^N), \text{ and hence} \\ D(v_n\overline{\theta}_n) \to D(v\theta) \text{ in } L^2(I;L^2(\Omega)^N), \text{ as } n\to\infty. \end{cases}$$

Besides, (6.1)–(6.2) and (6.5) enable to check:

$$\left|\int_I\int_\Omega |D(v_n\overline{\theta}_n)|^2dxdt - \int_I\int_\Omega |D(v_n\underline{\theta}_n)|^2\,dxdt\right| \le \frac{2F_*^\nu}{\delta_*}h_n \to 0, \text{ as } n\to\infty,$$

and the above convergence further enables to show that:

$$\begin{cases} D(v_n\underline{\theta}_n) \to D(v\theta) \text{ in } L^2(I;L^2(\Omega)^N), \text{ and hence} \\ (\overline{v}_n-\varpi)\cdot[\nabla\beta](\overline{v}_n)D(v_n\underline{\theta}_n) \\ \qquad \to (v-\varpi)\cdot[\nabla\beta](v)D(v\theta) \text{ in } L^2(I;L^2(\Omega)^N), \\ \text{for any } \varpi\in[H^1(\Omega)\cap L^\infty(\Omega)]^2, \text{ as } n\to\infty. \end{cases} \tag{6.14}$$

With (A2)–(A5), (♯1)–(♯3), (6.4)–(6.5), (6.7), (6.13)–(6.14) and lower semi-continuity of L^2-norm in mind, letting $n\to\infty$ in (6.9) yields that:

$$\begin{aligned}&\int_I (v_t(t), v(t)-\varpi)_{L^2(\Omega)^2}dt + \int_I (Dv(t), D(v(t)-\varpi))_{L^2(\Omega)^{N\times 2}}dt \\ &\quad + \int_I\int_\Omega \gamma(w(t))dxdt + \int_I ([\nabla G](u(t);v(t)), v(t)-\varpi)_{L^2(\Omega)^2}dt \\ &\quad + \int_I\int_\Omega d[(v(t)-\varpi)\cdot[\nabla\alpha](v(t))|D\theta(t)|]dt \\ &\quad + \int_I\int_\Omega [\nabla\beta](v(t))\cdot(v(t)-\varpi)|\nabla(v\theta)|^2dxdt \\ &\le \int_I\int_\Omega \gamma(\varphi)dxdt, \text{ for any } \varpi=[\varphi,\psi]\in[H^1(\Omega)\cap L^\infty(\Omega)]^2.\end{aligned} \tag{6.15}$$

Finally, taking the limit of (6.8), and applying (6.5)–(6.7), one can see that:

$$\begin{aligned}\int_I \langle u_t(t), z\rangle\, dt + \int_I (u(t), z)_V\, dt = \int_I (\lambda'(w(t))w_t(t), z)_{L^2(\Omega)}\, dt \\ + \int_I \langle f^*(t), z\rangle\, dt, \quad \text{for any } z \in V.\end{aligned} \tag{6.16}$$

Since the open interval $I \subset (0, T)$ is arbitrary, the conditions (S4)–(S5) will be verified by taking into account (6.4) and (6.15)–(6.16). □

Acknowledgments

This research was supported by JSPS KAKENHI Grant-in-Aid for Scientific Research (C), 16K05224, No. 26400138 and Young Scientists (B), No. 25800086. The authors express their gratitude to an anonymous referees for reviewing the original manuscript and for many valuable comments that helped clarify and refine this paper.

Conflict of Interest

All authors declare no conflicts of interest in this paper.

References

1. M. Amar, G. Bellettini, *A notion of total variation depending on a metric with discontinuous coefficients.* Ann. Inst. H. Poincaré Anal. Non Linéaire, **11** (1994), no. 1, 91–133.
2. L. Ambrosio, N. Fusco, D. Pallara, *Functions of Bounded Variation and Free Discontinuity Problems.* Oxford Mathematical Monographs. (2000).
3. H. Attouch, G. Buttazzo, G. Michaille, *Variational Analysis in Sobolev and BV Spaces.* Applications to PDEs and Optimization. MPS-SIAM Series on Optimization, **6**. SIAM and MPS, (2006).
4. V. Barbu, *Nonlinear semigroups and differential equations in Banach spaces.* Editura Academiei Republicii Socialiste România, Noordhoff International Publishing, (1976).
5. G. Bellettini, G. Bouchitté, I. Fragalà, *BV functions with respect to a measure and relaxation of metric integral functionals.* J. Convex Anal., **6** (1999), no. 2, 349–366.
6. H. Brézis, *Operateurs Maximaux Monotones et Semi-groupes de Contractions dans les Espaces de Hilbert.* North-Holland Mathematics Studies, **5**. Notas de Matemática (50). North-Holland Publishing and American Elsevier Publishing, (1973).
7. P. Colli, P. Laurençot, *Weak solutions to the Penrose-Fife phase field model for a class of admissible heat flux laws.* Phys. D, **111** (1998), 311–334.
8. P. Colli, J. Sprekels, *Glob al solution to the Penrose-Fife phase-field model with zero interfacial energy and Fourier law.* Adv. Math. Sci. Appl., **9** (1999), no. 1, 383–391.
9. G. Dal Maso, *An Introduction to* Γ*-convergence.* Progress in Nonlinear Differential Equations and their Applications, **8**. Birkhäuser Boston, Inc., Boston, Ma, (1993).

10. I. Ekeland, R. Temam, *Convex analysis and variational problems.* Translated from the French. Corrected reprint of the 1976 English edition. Classics in Applied Mathematics, **28**. SIAM, Philadelphia, (1999).
11. L. C. Evans, R. F. Gariepy, *Measure Theory and Fine Properties of Functions.* Studies in Advanced Mathematics. CRC Press, Boca Raton, (1992).
12. E. Giusti, *Minimal Surfaces and Functions of Bounded Variation.* Monographs in Mathematics, **80**. Birkhäuser, (1984).
13. M.-H. Giga, Y. Giga, *Very singular diffusion equations: second and fourth order problems.* Jpn. J. Ind. Appl. Math., **27** (2010), no. 3, 323–345.
14. W. Horn, J. Sprekels, S. Zheng, *Global existence of smooth solutions to the Penrose-Fife model for Ising ferromagnets.* Adv. Math. Sci. Appl., **6** (1996), no. 1, 227–241.
15. A. Ito, N. Kenmochi, N. Yamazaki, *A phase-field model of grain boundary motion.* Appl. Math., **53** (2008), no. 5, 433–454.
16. A. Ito, N. Kenmochi, N. Yamazaki, *Weak solutions of grain boundary motion model with singularity.* Rend. Mat. Appl. (7), **29** (2009), no. 1, 51–63.
17. A. Ito, N.Kenmochi, N. Yamazaki, *Global solvability of a model for grain boundary motion with constraint.* Discrete Contin. Dyn. Syst. Ser. S, **5** (2012), no. 1, 127–146.
18. N. Kenmochi, : *Systems of nonlinear PDEs arising from dynamical phase transitions.* In: *Phase transitions and hysteresis (Montecatini Terme, 1993)*, pp. 39–86, Lecture Notes in Math., **1584**, Springer, Berlin, (1994).
19. N. Kenmochi, M. Kubo, *Weak solutions of nonlinear systems for non-isothermal phase transitions.* Adv. Math. Sci. Appl., **9** (1999), no. 1, 499–521.
20. N. Kenmochi, N. Yamazaki, *Large-time behavior of solutions to a phase-field model of grain boundary motion with constraint.* In: *Current advances in nonlinear analysis and related topics*, pp. 389–403, GAKUTO Internat. Ser. Math. Sci. Appl., **32**, Gakkōtosho, Tokyo, (2010).
21. R. Kobayashi, Y. Giga, *Equations with singular diffusivity.* J. Statist. Phys., **95** (1999), 1187–1220.
22. R. Kobayashi, J. A. Warren, W. C. Carter, *A continuum model of grain boundary.* Phys. D, **140** (2000), no. 1-2, 141–150.
23. R. Kobayashi, J. A. Warren, W. C. Carter, *Grain boundary model and singular diffusivity.* In: *Free Boundary Problems: Theory and Applications*, pp. 283–294, GAKUTO Internat. Ser. Math. Sci. Appl., **14**, Gakkōtosho, Tokyo, (2000).
24. J. L. Lions, E. Magenes, *Non-Homogeneous Boundary Value Problems and Applications. Vol I.* Springer-Verlag, New York-Heidelberg, (1972).
25. S. Moll, K. Shirakawa, *Existence of solutions to the Kobayashi-Warren-Carter system.* Calc. Var. Partial Differential Equations, **51** (2014), 621–656. DOI:10.1007/s00526-013-0689-2
26. S. Moll, K. Shirakawa, H. Watanabe, *Energy dissipative solutions to the Kobayashi-Warren-Carter system.* submitted.

27. U. Mosco, *Convergence of convex sets and of solutions of variational inequalities.* Advances in Math., **3** (1969), 510–585.

28. K. Shirakawa, H. Watanabe, *Energy-dissipative solution to a one-dimensional phase field model of grain boundary motion.* Discrete Contin. Dyn. Syst. Ser. S, **7** (2014), no. 1, 139–159. DOI:10.3934/dcdss.2014.7.139

29. K. Shirakawa, H. Watanabe, *Large-time behavior of a PDE model of isothermal grain boundary motion with a constraint.* Discrete Contin. Dyn. Syst. 2015, Dynamical systems, differential equations and applications. 10th AIMS Conference. Suppl., 1009–1018.

30. K. Shirakawa, H. Watanabe, N. Yamazaki, *Solvability of one-dimensional phase field systems associated with grain boundary motion.* Math. Ann., **356** (2013), 301–330. DOI:10.1007/s00208-012-0849-2

31. K. Shirakawa, H. Watanabe, N. Yamazaki, *Phase-field systems for grain boundary motions under isothermal solidifications.* Adv. Math. Sci. Appl., **24** (2014), 353–400.

32. K. Shirakawa, H. Watanabe, N. Yamazaki, *Mathematical analysis for a Warren–Kobayashi–Lobkovsky–Carter type system.* RIMS Kôkyûroku, **1997** (2016), 64–85.

33. J. Simon, *Compact sets in the space $L^p(0,T;B)$.* Ann. Mat. Pura Appl. (4), **146** (1987), 65–96.

34. J. Sprekels, S. Zheng, *Global existence and asymptotic behaviour for a nonlocal phase-field model for non-isothermal phase transitions.* J. Math. Anal. Appl., **279** (2003), 97–110.

35. A. Visintin, *Models of phase transitions.* Progress in Nonlinear Differential Equations and their Applications, **28**, Birkhäuser, Boston, (1996).

36. J. A. Warren, R. Kobayashi, A. E. Lobkovsky, W. C. Carter, *Extending phase field models of solidification to polycrystalline materials.* Acta Materialia, **51** (2003), 6035–6058.

37. H. Watanabe, K. Shirakawa, *Qualitative properties of a one-dimensional phase-field system associated with grain boundary.* In: *Current Advances in Applied Nonlinear Analysis and Mathematical Modelling Issues*, pp. 301–328, GAKUTO Internat. Ser. Math. Sci. Appl., **36**, Gakkōtosho, Tokyo, (2013).

38. James A. Warren, Ryo Kobayashi, Alexander E. Lobkovsky, W. Craig Carter, *Extending phase field models of solidification to polycrystalline materials.* Acta Materialia, **51** (2003), 60356058.

39. H. Watanabe, K. Shirakawa, *Stability for approximation methods of the one-dimensional Kobayashi-Warren-Carter system.* Mathematica Bohemica, **139** (2014), special issue dedicated to Equadiff 13, no. 2, 381–389.

40. N. Yamazaki, *Global attractors for non-autonomous phase-field systems of grain boundary motion with constraint.* Adv. Math. Sci. Appl. **23** (2013), no. 1, 267–296.

6

On the property of bases of multiple systems in Sobolev-Liouville classes

Onur Alp Ilhan[1,*] **and Shakirbay G. Kasimov**[2]

[1] Faculty of Education,University of Erciyes Melikgazi 38039, Kayseri, Turkey
[2] Mechanics and Mathematics Faculty,National University of Uzbekistan ,Tashkent, Uzbekistan

[*] **Correspondence:** oailhan@erciyes.edu.tr

Abstract: In the present work we consider the question of preservation of the baseness property for the system of vectors $\varphi = \{\varphi_n\}_{n\in Z^N}$ in the Sobolev-Liouville and Besov classes at small perturbations with the purpose of the further application of obtained results to study decomposition on root vectors of differential operators.

Keywords: Banach Space; Hilbert Space; Sobolev Space; Normed Vector Space

1. Introduction

We say that a series $\sum_{n\in Z^N} c_n\varphi_n$ converges on rectangulars if there exists the limit of the partial sums $S_m = \sum_{|n_1|\le m_1}\sum_{|n_2|\le m_2}\cdots\sum_{|n_N|\le m_N} c_n\varphi_n$ as $\min\limits_{1\le j\le N} m_j \to \infty$.

Let us remind that a system of elements $\varphi = \{\varphi_n\}_{n\in Z^N}$ is called a basis of the Banach space E at summation on rectangulars if any vector $x \in E$ decomposes uniquely in the series

$$x = \sum_{n\in Z^N} c_n\varphi_n \tag{1.1}$$

which is convergent with respect to the norm of the space E at summation by rectangulars. Hence we exclude from consideration Banach spaces which do not possess the property of approximation (see [1] and [2]).

Factors c_n in (1.1) are linear functionals:

$$c_n = f_n(x)\ , \quad n \in Z^N$$

and, according to the well known Banach theorem (see, for example, [3], [4]), there is a constant C_φ such that

$$\left\|\varphi_n\right\|^{-1} \le \|f_n\| \le C_\varphi \|{-\varphi_n}\|^{-1} .$$

A system of elements $\psi = \{\psi_n\}_{n\in Z^N}$ from the Banach space is said to be ω-linear independent at summation by rectangulars if the equality $\sum_{n\in Z^N} c_n\psi_n = 0$ at summation on rectangulars is impossible at

$$\sum_{n=1}^{\infty} |c_n|^2 \cdot \| \psi_n \|^2 > 0.$$

2. Main Results

Theorem 2.1. *Let $\{\varphi_n\}_{n\in Z^N}$ be a normed basis in the Banach space E at summation by rectangulars. Further, let the system $\{\psi_n\}_{n\in Z^N}$ be ω–linear independent at summation by rectangulars and $\sum_{n\in Z^N} \|\varphi_n - \psi_n\| < \infty$. Then $\{\psi_n\}_{n\in Z^N}$ is also a basis in E at summation by rectangulars.*

Proof. Fix an N-dimensional vector $\beta = (\beta_1, \beta_2, \dots, \beta_N)$ with nonnegative integer components $\beta_1, \beta_2, \dots, \beta_N$ and define as,

$$\tilde{\psi}_n = \begin{cases} \varphi_n & as \quad |n_1| \le \beta_1, \quad |n_2| \le \beta_2, \dots, |n_N| \le \beta_N, \\ \psi_n & as \quad or\ |n_1| > \beta_1,\ or\ \ |n_2| > \beta_2, \ \dots,\ or\ |n_N| > \beta_N, \quad here\ n \in Z^N. \end{cases}$$

Let us introduce the operator $S : E \to E$ which compares to each element

$$x = \sum_{n\in Z^N} f_n(x)\,\varphi_n = \lim_{\substack{\min m_j \to\infty \\ 1\le j\le N}} \sum_{|n_1|\le m_1} \sum_{|n_2|\le m_2} \dots \sum_{|n_N|\le m_N} f_n(x)\,\varphi_n$$

to the element

$$S\,x = \sum_{n\in Z^N} f_n(x)\left(\varphi_n - \tilde{\psi}_n\right).$$

Obviously, for sufficiently large $\mu = \min_{1\le j\le N} \beta$, we have

$$\|S\,x\| \le C_\varphi \|x\| \sum_{or\ |n_1|>\beta_1} \ \sum_{or\ |n_2|>\beta_2} \dots \sum_{or\ |n_N|>\beta_N} \|\varphi_n - \psi_n\| < \varepsilon \|x\|.$$

Hence, for the operator U defined by equality

$$U\,x = x - S\,x = \sum_{n\in Z^N} f_n(x)\,\tilde{\psi}_n,$$

there is an inverse linear operator U^{-1}. Acting on both parts of the equality

$$U^{-1}x = \sum_{n\in Z^N} f_n\left(U^{-1}x\right)\varphi_n$$

with the operator U, we obtain

$$x = \sum_{n\in Z^N} f_n\left(U^{-1}x\right)\tilde{\psi}_n,$$

which implies that the system $\left\{\widetilde{\psi}_n\right\}_{n\in Z^N}$ forms a basis in E at summation on rectangulars, i.e. each vector $x \in E$ is decomposed uniquely in the series

$$x = \sum_{n\in Z^N} f_n\left(U^{-1}x\right)\widetilde{\psi}_n = \lim_{\substack{\min m_j\to\infty \\ 1\le j\le N}} \sum_{|n_1|\le m_1}\sum_{|n_2|\le m_2}\cdots\sum_{|n_N|\le m_N} f_n\left(U^{-1}x\right)\widetilde{\psi}_n$$

which is convergent with respect to the norm of the space E at summation on rectangulars.

Since the system $\left\{\widetilde{\psi}_n\right\}_{n\in Z^N}$ forms a basis in E at summation on rectangulars, then

$$\psi_k = \sum_{|n_1|\le\beta_1}\sum_{|n_2|\le\beta_2}\cdots\sum_{|n_N|\le\beta_N} f_n\left(U^{-1}\psi_k\right)\varphi_n + \sum_{or\ |n_1|>\beta_1}\sum_{or\ |n_2|>\beta_2}\cdots\sum_{or\ |n_N|>\beta_N} f_n\left(U^{-1}\psi_k\right)\psi_n = x_k^1 + x_k^2,$$

here $k = (k_1, k_2, \ldots, k_N)$ is a multi-index with components $|k_1| \le \beta_1, |k_2| \le \beta_2, \ldots, |k_N| \le \beta_N$, and

$$x_k^1 = \sum_{|n_1|\le\beta_1}\sum_{|n_2|\le\beta_2}\cdots\sum_{|n_N|\le\beta_N} f_n\left(U^{-1}\psi_k\right)\varphi_n,$$
$$x_k^2 = \sum_{or\ |n_1|>\beta_1}\sum_{or\ |n_2|>\beta_2}\cdots\sum_{or\ |n_N|>\beta_N} f_n\left(U^{-1}\psi_k\right)\psi_n$$

ω-linear independence of $\{\psi_n\}_{n\in Z^N}$ at summation on rectangulars implies linear independence of $\left\{x_k^1\right\}$. As concepts of linear independence and baseness are equivalent in finite dimensional space,

$$\varphi_n = \sum_{|k_1|\le\beta_1}\sum_{|k_2|\le\beta_2}\cdots\sum_{|k_N|\le\beta_N} \alpha_{nk}x_k^1$$

is a multi-index with components $|n_1| \le \beta_1,\ |n_2| \le \beta_2, \ldots, |n_N| \le \beta_N$ for $n = (n_1, n_2, \ldots, n_N) \in Z^N$.

Hence we have

$$x = \sum_{|n_1|\le\beta_1}\sum_{|n_2|\le\beta_2}\cdots\sum_{|n_N|\le\beta_N} f_n\left(U^{-1}x\right)\varphi_n + \sum_{or\ |n_1|>\beta_1}\sum_{or\ |n_2|>\beta_2}\cdots\sum_{or\ |n_N|>\beta_N} f_n\left(U^{-1}x\right)\psi_n =$$
$$= \sum_{|k_1|\le\beta_1}\sum_{|k_2|\le\beta_2}\cdots\sum_{|k_N|\le\beta_N} (T)\,\psi_k + \sum_{or\ |n_1|>\beta_1}\sum_{or\ |n_2|>\beta_2}\cdots\sum_{or\ |n_N|>\beta_N} f_n\left(U^{-1}x\right)\psi_n =$$
$$= \sum_{|k_1|\le\beta_1}\sum_{|k_2|\le\beta_2}\cdots\sum_{|k_N|\le\beta_N} (T)\,\psi_k + \sum_{or\ |s_1|>\beta_1}\sum_{or\ |s_2|>\beta_2}\cdots\sum_{or\ |s_N|>\beta_N}$$
$$\left(\sum_{|k_1|\le\beta_1}\sum_{|k_2|\le\beta_2}\cdots\sum_{|k_N|\le\beta_N} (T)\, f_s\left(U^{-1}\psi_k\right)\right)\psi_s + \sum_{or\ |n_1|>\beta_1}\sum_{or\ |n_2|>\beta_2}\cdots\sum_{or\ |n_N|>\beta_N} f_n\left(U^{-1}x\right)\psi_n$$

Here ,

$$T = \sum_{|n_1|\le\beta_1}\sum_{|n_2|\le\beta_2}\cdots\sum_{|n_N|\le\beta_N} f_n\left(U^{-1}x\right)a_{nk}$$

It means that the system $\{\psi_n\}_{n\in Z^N}$ is a basis in the Banach space E at summation on rectangulars. Hence, Theorem (2.1) is proved. □

When $N = 1$ Theorem (2.1) was proved in [6]

Remark 1. At absence of ω-linear independence of the system $\{\psi_n\}_{n\in Z^N}$ at summation on rectangulars, one states that the system $\{\psi_n\}_{n\in Z^N}$ is a basis (probably, overfilling) with the finite defect in the Banch space E.

A function $f(x) \in L_p\left(T^N\right)$ belongs to the space $W_p^s\left(T^N\right)$, if all its partial derivatives $D^\alpha f$ (in the sense of the theory of distributions) of the order $|\alpha| = s$ belong to $L_p\left(T^N\right)$, i.e. the norm

$$\|f\|_{W_p^s(T^N)} = \|f\|_{L_p(T^N)} + \sum_{|\alpha|=s} \|D^\alpha f\|_{L_p(T^N)},$$

where $1 \le p < \infty$, $s = 0, 1, 2, \ldots$, is finite.

In the case of $N = 1$ belonging of a function $f(x)$ to the class $W_p^s\left(T^N\right)$ it means that $f(x)$ has $s-1$ continuous derivatives, $f^{(s-1)}(x)$ is absolutely continuous, and $f^{(s)}(x)$ belongs to $L_p(T)$.

Corollary 1. *Let* $\psi_n(x) = (2\pi)^{-\frac{N}{p}} \cdot \left(1 + \sum_{|\alpha|=s} |n^\alpha|\right)^{-1} \cdot \exp(i\lambda_n x) + \alpha_n(x)$, *where* $\lambda_n \ne \lambda_m$ *as* $n \ne m$, *be an* ω*-linear independent system of functions satisfying the following conditions:*

1. $\sum_{n\in Z^N} |\lambda_n - n| < \infty$;
2. $\sum_{n\in Z^N} \|\alpha_n(x)\|_{W_p^s(T^N)} < \infty$.

Then the system of functions $\{\psi_n\}_{n\in Z^N}$ *forms at summation on rectangulars a basis in* $W_p^s\left(T^N\right)$, $1 < p < \infty$, $s = 0, 1, 2, \ldots$.

Theorem 2.2. *Let*

$$\psi_n(x) = (2\pi)^{-\frac{N}{2}} \cdot \left(1 + \sum_{|\alpha|=s} |n^\alpha|^2\right)^{-\frac{1}{2}} \cdot \exp(i\lambda_n x) + \alpha_n(x),$$

where $\lambda_n \ne \lambda_m$ *as* $n \ne m$, *be an* ω*-linear independent system of functions satisfying the following conditions:*

1. $k = \sqrt{\sup_n \frac{\theta^2 + \sum_{|\alpha|=s}\left(\theta^2|\lambda_n^\alpha|^2 + |\lambda_n^\alpha - n^\alpha|^2\right)}{1 + \sum_{|\alpha|=s}|n^\alpha|^2}} < 1$, *here* $\theta = \exp(MN\pi) - 1$,
2. $\sum_{n\in Z^N} \|\alpha_n(x)\|^2_{W_2^s(T^N)} < \infty$.

$$M = \sup_j \sup_{n_j} \left|\lambda_{n_j} - n_j\right|;$$

Then the system of functions $\{\psi_n(x)\}_{n\in Z^N}$ *forms the Riesz basis in the space* $W_2^s\left(T^N\right)$.

Proof. It is known that the system of functions $\varphi_n(x) = (2\pi)^{-\frac{N}{2}} \cdot \left(1 + \sum_{|\alpha|=s} |n^\alpha|^2\right)^{-\frac{1}{2}} \cdot \exp(inx)$ forms an orthonormal basis in the space $W_2^s\left(T^N\right)$. The norm in this space is introduced in such a way at following :

$$\|f\|^2_{W_2^s(T^N)} = \|f\|^2_{L_2(T^N)} + \sum_{|\alpha|=s} \|D^\alpha f\|^2_{L_2(T^N)}.$$

Let

$$\tilde{\psi}_n(x) = (2\pi)^{-\frac{N}{2}} \cdot \left(1 + \sum_{|\alpha|=s} |n^\alpha|^2\right)^{-\frac{1}{2}} \cdot \exp(i\lambda_n x),$$

where $\lambda_n \neq \lambda_m$ as $n \neq m$, be an ω-linear independent system of functions satisfying the condition:

$$k = \sqrt{\sup_n \frac{\theta^2 + \sum_{|\alpha|=s}\left(\theta^2 \left|\lambda_n^\alpha\right|^2 + \left|\lambda_n^\alpha - n^\alpha\right|^2\right)}{1 + \sum_{|\alpha|=s} |n^\alpha|^2}} < 1,$$

here $\theta = \exp(MN\pi) - 1$, $M = \sup_j \sup_{n_j} \left|\lambda_{n_j} - n_j\right|$.

Further, let $\{a_n\}$ be a finite system of complex numbers. Then

$$\left\|\sum_n a_n\left(\tilde{\psi}_n - \varphi_n\right)\right\|^2_{W_2^s(T^N)} = \left\|\sum_n a_n\left(\tilde{\psi}_n - \varphi_n\right)\right\|^2_{L_2(T^N)} + \sum_{|\alpha|=s}\left\|D^\alpha\left(\sum_n a_n\left(\tilde{\psi}_n - \varphi_n\right)\right)\right\|^2_{L_2(T^N)} =$$

$$= (2\pi)^{-N}\left[\left\|\sum_n a_n \cdot \left(\left(1 + \sum_{|\alpha|=s}|n^\alpha|^2\right)^{-\frac{1}{2}}(\exp(i\lambda_n x) - \exp(inx)\right)\right\|^2_{L_2(T^N)} + \right.$$

$$\left. + \sum_{|\alpha|=s}\left\|D^\alpha\left(\sum_n a_n \cdot \left(1 + \sum_{|\alpha|=s}|n^\alpha|^2\right)^{-\frac{1}{2}} \cdot (\exp(i\lambda_n x) - \exp(inx))\right)\right\|^2_{L_2(T^N)}\right]$$

As we have,

$$\left\|\sum_n a_n \cdot \left(1 + \sum_{|\alpha|=s}|n^\alpha|^2\right)^{-\frac{1}{2}} \cdot (\exp(i\lambda_n x) - \exp(inx))\right\|_{L_2(T^N)} \leq$$

$$\leq \sum_{k=1}^{\infty}\frac{1}{k!}\left\|\sum_n a_n \cdot \left(1 + \sum_{|\alpha|=s}|n^\alpha|^2\right)^{-\frac{1}{2}} \cdot [i(\lambda_n - n)x]^k \cdot \exp(inx)\right\|_{L_2(T^N)}.$$

Further,

$$\left\|\sum_n a_n \cdot \left(1 + \sum_{|\alpha|=s}|n^\alpha|^2\right)^{-\frac{1}{2}} \cdot [i(\lambda_n - n)x]^k \cdot \exp(inx)\right\|_{L_2(T^N)} =$$

$$= \left\| \sum_n a_n \cdot \left(1 + \sum_{|\alpha|=s}|n^\alpha|^2\right)^{-\frac{1}{2}}\left(\sum_{\beta_1+\beta_2+\ldots+\beta_N=k}\frac{k!}{\beta_1!\beta_2!\ldots\beta_N!} \cdot \right.\right.$$

$$\left.\left. \cdot \prod_{j=1}^N\left(\lambda_{n_j} - n_j\right)^{\beta_j} \cdot x_j^{\beta_j}\right) \cdot \exp(inx)\right\|_{L_2(T^N)} =$$

$$= \left\|\sum_{\beta_1+\beta_2+\ldots+\beta_N=k}\frac{k!}{\beta_1!\beta_2!\ldots\beta_N!} \cdot \prod_{j=1}^N x_j^{\beta_j} \cdot \left(\sum_n a_n \cdot \left(1 + \sum_{|\alpha|=s}|n^\alpha|^2\right)^{-\frac{1}{2}} \cdot \right.\right.$$

$$\cdot \prod_{j=1}^{N} \left(\lambda_{n_j} - n_j\right)^{\beta_j} \cdot \exp(inx) \Bigg) \Bigg\|_{L_2(T^N)} \leq \sum_{\beta_1+\beta_2+\ldots+\beta_N=k} \frac{k!}{\beta_1!\beta_2!\ldots\beta_N!} \cdot \pi^k \cdot$$

$$\cdot \left\| \sum_n a_n \cdot \left(1 + \sum_{|\alpha|=s} |n^\alpha|^2\right)^{-\frac{1}{2}} \cdot \prod_{j=1}^{N} \left(\lambda_{n_j} - n_j\right)^{\beta_j} \cdot \exp(inx) \right\|_{L_2(T^N)} \leq$$

$$\sum_{\beta_1+\beta_2+\ldots+\beta_N=k} \frac{k!}{\beta_1!\beta_2!\ldots\beta_N!} \cdot \pi^k \cdot \left(\sum_n |a_n|^2 \cdot \left(1 + \sum_{|\alpha|=s} |n^\alpha|^2\right)^{-1} \cdot \prod_{j=1}^{N} \left|\lambda_{n_j} - n_j\right|^{2\beta_j} \cdot (2\pi)^N \right)^{\frac{1}{2}} \leq$$

$$\leq (2\pi)^{\frac{N}{2}} \sum_{\beta_1+\beta_2+\ldots+\beta_N=k} \frac{k!}{\beta_1!\beta_2!\ldots\beta_N!} \cdot \pi^k \cdot M^k \cdot \left(\sum_n |a_n|^2 \cdot \left(1 + \sum_{|\alpha|=s} |n^\alpha|^2\right)^{-1} \right)^{\frac{1}{2}} =$$

$$= (2\pi)^{\frac{N}{2}} \cdot \pi^k \cdot M^k \cdot \left(\sum_n |a_n|^2 \cdot \left(1 + \sum_{|\alpha|=s} |n^\alpha|^2\right)^{-1} \right)^{\frac{1}{2}} \cdot \sum_{\beta_1+\beta_2+\ldots+\beta_N=k} \frac{k!}{\beta_1!\beta_2!\ldots\beta_N!} =$$

$$= (2\pi)^{\frac{N}{2}} \cdot \pi^k \cdot M^k \cdot N^k \cdot \left(\sum_n |a_n|^2 \cdot \left(1 + \sum_{|\alpha|=s} |n^\alpha|^2\right)^{-1} \right)^{\frac{1}{2}},$$

where summation is carried out on all integer nonnegative $\beta_1, \beta_2, \ldots, \beta_N$ such that $\beta_1 + \beta_2 + \ldots + \beta_N = k$, $M = \sup_j \sup_{n_j} \left|\lambda_{n_j} - n_j\right|$, we get

$$\left\| \sum_n a_n \cdot \left(1 + \sum_{|\alpha|=s} |n^\alpha|^2\right)^{-\frac{1}{2}} \cdot (\exp(i\lambda_n x) - \exp(inx)) \right\|_{L_2(T^N)} \leq$$

$$\leq (2\pi)^{\frac{N}{2}} \cdot \sum_{k=1}^{\infty} \frac{1}{k!} \cdot \pi^k \cdot M^k \cdot N^k \cdot \left(\sum_n |a_n|^2 \cdot \left(1 + \sum_{|\alpha|=s} |n^\alpha|^2\right)^{-1} \right)^{\frac{1}{2}} =$$

$$= (2\pi)^{\frac{N}{2}} (\exp(MN\pi) - 1) \cdot \left(\sum_n |a_n|^2 \cdot \left(1 + \sum_{|\alpha|=s} |n^\alpha|^2\right)^{-1} \right)^{\frac{1}{2}}.$$

Further,

$$\sum_{|\alpha|=s} \left\| D^\alpha \left(\sum_n a_n \cdot \left(1 + \sum_{|\alpha|=s} |n^\alpha|^2\right)^{-\frac{1}{2}} \cdot (\exp(i\lambda_n x) - \exp(inx)) \right) \right\|_{L_2(T^N)} =$$

$$= \sum_{|\alpha|=s} \left\| \sum_n a_n \cdot \left(1 + \sum_{|\alpha|=s} |n^\alpha|^2\right)^{-\frac{1}{2}} \cdot D^\alpha (\exp(i\lambda_n x) - \exp(inx)) \right\|_{L_2(T^N)}.$$

Therefore,

$$\left\|\sum_n a_n\left(\widetilde{\psi}_n-\varphi_n\right)\right\|_{W_2^s(T^N)} \leq (\exp(MN\pi)-1)^2\cdot\left(\sum_n |a_n|^2\cdot\left(1+\sum_{|\alpha|=s}|n^\alpha|^2\right)^{-1}\right)+$$

$$+\sum_{|\alpha|=s}\left((\exp(MN\pi)-1)^2\cdot\left(\sum_n |a_n|^2\cdot\left|\lambda_n^\alpha\right|^2\cdot\left(1+\sum_{|\alpha|=s}|n^\alpha|^2\right)^{-1}\right)+\right.$$

$$\left.+\sum_n |a_n|^2\cdot\left|\lambda_n^\alpha-n^\alpha\right|^2\cdot\left(1+\sum_{|\alpha|=s}|n^\alpha|^2\right)^{-1}\right).$$

Hence, we have

$$\left\|\sum_n a_n\left(\widetilde{\psi}_n-\varphi_n\right)\right\|_{W_2^s(T^N)} \leq k\cdot\left(\sum_n |a_n|^2\right)^{\frac{1}{2}}.$$

Since $k<1$, then by theorem by R. Paly and N. Winner ([5], p.224) the system of functions $\left\{\widetilde{\psi}_n(x)\right\}_{n\in Z^N}$ forms a basis in the space $W_2^s\left(T^N\right)$. On the other hand, the theorem by N.K. Bary (see [3], p. 382) implies that the ω-linear system of functions $\{\psi_n(x)\}_{n\in Z^N}$, quadratically close to the Riesz basis $\left\{\widetilde{\psi}_n(x)\right\}_{n\in Z^N}$ in $W_2^s\left(T^N\right)$, is a Riesz basis in $W_2^s\left(T^N\right)$. Hence, Theorem (2.2) is proved. □

Theorem 2.3. *Let* $\psi_n(x)=(2\pi)^{-\frac{N}{p}}\cdot\left(1+|n|^2\right)^{-\frac{s}{2}}\cdot\exp(i\lambda_n x)+\alpha_n(x)$, $n\in Z^N$, *where* $\lambda_n\neq\lambda_m$, *as* $n\neq m$, ω *be an linear independent system of functions at summation on rectangles that satisfies the following conditions:*

1. $\sum_{n\in Z^N}\left\|\sum_{k\in Z^N}\left(\frac{1+|k|^2}{1+|n|^2}\right)^{\frac{s}{2}}\left(\prod_{j=1}^N\frac{\sin\left(\lambda_{n_j}-k_j\right)\pi}{\left(\lambda_{n_j}-k_j\right)\pi}-\delta_{nk}\right)\cdot\exp(ikx)\right\|_{L_p(T^N)}<\infty;$
2. $\sum_{n\in Z^N}\|\alpha_n(x)\|_{L_p^s(T^N)}<\infty.$

Then, the summation on rectangles system functions $\{\psi_n\}_{n\in Z^N}$ *forms a basis* $L_p^s\left(T^N\right)$, $1<p<\infty$.

Proof. By Theorem Sokol-Sokolowski, the system functions $\varphi_n(x)=(2\pi)^{-\frac{N}{p}}\exp(inx)$ forms a normalized basis in $L_p\left(T^N\right)$ at summation on rectangles, i.e, for every $f\in L_p\left(T^N\right)$, there is a single row $\sum_{n\in Z^N} f_n\exp(inx)$ such that

$$S_m(x)=\sum_{|n_1|\leq m_1}\sum_{|n_2|\leq m_2}\dots\sum_{|n_N|\leq m_N} f_n\exp(inx)$$

which partial sums converges (on rectangles) to function $f(x)$ in $L_p\left(T^N\right)$ with respect to norm topology ,while $\min\limits_{1\leq j\leq N} m_j\to\infty$.

Similarly, the system functions

$$\varphi_n(x)=(2\pi)^{-\frac{N}{p}}\cdot\left(1+|n|^2\right)^{-\frac{s}{2}}\cdot\exp(inx)$$

forms a normalized basis in $L_p^s\left(T^N\right)$ at summation on rectangles, ie, for every $f\in L_p^s\left(T^N\right)$, there is a single row

$$\sum_{n\in Z^N}\widetilde{f}_n\cdot\varphi_n(x)$$

such that

$$S_m(x) = \sum_{|n_1|\le m_1} \sum_{|n_2|\le m_2} \cdots \sum_{|n_N|\le m_N} \widetilde{f_n} \cdot \varphi_n(x)$$

which partial sums converges (on rectangles) to function $f(x)$ in $L_p^s\left(T^N\right)$ with respect to norm topology, while $\min\limits_{1\le j\le N} m_j \to \infty$.
Consequently,

$$\|f(x) - S_m(x)\|_{L_p^s(T^N)} = \Big\| \textstyle\sum_{n\in Z^N} (2\pi)^{-\frac{N}{p}} \cdot \widetilde{f_n} \cdot \exp(inx)$$
$$- \textstyle\sum_{|n_1|\le m_1} \sum_{|n_2|\le m_2} \cdots \sum_{|n_N|\le m_N} (2\pi)^{-\frac{N}{p}} \cdot \widetilde{f_n} \cdot \exp(inx) \Big\|_{L_p(T^N)} \to 0$$

while $\min\limits_{1\le j\le N} m_j \to \infty$ where $p \ge 1,\ s \ge 0, \widetilde{f_n} = (2\pi)^{-\frac{N}{q}} \cdot \left(1 + |n|^2\right)^{\frac{s}{2}} \cdot \int_{T^N} f(x) \cdot \exp(-inx)dx,\ \frac{1}{p} + \frac{1}{q} = 1.$
We have

$$\widetilde{\psi}_n(x) = (2\pi)^{-\frac{N}{p}} \cdot \left(1 + |n|^2\right)^{-\frac{s}{2}} \cdot \exp(i\lambda_n x)$$

where $\lambda_n \ne \lambda_m$, while $n \ne m$,be an ω - linear independent system of functions that satisfies the following conditions:

$$\sum_{n\in Z^N} \left\| \sum_{k\in Z^N} \left(\frac{1 + |k|^2}{1 + |n|^2}\right)^{\frac{s}{2}} \left(\prod_{j=1}^{N} \frac{\sin\left(\lambda_{n_j} - k_j\right)\pi}{\left(\lambda_{n_j} - k_j\right)\pi} - \delta_{nk} \right) \cdot \exp(ikx) \right\|_{L_p(T^N)} < \infty.$$

as

$$\left\|\varphi_n - \widetilde{\psi}_n\right\|_{L_p^s(T^N)} = \left\| \sum_{k\in Z^N} \left(1 + |k|^2\right)^{\frac{s}{2}} \left(\varphi_n - \widetilde{\psi}_n\right)_k \cdot \exp(ikx) \right\|_{L_p(T^N)}$$

where

$$\left(\varphi_n - \widetilde{\psi}_n\right)_k = (2\pi)^{-N} \int_{T^N} \left[\varphi_n(x) - \widetilde{\psi}_n(x)\right] \cdot \exp(-ikx)dx$$

are Fourier coefficients. Hence we get

$$\left(\varphi_n - \widetilde{\psi}_n\right)_k = (2\pi)^{-N} \textstyle\int_{T^N} \left[\varphi_n(x) - \widetilde{\psi}_n(x)\right] \cdot \exp(-ikx)dx$$
$$= (2\pi)^{-N-\frac{N}{p}} \left(1 + |n|^2\right)^{-\frac{s}{2}} \textstyle\int_{T^N} \left[\exp(inx) - \exp(i\lambda_n x)\right] \cdot \exp(-ikx)dx$$

$$= (2\pi)^{-N-\frac{N}{p}} \left(1 + |n|^2\right)^{-\frac{s}{2}} \left[\int_{T^N} \exp\left(i\left(n - k\right)x\right)dx - \int_{T^N} \exp\left(i\left(\lambda_n - k\right)x\right)dx \right] =$$

$$= (2\pi)^{-\frac{N}{p}} \left(1 + |n|^2\right)^{-\frac{s}{2}} \left[\delta_{nk} - (2\pi)^{-N} \int_{T^N} \exp\left(i\left(\lambda_n - k\right)x\right)dx \right] =$$

$$= (2\pi)^{-\frac{N}{p}} \left(1 + |n|^2\right)^{-\frac{s}{2}} \left[\delta_{nk} - (2\pi)^{-N} \prod_{j=1}^{N} \int_{-\pi}^{\pi} \exp\left(i\left(\lambda_{n_j} - k_j\right) x_j\right) dx_j\right] =$$

$$= (2\pi)^{-\frac{N}{p}} \left(1 + |n|^2\right)^{-\frac{s}{2}} \left[\delta_{nk} - (2\pi)^{-N} \prod_{j=1}^{N} \frac{1}{i(\lambda_{n_j} - k_j)} \exp\left(i\left(\lambda_{n_j} - k_j\right) x_j\right)\Big|_{-\pi}^{\pi}\right] =$$

$$= (2\pi)^{-\frac{N}{p}} \left(1 + |n|^2\right)^{-\frac{s}{2}} \left[\delta_{nk} - \prod_{j=1}^{N} \frac{\sin\left(\lambda_{n_j} - k_j\right)\pi}{\left(\lambda_{n_j} - k_j\right)\pi}\right]$$

in this way,

$$\left(\varphi_n - \widetilde{\psi}_n\right)_k = (2\pi)^{-\frac{N}{p}} \left(1 + |n|^2\right)^{-\frac{s}{2}} \left[\delta_{nk} - \prod_{j=1}^{N} \frac{\sin\left(\lambda_{n_j} - k_j\right)\pi}{\left(\lambda_{n_j} - k_j\right)\pi}\right]$$

hence

$$\sum_{n\in Z^N} \|\varphi_n - \psi_n\|_{L_p^s(T^N)} = \sum_{n\in Z^N} \left\|\varphi_n - \widetilde{\psi}_n - \alpha_n(x)\right\|_{L_p^s(T^N)} \le \sum_{n\in Z^N} \left\|\varphi_n - \widetilde{\psi}_n\right\|_{L_p^s(T^N)}$$

$$+ \sum_{n\in Z^N} \|\alpha_n(x)\|_{L_p^s(T^N)} =$$

$$= \sum_{n\in Z^N} \left\| \sum_{k\in Z^N} \left(\tfrac{1+|k|^2}{1+|n|^2}\right)^{\frac{s}{2}} \cdot (2\pi)^{-\frac{N}{p}} \left[\delta_{nk} - \prod_{j=1}^{N} \tfrac{\sin\left(\lambda_{n_j}-k_j\right)\pi}{\left(\lambda_{n_j}-k_j\right)\pi}\right] \cdot \exp(ikx) \right\|_{L_p(T^N)}$$

$$+ \sum_{n\in Z^N} \|\alpha_n(x)\|_{L_p^s(T^N)} < \infty.$$

□

By Theorem (2.2) we have the proof of the Theorem (2.3).

References

1. Grothendieck A., *Produits tensoriels topologiques et espaces nucleaires,* Mem. Amer. Math. Soc., **16** (1955).
2. Enflo P., *A counterexample to the approximation problem in Banach spaces,* Acta Math., **130** (1973), 309-317.
3. Gokhberg I.C., Kreyn M.G., *Introduction to the Theory of Linear non Self-adjoint Operators in the Hilbert Space,* Moscow: Nauka, 1969.

4. Gokhberg I.C., Markus A.S., *Stability for bases of Banach and Hilbert spaces,* Izvestiya AN MSSR., **5** (1962), 17-35.

5. Riesz F., Sekyofalvi-Nad B., *Lectures on Functional Analysis,* Moscow, "Mir", 1979.

6. Shakirbay G. Kasimov., *On a Property of Bases in Banach and Hilbert Spaces,* Malaysian Journal of Mathematical Sciences, **5** (2011), 229-240.

7

On the approximate controllability for some impulsive fractional evolution hemivariational inequalities

Yanfang Li*, **Yanmin Liu, Xianghu Liu and He Jun**

Mathematics Department, Zunyi Normal College, 563006, Guizhou, P. R. China

* **Correspondence:** liyanfang998@163.com

Abstract: In this paper, we study the approximate controllability for some impulsive fractional evolution hemivariational inequalities. We show the concept of mild solutions for these problems. The approximate controllability results are formulated and proved by utilizing fractional calculus, fixed points theorem of multivalued maps and properties of generalized Clarke subgradient under some certain conditions.

Keywords: Approximate controllability; hemivariational inequalities; fractional differential; mild solutions; generalized Clarke subdifferential

Mathematics Subject Classification:20M05, 49N25, 34G25, 65L05

---

1. Introduction

In this paper,we will study the approximate controllability results of the following impulsive fractional evolution hemivariational inequalities:

$$\begin{cases} {}^cD_t^\alpha x(t) \in Ax(t) + Bu(t) + \partial F(t, x(t)), & t \in J,\ t \neq t_k,\ \frac{1}{2} < \alpha \leq 1, \\ \Delta x(t_k) \in I_k(x(t_k^-)), & k = 1, 2, \ldots, m, \\ x(t) = x_0, & \end{cases} \tag{1}$$

where ${}^CD_t^\alpha$ denotes the Caputo fractional derivative of order α with the lower limit zero. $A : D(A) \subseteq X \to X$ is the infinitesimal generator of a C_0−semigroup $T(t)(t \geq 0)$ on a separable Hilbert space X. The notation ∂F stands for the generalized Clarke subgradient (cf. [5]) of a locally Lipschitz function $F(t, \cdot) : X \to R$. The control function $u(t)$ takes value in $L^2([0, b]; U)$ and U is a Hilbert space, B is a linear operator from U into X. The function $I_k : X \to X$ is continous, and $0 = t_0 < t_1 < t_2 < \cdots < t_k < \cdots < t_m = T, \Delta x(t_k) = x(t_k^+) - x(t_k^-)$, $x(t_k^+)$ and $x(t_k^-)$ denote the right and the left limits of $x(t)$ at $t = t_k (k = 1, 2, \cdots, m)$.

Since the hemivariational inequality was introduced by Panagiotopoulos in [23] to solve the mechanical problems with nonconvex and nonsmooth superpotentials, an extensive attention has been paid to this field and the great progress has been made in the last two decades. As a natural generalization of variational inequality, the notion of hemivariational inequality plays an very important role in both the qualitative and numerical analysis of nonlinear boundary value problems arising in mechanics, physics, engineering sciences and so on. For more details, one can see, Carl and Motreanu [4], Liu [12,13], Migórski and Ochal [18,19], Panagiotopoulos [24,25]. The theory of the fractional derivatives and integrals is an expanding and vibrant branch of applied mathematics that has found numerous applications. Recently, both the ordinary and the partial differential equations of fractional order have been used within the last few decades for modeling of many physical and chemical processes and in engineering, see e.g. [8, 11, 14–16, 26, 28] and references therein.

It is well known that the controllability, introduced firstly by R.Kalman in 1960, plays an important role in control theory and engineering. It lies in the fact that they have close connections to pole assignment, structural decomposition, quadratic optimal control, observer design, etc. For this reason, the controllability has become an active area of investigation by many researchers and an impressive progress has been made in recent years [1,3,11,15–17,27,29]. However, to the best of our knowledge, the approximate controllability of some impulsive fractional evolution hemivariational inequalities is still an untreated topic, so it is more interesting and necessary to study it.

Motivated by the above mentioned works, the rest of this paper is organized as follows: In Section 2, we will show some definitions and preliminaries which will be used in the following parts. By applying the fixed point theorem of multivalued maps, the approximate controllability of the control system (1) is given in Section 3 under some appropriate conditions.

2. Preliminaries

In this section, we will give some definitions and preliminaries which will be used in the paper. For the uniformly bounded C_0-semigroup $T(t)(t \geq 0)$, we set $M := \sup_{t\in[0,\infty)}\|T(t)\|_{L_b(X)} < \infty$. The norm of the space X will be defined by $\|\cdot\|_X$. Let $C(J, X)$ denote the Banach space of all X-value continous functions from $J = [0, T]$ into X, the norm $\|\cdot\|_c$=sup$\|\cdot\|_X$. Let the another Banach space $PC(J, X)$=$\{x : J \to X, x \in C((t_k, t_{k+1}], X), k = 0, 1, 2, \cdots, n,$there exist$x(t_k^-), x(t_k^+), k = 1, 2, \cdots, n,$ and $x(t_k^-) = x(t_k)\}$, the norm $\|x\|_{PC} = max\{sup\|x(t + 0)\|, sup\|x(t - 0)\|\}$. We can use $L^p(J, R)$ denote the Banach space of all Lebesgue measurable functions from J to R with $\|f\|_{L^p(J,R)} = (\int_J |f(t)|^p dt)^{\frac{1}{p}}$,$L^p(J, X)$ denote the Banach space of functions $f : J \to X$ which are Bochner integroble normed by $\|f\|_{L^p(J,X)}$, $u \in L^p(J, R)$.

Let us recall some known definitions, for more details, one can see [8] and [26].

Definition 2.1 For a given function $f : [0, +\infty) \to R$, the integral

$$I_t^\alpha f(t) = \frac{1}{\Gamma(\alpha)} \int_0^t (t - s)^{\alpha-1} f(s)ds, \qquad \alpha > 0,$$

is called Riemann-Liouville fractional integral of order α, where Γ is the gamma function.

The expression

$${}^L D_t^\alpha f(t) = \frac{1}{\Gamma(n - \alpha)} (\frac{d}{dt})^{(n)} \int_0^t (t - s)^{n-\alpha-1} f(s)dt,$$

where $n = [\alpha] + 1$, $[\alpha]$ denotes the integer part of number α, is called the Riemann-Liouville fractional derivative of order $\alpha > 0$.

Definition 2.2 Caputo's derivative for a function $f : [0, \infty) \to R$ can be defined as

$$^{c}D_t^{\alpha} f(t) = {}^{L}D_t^{\alpha}[f(t) - \sum_{k=0}^{n-1} \frac{t^k}{k!} f^{(k)}(0)], \quad n = [\alpha] + 1,$$

where $[\alpha]$ denotes the integer part of real number α.

Now, let us recall the definition of the generalized gradient of Clarke for a locally Lipschitz functional $h : E \to R$ (where E is a Banach space), cf. [5]. We denote by $h^0(y, z)$ the Clarke generalized directional derivative of h at y in the direction z, that is

$$h^0(y, z) := \limsup_{\lambda \to 0^+,\ \xi \to y} \frac{h(\xi + \lambda z) - h(\xi)}{\lambda}.$$

Recall also that the generalized Clarke subgradient of h at y, denote by $\partial h(y)$, is a subset of E^* is given by

$$\partial h(y) := \{y^* \in E^* : h^0(y, z) \geq \langle y^*, z \rangle_X, \ \forall z \in E\}.$$

The following basic properties of the generalized subgradient play important role in our main results.

Lemma 2.3 (see Proposition 2.1.2 of [5]). Let h be locally Lipschitz of rank K near y. Then

(a) $\partial h(y)$ is a nonempty, convex, weak*-compact subset of E^* and $\|y^*\|_{E^*} \leq K$ for every y^* in $\partial h(y)$;

(b) for every $z \in E$, one has $h^0(y, z) = \max\{\langle y^*, z \rangle : \text{ for all } y^* \in \partial h(y)\}$.

Lemma 2.4 (see Proposition 5.6.10 of [6]). If $h : E \to R$ is locally Lipschitz, then the multifunction $y \to \partial h(y)$ is upper semicontinuous (u.s.c. for short) from E into $E^*_{w^*}$ (where $E^*_{w^*}$ denotes the Banach space E^* furnished with the w^*-topology).

Next, we present a result on measurability of the multifunction of the subgradient type whose proofs can be found in Kulig [10].

Lemma 2.5 (Proposition 3.44 of [20], page 66). Let E be a separable reflexive Banach space, $0 < b < \infty$ and $h : (0, b) \times E \to R$ be a function such that $h(\cdot, x)$ is measurable for all $x \in E$ and $h(t, \cdot)$ is locally Lipschitz for all $t \in (0, b)$. Then the multifunction $(0, b) \times E \ni (t, x) \mapsto \partial h(t, x)$ is measurable, where ∂h denotes the Clarke generalized gradient of $h(t, \cdot)$.

Now, we also introduce some basic definitions and results from multivalued analysis. For more details ,one can see the book [7]:

- In a Banach space E, a multivalued map $F : E \to 2^E \setminus \{\emptyset\} := \mathcal{P}(E)$ is convex (closed) valued, if $F(x)$ is convex (closed) for all $x \in E$. F is bounded on bounded sets if $F(V) = \bigcup_{x \in V} f(x)$ is bounded in E, for any bounded set V of E (i.e., $\sup_{x \in V}\{\sup\{\|y\| : y \in F(x)\}\} < \infty$).
- F is called u.s.c on E, if for each $x \in E$, the set $F(x)$ is a nonempty, closed subset of E, and if for each open set V of E containing $F(x)$, there exists an open neighborhood N of x such that $F(N) \subseteq V$.
- F is said to be completely continuous if $F(V)$ is relatively compact, for every bounded subset $V \subseteq E$.
- If the multivalued map F is completely continuous with nonempty compact values, then F is u.s.c. if and only if F has a closed graph (i.e., $x_n \to x,\ y_n \to y,\ y_n \in F(x_n)$ imply $y \in F(x)$).
- F has a fixed point if there is $x \in E$, such that $x \in F(x)$.

• A multivalued map $F : J \to \mathcal{P}(E)$ is measurable if $F^{-1}(C) = \{t \in J : F(t) \cap C \neq \emptyset\} \in \Sigma$ for every closed set $C \subseteq E$. If $F : J \times E \to \mathcal{P}(E)$, then measurability of F means that $F^{-1}(C) \in \Sigma \otimes \mathcal{B}_E$, where $\Sigma \otimes \mathcal{B}_E$ is the $\sigma-$algebra of subsets in $J \times E$ generated by the sets $A \times B$, $A \in \Sigma$, $B \in \mathcal{B}_E$, and $\mathcal{B}_E$ is the $\sigma-$algebra of the Borel sets in E.

Now, according to the paper [15, 16, 28, 30], we shall recall the following definitions:

Definition 2.6 For each $u \in L^2(J, U)$, a function $x \in C(J, X)$ is a solution (mild solution) of the system (1) if $x(0) = x_0$ and there exists $f \in L^p(J, X)$ $(p > \frac{1}{\alpha})$ such that $f(t) \in \partial F(t, x(t))$, $\mathcal{I}_i \in I_i(x(t_i^-))$, without loss of generality, let $t \in (t_k, t_{k+1}]$, $1 \leq k \leq m-1$.

$$x(t) = S_\alpha(t)x_0 + \sum_{i=1}^{k} S_\alpha(t-t_i)\mathcal{I}_i(x(t_i^-)) + \int_0^t (t-s)^{\alpha-1}T_\alpha(t-s)f(s)ds + \int_0^t (t-s)^{\alpha-1}T_\alpha(t-s)Bu(s)ds. \quad (2.4)$$

where

$$S_\alpha(t) = \int_0^\infty \xi_\alpha(\theta)T(t^\alpha\theta)d\theta, \qquad T_\alpha(t) = \alpha\int_0^\infty \theta\xi_\alpha(\theta)T(t^\alpha\theta)d\theta,$$

and

$$\xi_\alpha(\theta) = \frac{1}{\alpha}\theta^{-1-\frac{1}{\alpha}}\varpi_\alpha(\theta^{-\frac{1}{\alpha}}) \geq 0,$$

$$\varpi_\alpha(\theta) = \frac{1}{\pi}\sum_{n=1}^{\infty}(-1)^{n-1}\theta^{-n\alpha-1}\frac{\Gamma(n\alpha+1)}{n!}\sin(n\pi\alpha), \quad \theta \in (0, \infty),$$

ξ_α is a probability density function defined on $(0, \infty)$, that is

$$\xi_\alpha(\theta) \geq 0, \ \theta \in (0, \infty), \quad \text{and} \quad \int_0^\infty \xi_\alpha(\theta)d\theta = 1.$$

Lemma 2.7 (Lemma 3.2-3.4 in [28]) The operators $S_\alpha(t)$ and $T_\alpha(t)$ have the following properties:

(i) For any fixed $t \geq 0$, $S_\alpha(t)$ and $T_\alpha(t)$ are linear and bounded operators, i.e., for any $x \in X$,

$$\|S_\alpha(t)x\| \leq M\|x\|, \quad \text{and} \quad \|T_\alpha(t)x\| \leq \frac{M}{\Gamma(\alpha)}\|x\|.$$

(ii) $\{S_\alpha(t), t \geq 0\}$ and $\{T_\alpha(t), t \geq 0\}$ are strongly continuous.

(iii) For any $t > 0$, $S_\alpha(t)$ and $T_\alpha(t)$ are also compact operators if $T(t)$ is compact.

The key tool in our main results is the following fixed point theorem stated in [2].

Theorem 2.8 (Bohnenblust-Karlin [2]). Let Ω be a nonempty subset of a Banach space E, which is bounded, closed and convex. Suppose that $F : \Omega \to 2^E \setminus \{\emptyset\}$ is u.s.c. with closed, convex values such that $F(\Omega) \subseteq \Omega$ and $F(\Omega)$ is compact. Then F has a fixed point.

3. Approximate controllability results

In this section, we investigate the approximate controllability of the control systems described by impulsive fractional evolution hemivariational inequalities.

Let $x(t; 0, x_0, u)$ be a solution of system (1) at time t corresponding to the control $u(\cdot) \in L^2(J, U)$ and the initial value $x_0 \in X$. The set $\Re(b, x_0) = \{x(b; 0, x_0, u) : u(\cdot) \in L^2(J, U)\}$ is called the reachable set of system (1) at terminal time b. Then, the following definition of the approximate controllability is standard.

Definition 3.1 The control system (1) is said to be approximately controllable on the interval J, if for every initial function $x_0 \in X$, we have $\overline{\Re(b, x_0)} = X$.

Now, we consider the following linear fractional differential system:

$$\begin{cases} {}^C D_t^\alpha x(t) = Ax(t) + Bu(t), \quad t \in J = [0, b], \\ x(0) = x_0. \end{cases} \tag{3.1}$$

It is convenient at this point to introduce the controllability operator associated with (3.1) as follows:

$$\Gamma_0^b = \int_0^b (b-s)^{\alpha-1} T_\alpha(b-s) BB^* T_\alpha^*(b-s) ds,$$

$$R(\varepsilon, \Gamma_0^b) = (\varepsilon I + \Gamma_0^b)^{-1}, \quad \varepsilon > 0,$$

respectively, where B^* denotes the adjoint of B and $T_\alpha^*(t)$ is the adjoint of $T_\alpha(t)$. It is straightforward to see that the operator Γ_0^b is a linear bounded operator.

The following Lemma is of great importance for our main results.

Lemma 3.2 [1,16]The linear fractional control system (3.1) is approximately controllable on J if and only if $\varepsilon R(\varepsilon, \Gamma_0^b) \to 0$ as $\varepsilon \to 0^+$ in the strong operator topology.

To obtain the approximate controllability result, we impose the following hypotheses:

$H(1)$: The C_0-semigroup $T(t)$ is compact and $\sup_{t\in[0,\infty)} \|T(t)\|_{L_b(X)} \leq M$.

$H(2)$: $F : J \times X \to \mathbb{R}$ is a function such that:

(i) for all $x \in X$, the function $t \mapsto F(t, x)$ is measurable;

(ii) the function $x \mapsto F(t, x)$ is locally Lipschitz for a.e. $t \in J$;

(iii) there exists a function $a(t) \in L^p(J, R^+)(p > \frac{1}{\alpha})$ and a constant $c > 0$ such that

$$\|\partial F(t, x)\|_{X^*} = \sup\{\|f\|_{X^*} : f(t) \in \partial F(t; x)\} \leq a(t) + c\|x\|_X, \text{ for a.e. } t \in J, \text{ all } x \in X.$$

$H(3)$: $\mathcal{I}_i : X \to X (i = 1, 2, \cdots, m)$ satisfies:

(i) $\mathcal{I}_i$ maps a bounded set to a bounded set;

(ii) There exist constants $d_i > 0 (i = 1, 2, \cdots, m)$ such that

$$\|\mathcal{I}_i(x) - \mathcal{I}_i(y)\| \leq d_i \|x - y\|, \quad x, y \in X.$$

$$(iii) \|\mathcal{I}(0)\| = max(\|\mathcal{I}_1(0)\|, \|\mathcal{I}_2(0)\|, \cdots, \|\mathcal{I}_m(0)\|).$$

Next, we define an operator $\mathcal{N} : L^{\frac{p}{p-1}}(J, X) \to 2^{L^p(J,X)}$ as follows

$$\mathcal{N}(x) = \{w \in L^p(J, X) : w(t) \in \partial F(t; x(t)) \text{ a.e. } t \in J\}, \ x \in L^{\frac{p}{p-1}}(J, X).$$

The following Lemma due to Migórski and Ochal [20] is crucial in our main results.

Lemma 3.3 If the assumption $H(2)$ holds, then the set $\mathcal{N}(x)$ has nonempty, convex and weakly compact values for $x \in L^{\frac{p}{p-1}}(J, X)$, that is the multifunction $t \mapsto \partial F(t, x(t))$ has a measurable X^* selection.

Proof. Our main idea comes from Lemma 5.3 of [20]. Firstly, it is easy to see that $\mathcal{N}(x)$ has convex and weakly compact values from Lemma 2.3. Now, we only show that its values are nonempty. Let

$x \in L^{\frac{p}{p-1}}(J,X)$. Then, by Theorem 2.35 (ii) of [20], there exists a sequence $\{\varphi_n\} \in L^{\frac{p}{p-1}}(J,X)$ of simple functions such that

$$\varphi_n(t) \to x(t), \quad \text{in } L^{\frac{p}{p-1}}(J,X). \tag{3.2}$$

From Lemma 2.5 and hypotheses $H(2)$(i), (ii), the multifunction $t \mapsto \partial F(t,x)$ is measurable from J into $\mathcal{P}_{fc}(X^*)$ (where $\mathcal{P}_{fc}(X^*) = \{\Omega \subseteq X^* : \Omega$ is nonempty, convex and closed $\}$) (since the weak and weak*-topologies on the dual space of a reflexive Banach space coincide (cf. e.g. p7 of [9]), the multifunction ∂F is $P_{fc}(X^*)$-valued). Applying Theorem 3.18 of [20], for every $n \geq 1$, there exists a measurable function $\zeta_n : J \to X^*$ such that $\zeta_n(t) \in \partial F_n(t, \varphi_n(t))$ a.e. $t \in J$. Next, from hypothesis $H(F)(iii)$, we have

$$\|\zeta_n\|_{X^*} \leq a(t) + c\|\varphi_n\|_X.$$

Hence, $\{\zeta_n\}$ remains in a bounded subset of X^*. Thus, by passing to a subsequence, if necessary, we may suppose, by Theorem 1.36 of [20], that $\zeta_n \to \zeta$ weakly in X^* with $\zeta \in X^*$. From Proposition 3.16 of [20], it follows that

$$\zeta(t) \in \overline{\text{conv}}\Big((w - X^*) - \limsup\{\zeta_n(t)\}_{n\geq 1}\Big) \quad \text{a.e. } t \in J, \tag{3.3}$$

where $\overline{\text{conv}}$ denotes the closed convex hull of a set. From hypothesis $H(2)$(ii) and Lemma 2.4, we know that the multifunction $x \mapsto \partial F(t, x(t))$ is u.s.c from X into $X^*_{w^*}$. Recalling that the graph of an u.s.c multifunction with closed values is closed (see Proposition 3.12 of [20]), we get for a.e. $t \in J$, if $f_n \in \partial F(t, \zeta_n)$, $f_n \in X^*$, $f_n \to f$ weakly in X^*, $\zeta_n \in L^{\frac{p}{p-1}}(J,X)$, $\zeta_n \to \zeta$ in $L^{\frac{p}{p-1}}(J,X)$, then $f \in \partial F(t,\zeta)$. Therefore, by (3.2), we have

$$(w - X^*) - \limsup \partial F(t, \zeta_n(t)) \subset \partial F(t, x(t)) \quad \text{a.e. } t \in J, \tag{3.4}$$

where the Kuratowski limit superior is given by

$$\begin{aligned} & (w - X^*) - \limsup \partial F(t, \varphi_n(t)) \\ = \ & \{\zeta^* \in X^* : \zeta^* = (w - X^*) - \limsup \partial \zeta^*_{n_k} \ \zeta^*_{n_k} \in \partial F(t, \varphi_n(t)), \ n_1 < n_2 < \cdots < n_k < \cdots\} \end{aligned}$$

(see Definition 3.14 of [20]). So, from (3.3) and (3.4), we deduce that

$$\begin{aligned} \zeta(t) \ & \subset \ \overline{\text{conv}}((w - X^*) - \limsup\{\zeta_n(t)\}_{n\geq 1}) \\ & \subset \ \overline{\text{conv}}((w - X^*) - \limsup \partial F(t, \varphi_n(t)) \\ & \subset \ \partial F(t, x(t)), \quad \text{a.e. } t \in J. \end{aligned}$$

Since $\zeta \in X^*$ and $\zeta(t) \in \partial F(t, x(t))$ a.e. $t \in J$, it is clear that $\zeta \in \mathcal{N}(x)$. This proves that $\mathcal{N}(x)$ has nonempty values and completes the proof .

we prove that there exists $f \in L^p(J,X)$ $(p > \frac{1}{\alpha})$ such that $f(t) \in \partial F(t, x(t))$, so the $\mathcal{I}_i \in I_i(x(t_i^-))$, we omit the same kind of arguement.

The following Lemma is of great importance in our main results.

Lemma 3.4 (see Lemma 11 in [19]). If $H(2)$ holds, the operator $\mathcal{N}$ satisfies: if $z_n \to z$ in $L^{\frac{p}{p-1}}(J,X)$, $w_n \rightharpoonup w$ in $L^p(J,X)$ and $w_n \in \mathcal{N}(z_n)$, then we have $w \in \mathcal{N}(z)$. (Where $\rightharpoonup$ means weak convergence).

Now, we are in the position to prove the existence results of this paper.

Theorem 3.5 Suppose that the hypotheses $H(1)$ and $H(2)$ are satisfied, then the system (1.1) has a mild solution on J provided that

$$\left[1+\frac{M^2M_B^2b^\alpha}{\varepsilon\alpha[\Gamma(\alpha)]^2}\right]\left[\sum_{i=1}^k Md_i+\frac{Mcb^\alpha}{\Gamma(1+\alpha)}\right]<\frac{1}{2},\quad \text{where}\quad M_B:=\|B\|.$$

Proof. For any $\varepsilon>0$, we consider the multivalued map $F_\varepsilon : C(J,X)\to 2^{C(J,X)}$ as follows

$$\begin{aligned}F_\varepsilon(x) &= \Big\{h\in C(J,X):\ h(t)=S_\alpha(t)x_0+\sum_{i=1}^k S_\alpha(t-t_i)\mathcal{I}_i(x(t_i^-))+\int_0^t(t-s)^{\alpha-1}T_\alpha(t-s)f(s)ds\\ &\quad+\int_0^t(t-s)^{\alpha-1}T_\alpha(t-s)Bu_\varepsilon(s)ds,\ \text{with } f\in\mathcal{N}(x)\Big\},\quad \text{for } x\in C(J,X),\end{aligned}$$

where

$$u_\varepsilon(t)=B^*T_\alpha^*(b-t)R(\varepsilon,\Gamma_0^b)\Big(x_1-S_\alpha(b)x_0-\sum_{i=1}^k S_\alpha(b-t_i)\mathcal{I}_i(x(t_i^-))-\int_0^b(b-\tau)^{\alpha-1}T_\alpha(b-\tau)f(\tau)d\tau\Big).$$

It is clear that the problem of finding mild solutions of (1) is reduced to find the fixed point of F_ε. We prove the operator F_ε satisfies all the conditions of the Theorem 2.8 and we divide the proof into several steps.

Step 1: F_ε is convex for each $x\in C(J,X)$.

In fact, for any ρ_1, ρ_2 belong to F_ε, then there exist $f_1,f_2\in\mathcal{N}(x)$ such that

$$\begin{aligned}\rho_i(t) &= S_\alpha(t)x_0+\sum_{i=1}^k S_\alpha(t-t_i)\mathcal{I}_i(x(t_i^-))+\int_0^t(t-s)^{\alpha-1}T_\alpha(t-s)f_i(s)ds \qquad (3.1)\\ &\quad+\int_0^t(t-s)^{\alpha-1}T_\alpha(t-s)BB^*T_\alpha^*(b-s)\\ &\quad\times R(\varepsilon,\Gamma_0^b)\Big(x_1-S_\alpha(b)x_0-\sum_{i=1}^k S_\alpha(b-t_i)\mathcal{I}_i(x(t_i^-))-\int_0^b(b-\tau)^{\alpha-1}T_\alpha(b-\tau)f(\tau)d\tau\Big)ds,\\ &\quad i=1,2,\ t\in J.\end{aligned}$$

Let $\lambda\in[0,1]$, then for each $t\in J$, we have

$$\begin{aligned}&[\lambda\rho_1+(1-\lambda)\rho_2](t) \qquad (3.2)\\ &= S_\alpha(t)x_0+\sum_{i=1}^k S_\alpha(t-t_i)\mathcal{I}_i(x(t_i^-))+\int_0^t(t-s)^{\alpha-1}T_\alpha(t-s)[\lambda f_1+(1-\lambda)f_2](s)ds\\ &\quad+\int_0^t(t-s)^{\alpha-1}T_\alpha(t-s)BB^*T_\alpha^*(b-s)R(\varepsilon,\Gamma_0^b)\Big(x_1-S_\alpha(b)x_0\\ &\quad-\sum_{i=1}^k S_\alpha(b-t_i)\mathcal{I}_i(x(t_i^-))-\int_0^b(b-\tau)^{\alpha-1}T_\alpha(b-\tau)[\lambda f_1+(1-\lambda)f_2](\tau)d\tau\Big)ds.\end{aligned}$$

From Lemma 2.3, we know that $\partial F(t,x(t))$ is convex, hence for $\lambda \in [0,1]$, $\lambda f_1 + (1-\lambda)f_2 \in \mathcal{N}(x)$, then $\lambda\rho_1(t) + (1-\lambda)\rho_2(t) \in F_\varepsilon$, which implies that F_ε is convex for each $x \in C(J,X)$.

Step 2: There exists a nonempty, bounded, closed and convex subset $B_r \subseteq C(J,X)$ such that $F_\varepsilon(B_r) \subseteq B_r$.

Take

$$
\begin{aligned}
r &= 2\Big[M\|x_0\| + \sum_{i=1}^{k} M\|\mathcal{I}(0)\| + (1 + \frac{M^2M_B^2b^\alpha}{\varepsilon\alpha[\Gamma(\alpha)]^2})\frac{M}{\Gamma(\alpha)}(\frac{p-1}{p\alpha-1})^{1-\frac{1}{p}}b^{\alpha-\frac{1}{p}}\|a\|_{L^p} \\
&\quad + \frac{M^2M_B^2b^\alpha}{\varepsilon\alpha[\Gamma(\alpha)]^2}(\|x_1\| + M\|x_0\| + \sum_{i=1}^{k} M\|\mathcal{I}(0)\|)\Big],
\end{aligned}
$$

and denote $B_r = \{x \in C(J,X) : \|x(t)\|_X \le r\}$. Obviously, B_r is a bounded, closed and convex subset of $C(J,X)$. In fact, for any $x \in B_r$, $\varphi \in F_\varepsilon(x)$, there exists $f \in \mathcal{N}(x)$ such that

$$
\begin{aligned}
\varphi(t) &= S_\alpha(t)x_0 + \sum_{i=1}^{k} S_\alpha(t-t_i)\mathcal{I}_i(x(t_i^-)) + \int_0^t (t-s)^{\alpha-1}T_\alpha(t-s)f(s)ds \\
&\quad + \int_0^t (t-s)^{\alpha-1}T_\alpha(t-s)BB^*T_\alpha^*(b-s) \times R(\varepsilon,\Gamma_0^b)\Big(x_1 - S_\alpha(b)x_0 \\
&\quad - \sum_{i=1}^{k} S_\alpha(b-t_i)\mathcal{I}_i(x(t_i^-)) - \int_0^b (b-\tau)^{\alpha-1}T_\alpha(b-\tau)f(\tau)d\tau\Big)ds, \qquad t \in J.
\end{aligned}
$$

Taking the assumptions $H(1)$ and Hölder inequality into account, we obtain

$$
\begin{aligned}
\|\varphi(t)\| &\le \|S_\alpha(t)x_0\| + \|\sum_{i=1}^{k} S_\alpha(t-t_i)\mathcal{I}_i(x(t_i^-))\| + \int_0^t (t-s)^{\alpha-1}\|T_\alpha(t-s)f(s)\|ds \\
&\quad + \int_0^t (t-s)^{\alpha-1}\|T_\alpha(t-s)B \times B^*T_\alpha^*(b-s)R(\varepsilon,\Gamma_0^b)(x_1 - S_\alpha(b)x_0 \\
&\quad - \sum_{i=1}^{k} S_\alpha(b-t_i)\mathcal{I}_i(x(t_i^-)) - \int_0^b (b-\tau)^{\alpha-1}T_\alpha(b-\tau)f(\tau)d\tau)\|ds \\
&\le M\|x_0\| + \sum_{i=1}^{k} M(d_i\|x(t_i^-)\| + \|\mathcal{I}_i(0)\|) + \frac{M}{\Gamma(\alpha)}\int_0^t (t-s)^{\alpha-1}[a(s) + c\|x(s)\|_X]ds \\
&\quad + \frac{M^2M_B^2b^\alpha}{\varepsilon\alpha[\Gamma(\alpha)]^2}\Big[\|x_1\| + M\|x_0\| + \sum_{i=1}^{k} M(d_i\|x(t_i^-)\| + \|\mathcal{I}_i(0)\|) \\
&\quad + \frac{M}{\Gamma(\alpha)}\int_0^b (b-\tau)^{\alpha-1}[a(\tau) + c\|x(\tau)\|_X\Big]d\tau \\
&\le M\|x_0\| + \sum_{i=1}^{k} M(d_ir + \|\mathcal{I}_i(0)\|) + \frac{M}{\Gamma(\alpha)}(\frac{p-1}{p\alpha-1})^{1-\frac{1}{p}}b^{\alpha-\frac{1}{p}}\|a\|_{L^p} + \frac{Mcb^\alpha}{\Gamma(1+\alpha)}r \\
&\quad + \frac{M^2M_B^2b^\alpha}{\varepsilon\alpha[\Gamma(\alpha)]^2}\Big[\|x_1\| + M\|x_0\| + \sum_{i=1}^{k} M(d_ir + \|\mathcal{I}_i(0)\|)
\end{aligned}
$$

$$+\frac{M}{\Gamma(\alpha)}(\frac{p-1}{p\alpha-1})^{1-\frac{1}{p}}b^{\alpha-\frac{1}{p}}\|a\|_{L^p}+\frac{Mcb^{\alpha}}{\Gamma(1+\alpha)}r\Big]$$
$$\leq r.$$

Thus, we obtain that $F_\varepsilon(B_r)\subseteq B_r$.

Step 3. F_ε is equicontinuous on B_r.

Firstly, for any $x\in B_r$, $\varphi\in F_\varepsilon(x)$, there exists $f\in\mathcal{N}(x)$ such that

$$\varphi(t) = S_\alpha(t)x_0+\sum_{i=1}^{k}S_\alpha(t-t_i)\mathcal{I}_i(x(t_i^-))+\int_0^t(t-s)^{\alpha-1}T_\alpha(t-s)[f(s)+Bu_\varepsilon(s)]ds,\quad t\in J.$$

For any $\epsilon>0$, when $\tau_1=0,\ 0<\tau_2\leq\delta_0$, we obtain

$$\begin{aligned}
&\|\varphi(\tau_2)-\varphi(\tau_1)\|=\|\varphi(\tau_2)-x_0\|\\
\leq\ &\|S_\alpha(\tau_2)x_0-x_0\|+\|\sum_{i=1}^{k}S_\alpha(t-t_i)\mathcal{I}_i(x(t_i^-))\|+\|\int_0^{\tau_2}(\tau_2-s)^{\alpha-1}T_\alpha(\tau_2-s)f(s)ds\|\\
&+\|\int_0^{\tau_2}(\tau_2-s)^{\alpha-1}T_\alpha(\tau_2-s)Bu_\varepsilon(s)ds\|\\
\leq\ &\|S_\alpha(\tau_2)x_0-x_0\|+\frac{M}{\Gamma(\alpha)}(\frac{p-1}{p\alpha-1})^{1-\frac{1}{p}}\|a\|_{L^p}\tau_2^{\alpha-\frac{1}{p}}+\frac{Mc\tau_2^{\alpha}}{\Gamma(1+\alpha)}r+\frac{MM_B}{\sqrt{2\alpha-1}\Gamma(\alpha)}\|u_\varepsilon\|_{L^2}\tau_2^{\alpha-\frac{1}{2}}.
\end{aligned}$$

Hence, we can choose $\delta_0>0$ is small enough so that for all $0<\tau_2\leq\delta_0$, the impulsive term is 0, $\|\varphi(\tau_2)-\varphi(\tau_1)\|<\frac{\epsilon}{2}$. Thus, for $\forall\epsilon>0$, $\forall\tau_1,\ \tau_2\in[0,\delta_0]$, $\forall\varphi\in F_\varepsilon(x)$, we have $\|\varphi(\tau_2)-\varphi(\tau_1)\|<\epsilon$ independently of $x\in B_r$.

Next, for any $x\in B_r$ and $\frac{\delta_0}{2}\leq\tau_1<\tau_2\leq b$, we obtain

$$\begin{aligned}
&\|\varphi(\tau_2)-\varphi(\tau_1)\|\\
\leq\ &\|S_\alpha(\tau_2)x_0-S_\alpha(\tau_1)x_0\|+\|\int_0^{\tau_1}[(\tau_2-s)^{\alpha-1}-\tau_1-s)^{\alpha-1}]T_\alpha(\tau_2-s)f(s)ds\\
&+\|\int_0^{\tau_1}(\tau_1-s)^{\alpha-1}[T_\alpha(\tau_2-s)-T_\alpha(\tau_1-s)]f(s)ds\|\\
&+\|\int_{\tau_1}^{\tau_2}(\tau_2-s)^{\alpha-1}T_\alpha(\tau_2-s)f(s)ds\|\\
&+\|\int_0^{\tau_1}[(\tau_1-s)^{\alpha-1}-(\tau_2-s)^{\alpha-1}]T_\alpha(\tau_1-s)Bu_\varepsilon(s)ds\|\\
&+\|\int_0^{\tau_1}(\tau_1-s)^{\alpha-1}[T_\alpha(\tau_2-s)-T_\alpha(\tau_1-s)]Bu_\varepsilon(s)ds\|\\
&+\|\int_{\tau_1}^{\tau_2}(\tau_2-s)^{\alpha-1}T_\alpha(\tau_2-s)Bu_\varepsilon(s)ds\|\\
&\leq Q_1+Q_2+Q_3+Q_4+Q_5+Q_6+Q_7.
\end{aligned}$$

By the assumptions and Hölder's inequality, we have

$$Q_2 \leq \frac{M}{\Gamma(\alpha)}\int_0^{\tau_1}[(\tau_1-s)^{\alpha-1}-(\tau_2-s)^{\alpha-1}]\|f(s)\|ds$$

$$
\begin{aligned}
&\leq \frac{M}{\Gamma(\alpha)}(\frac{p-1}{p\alpha-1})^{1-\frac{1}{p}}||a||_{L^p}[\tau_2^{\alpha-\frac{1}{p}}-\tau_1^{\alpha-\frac{1}{p}}+(\tau_2-\tau_1)^{\alpha-\frac{1}{p}}]\\
&\quad+\frac{Mcr}{\Gamma(1+\alpha)}[\tau_2^{\alpha}-\tau_1^{\alpha}+(\tau_2-\tau_1)^{\alpha}]\\
&\leq \frac{2M}{\Gamma(\alpha)}(\frac{p-1}{p\alpha-1})^{1-\frac{1}{p}}||a||_{L^p}(\tau_2-\tau_1)^{\alpha-\frac{1}{p}}+\frac{2Mcr}{\Gamma(1+\alpha)}(\tau_2-\tau_1)^{\alpha},
\end{aligned}
$$

Similarly, we obtain

$$
\begin{aligned}
Q_4 &\leq \frac{M}{\Gamma(\alpha)}(\frac{p-1}{p\alpha-1})^{1-\frac{1}{p}}||a||_{L^p}(\tau_2-\tau_1)^{\alpha-\frac{1}{p}}+\frac{Mcr}{\Gamma(1+\alpha)}(\tau_2-\tau_1)^{\alpha},\\
Q_5 &\leq \frac{MM_B}{\sqrt{2\alpha-1}\Gamma(\alpha)}||u_\varepsilon||_{L^2}[\tau_2^{\alpha-\frac{1}{2}}-\tau_1^{\alpha-\frac{1}{2}}+(\tau_2-\tau_1)^{\alpha-\frac{1}{2}}],\\
Q_7 &\leq \frac{MM_B}{\sqrt{2\alpha-1}\Gamma(\alpha)}||u_\varepsilon||_{L^2}(\tau_2-\tau_1)^{\alpha-\frac{1}{2}}.
\end{aligned}
$$

For $\tau_1 \geq \frac{\delta_0}{2} > 0$ and $\delta_1 > 0$ small enough, we obtain

$$
\begin{aligned}
Q_3 &\leq \Big[||\int_0^{\tau_1-\delta_1}(\tau_1-s)^{\alpha-1}[T_\alpha(\tau_2-s)-T_\alpha(\tau_1-s)]f(s)ds||\\
&\quad+||\int_{\tau_1-\delta_1}^{\tau_1}(\tau_1-s)^{\alpha-1}[T_\alpha(\tau_2-s)-T_\alpha(\tau_1-s)]f(s)ds||\Big]\\
&\leq \sup_{s\in[0,\tau_1-\delta_1]}||T_\alpha(\tau_2-s)-T_\alpha(\tau_1-s)||\Big[(\frac{p-1}{p\alpha-1})^{1-\frac{1}{p}}||a||_{L^p}(\tau_1^{\alpha-\frac{1}{p}}-\delta_1^{\alpha-\frac{1}{p}})\\
&\quad+\frac{cr}{\alpha}(\tau_1^{\alpha}-\delta_1^{\alpha})\Big]+\frac{2M}{\Gamma(\alpha)}(\frac{p-1}{p\alpha-1})^{1-\frac{1}{p}}||a||_{L^p}\delta_1^{\alpha-\frac{1}{p}}+\frac{2Mcr}{\Gamma(1+\alpha)}\delta_1^{\alpha},\\
Q_6 &\leq \sup_{s\in[0,\tau_1-\delta_1]}||T_\alpha(\tau_2-s)-T_\alpha(\tau_1-s)||\\
&\quad\times\sqrt{\frac{1}{2\alpha-1}}||u_\varepsilon||_{L^2}(\tau_1^{\alpha-\frac{1}{2}}-\delta_1^{\alpha-\frac{1}{2}})+\frac{2MM_B}{\sqrt{2\alpha-1}\Gamma(\alpha)}||u_\varepsilon||_{L^2}\delta_1^{\alpha-\frac{1}{2}}.
\end{aligned}
$$

Since the compactness of $T(t)(t>0)$ and Lemma 2.7 imply the continuity of $T_\alpha(t)(t>0)$ in t in the uniform operator topology, it can be easily seen that Q_3 and Q_6 tend to zero independently of $x \in B_r$ as $\tau_2 \to \tau_1$, $\delta_1 \to 0$. And it is clear that $Q_i(i=1,2,4,5,7)$ tend to zero as $\tau_2 \to \tau_1$ does not depend on particular choice of x. Thus, one can choose $\delta = \min\{\delta_0, \delta_1\}$, then it is easy to get that $||\varphi(\tau_2)-\varphi(\tau_1)||$ tends to zero independently of $x \in B_r$ as $\delta \to 0$ which implies $\{(F_\varepsilon x)(t) : x \in B_r\}$ is an equicontinuous set in $C(J,X)$.

Step 4: F_ε is a compact multivalued map.

Let $t \in J$ be fixed, we show that the set $\Pi(t) = \{(F_\varepsilon x)(t) : x \in B_r\}$ is relatively compact in X.

Clearly, $\Pi(0) = \{x_0\}$ is compact, so it is only necessary to consider $t > 0$. Let $0 < t \leq b$ be fixed. For any $x \in B_r$, $\varphi \in F_\varepsilon(x)$, there exists $f \in \mathcal{N}(x)$ such that

$$
\varphi(t) = S_\alpha(t)x_0+\sum_{i=1}^{k}S_\alpha(t-t_i)\mathcal{I}_i(x(t_i^-))+\int_0^t(t-s)^{\alpha-1}T_\alpha(t-s)[f(s)+Bu_\varepsilon(s)]ds, \quad t\in J.
$$

For each $\epsilon \in (0,t)$, $t \in (0,b]$, $x \in B_r$ and any $\delta > 0$, we define

$$\begin{aligned}
\varphi^{\epsilon,\delta}(t) &= S_\alpha(t)x_0 + \sum_{i=1}^{k} S_\alpha(t-t_i)\mathcal{I}_i(x(t_i^-)) \\
&\quad +\alpha \int_0^{t-\epsilon}\int_\delta^\infty \theta(t-s)^{\alpha-1}\xi_\alpha(\theta)T((t-s)^\alpha\theta)[f(s)+Bu_\varepsilon(s)]d\theta ds. \\
&= S_\alpha(t)x_0 + \sum_{i=1}^{k} S_\alpha(t-t_i)\mathcal{I}_i(x(t_i^-)) \\
&\quad +\alpha T(\epsilon^\alpha\delta) \int_0^{t-\epsilon}\int_\delta^\infty \theta(t-s)^{\alpha-1}\xi_\alpha(\theta)T((t-s)^\alpha\theta-\epsilon^\alpha\delta)[f(s)+Bu_\varepsilon(s)]d\theta ds.
\end{aligned}$$

From the compactness of $S_\alpha(t)(t>0)$ and $T(\epsilon^\alpha\delta)$ $(\epsilon^\alpha\delta>0)$, we obtain that the set

$$\Pi_{\epsilon,\delta}(t) = \{F_\varepsilon^{\epsilon,\delta}(x)(t) : x \in B_r\},$$

is relatively compact set in X for each $\epsilon \in (0,t)$ and $\delta > 0$. Moreover, we have

$$\begin{aligned}
&\|\varphi(t) - \varphi^{\epsilon,\delta}(x)(t)\| \\
&= \|\alpha \int_0^t\int_0^\infty \theta(t-s)^{\alpha-1}\xi_\alpha(\theta)T((t-s)^\alpha\theta)[f(s)+Bu_\varepsilon(s)]d\theta ds \\
&\quad -\alpha \int_0^{t-\epsilon}\int_\delta^\infty \theta(t-s)^{\alpha-1}\xi_\alpha(\theta)T((t-s)^\alpha\theta)[f(s)+Bu_\varepsilon(s)]d\theta ds\| \\
&\le \alpha M(\frac{p-1}{p\alpha-1})^{1-\frac{1}{p}}\|a\|_{L^p}\Big[b^{\alpha-\frac{1}{p}}\int_0^\delta \theta\xi_\alpha(\theta)d\theta + \frac{1}{\Gamma(1+\alpha)}\epsilon^{\alpha-\frac{1}{p}}\Big] + Mcr\Big[\frac{1}{\Gamma(1+\alpha)}\epsilon^\alpha \\
&\quad +b^\alpha\int_0^\delta \theta\xi_\alpha(\theta)d\theta\Big] + \alpha M\sqrt{\frac{1}{2\alpha-1}}\|u_\varepsilon\|_{L^2}\Big[b^{\alpha-\frac{1}{2}}\int_0^\delta \theta\xi_\alpha(\theta)d\theta + \frac{b^{\frac{1}{2}}}{\Gamma(1+\alpha)}\epsilon^{\alpha-\frac{1}{2}}\Big].
\end{aligned}$$

Since $\int_0^\infty \xi_\alpha(\theta)d\theta = 1$, the last inequality tends to zero when $\epsilon \to 0$ and $\delta \to 0$. Therefore, there are relatively compact sets arbitrarily close to the set $\Pi(t)$ $(t>0)$. Hence the set $\Pi(t)$ $(t>0)$ is also relatively compact in X.

Step5: F_ε has a closed graph.

Let $x_n \to x_*$ in $C(J,X)$, $\varphi_n \in F_\varepsilon(x_n)$ and $\varphi_n \to \varphi_*$ in $C(J,X)$. we will show that $\varphi_* \in F_\varepsilon(x_*)$. Indeed, $\varphi_n \in F_\varepsilon(x_n)$ means that there exists $f_n \in \mathcal{N}(x_n)$ such that

$$\begin{aligned}
\varphi_n(t) &= S_\alpha(t)x_0 + \sum_{i=1}^{k} S_\alpha(t-t_i)\mathcal{I}_i(x(t_i^-)) + \int_0^t (t-s)^{\alpha-1}T_\alpha(t-s)f_n(s)ds \\
&\quad + \int_0^t (t-s)^{\alpha-1}T_\alpha(t-s)\times BB^*T_\alpha^*(b-s)R(\varepsilon,\Gamma_0^b)\Big(x_1 - S_\alpha(b)x_0 \\
&\quad -\sum_{i=1}^{k} S_\alpha(b-t_i)\mathcal{I}_i(x(t_i^-)) - \int_0^b (b-\tau)^{\alpha-1}T_\alpha(b-\tau)f_n(\tau)d\tau\Big)ds. \quad (3.5)
\end{aligned}$$

From Step 2, we know that $\{f_n\}_{n\ge1} \subseteq L^p(J,X)$ is bounded. Hence we may assume, passing to a subsequence if necessary, that

$$f_n \rightharpoonup f_*, \quad \text{for some } f_* \in L^p(J,X), \qquad (3.6)$$

It follows from (3.5), (3.6) and Lemma 3.4 that

$$\begin{aligned}\varphi_*(t) &= S_\alpha(t)x_0 + \sum_{i=1}^{k} S_\alpha(t-t_i)\mathcal{I}_i(x(t_i^-)) + \int_0^t (t-s)^{\alpha-1}T_\alpha(t-s)f_*(s)ds \\ &\quad + \int_0^t (t-s)^{\alpha-1}T_\alpha(t-s)\times BB^*T_\alpha^*(b-s)R(\varepsilon,\Gamma_0^b)\Big(x_1 - S_\alpha(b)x_0 \\ &\quad - \sum_{i=1}^{k} S_\alpha(b-t_i)\mathcal{I}_i(x(t_i^-)) - \int_0^b (b-\tau)^{\alpha-1}T_\alpha(b-\tau)f_*(\tau)d\tau\Big)ds. \quad (3.7)\end{aligned}$$

Note that $x_n \to x_*$ in $C(J,X)$ and $f_n \in \mathcal{N}(x_n)$. From Lemma 3.4 and (3.6), we obtain $f_* \in \mathcal{N}(x_*)$ Hence, we prove that $\varphi_* \in F_\varepsilon(x_*)$, which implies that F_ε has a closed graph.

Hence by Steps 1-5 and Arzelà-Ascoli theorem, we obtain that F_ε is a completely continuous multivalued map, u.s.c. with convex closed values and satisfies all the assumptions of Theorem 2.8. Thus F_ε has a fixed point which is a mild solution of problem (1). This is the end of the proof.

The following result concerns the approximately controllable of the problem (1). We need the following assumption.

$H(2)$(iii)': There exists a function $\eta \in L^\infty(J,R^+)$ such that

$$\|\partial F(t,x)\|_{X^*} = \sup\{\|f\|_{X^*} : f(t) \in \partial F(t,x)\} \le \eta(t), \quad \text{for a.e. } t \in J, \text{ all } x \in X.$$

Now, we are now in a position to prove the main result of this paper.

Theorem 3.6 Assume that assumptions of Theorem 3.5 and $H(2)$(iii)' are satisfied, and the linear system (3.1) is approximately controllable on J, then system (1) is approximately controllable on J.

Proof. By employing the technique used in Theorem 3.5, we can easily show that, for all $\varepsilon > 0$, the operator F_ε has a fixed point in B_{r_0}, where $r_0 = r(\varepsilon)$. Let $x^\varepsilon(\cdot)$ be a fixed point of F_ε in B_{r_0}. Any fixed point of F_ε is a mild solution of (1.1), this means that there exists $f^\varepsilon \in \mathcal{N}(x^\varepsilon)$ such that for each $t \in J$,

$$\begin{aligned}x^\varepsilon(t) &\in S_\alpha(t)x_0 + \sum_{i=1}^{k} S_\alpha(t-t_i)\mathcal{I}_i(x(t_i^-)) + \int_0^t (t-s)^{\alpha-1}T_\alpha(t-s)f^\varepsilon(s)ds \\ &\quad + \int_0^t (t-s)^{\alpha-1}T_\alpha(t-s)BB^*T_\alpha^*(b-s)\times R(\varepsilon,\Gamma_0^b)\Big(x_1 - S_\alpha(b)x_0 \\ &\quad - \sum_{i=1}^{k} S_\alpha(b-t_i)\mathcal{I}_i(x(t_i^-)) - \int_0^b (b-\tau)^{\alpha-1}T_\alpha(b-\tau)f^\varepsilon(\tau)d\tau\Big)ds.\end{aligned}$$

Define $G(f^\varepsilon) = x_1 - S_\alpha(b)x_0 - \sum_{i=1}^k S_\alpha(b-t_i)\mathcal{I}_i(x(t_i^-)) - \int_0^b (b-\tau)^{\alpha-1}T_\alpha(b-\tau)f^\varepsilon(\tau)d\tau$.

Noting that $I - \Gamma_0^b R(\varepsilon,\Gamma_0^b) = \varepsilon R(\varepsilon,\Gamma_0^b)$, we get $x(b) = x_1 - \varepsilon R(\varepsilon,\Gamma_0^b)G(f^\varepsilon)$.

By assumption $H(2)$(iii)', we have $\int_0^b \|f^\varepsilon(s)\|^2 ds \le \|\eta\|_{L^2(J,R)}\sqrt{b}$.

Consequently the sequence $\{f^\varepsilon\}$ is uniformly bounded in $L^2(J,X)$. Thus, there is a subsequence, still denoted by $\{f^\varepsilon\}$, that converges weakly to say f in $L^2(J,X)$. Denoting

$$h = x_1 - S_\alpha(b)x_0 - \sum_{i=1}^{k} S_\alpha(b-t_i)\mathcal{I}_i(x(t_i^-)) - \int_0^b (b-\tau)^{\alpha-1}T_\alpha(b-\tau)f(\tau)d\tau.$$

we see that

$$\begin{aligned}||G(f^{\varepsilon}) - h|| &= ||\int_0^b (b-\tau)^{\alpha-1}T_\alpha(b-\tau)[f^{\varepsilon}(\tau) - f(\tau)]d\tau \\ &\le \sup_{0\le t\le b} ||\int_0^t (t-\tau)^{\alpha-1}T_\alpha(t-\tau)[f^{\varepsilon}(\tau) - f(\tau)]d\tau. \qquad (3.8)\end{aligned}$$

Using the Ascoli-Arzela theorem one can show that the linear operator $g \mapsto \int_0^{\cdot}(\cdot-\tau)^{\alpha-1}T_\alpha(\cdot-\tau)g(\tau)d\tau :$ $L^2(J,X) \to C(J,X)$ is compact, consequently the right-hand side of (3.8) tends to zero as $\varepsilon \to 0^+$. This implies

$$\begin{aligned}||x^{\varepsilon}(b) - x_1|| &= ||\varepsilon R(\varepsilon, \Gamma_0^b)G(f^{\varepsilon})|| \\ &\le ||\varepsilon R(\varepsilon, \Gamma_0^b)(h)|| + ||\varepsilon R(\varepsilon, \Gamma_0^b)[G(f^{\varepsilon}) - h]|| \\ &\le ||\varepsilon R(\varepsilon, \Gamma_0^b)(h)|| + ||G(f^{\varepsilon}) - h|| \to 0, \quad \text{as } \varepsilon \to 0^+.\end{aligned}$$

This proves the approximate controllability of system (1).

Acknowledgments

Project supported by the National Natural Science Foundation of China (Grants nos.71461027,71471158),the Zunyi Normal College Doctoral Scientific Research Fund BS[2014]19, BS[2015]09, Guizhou Province Mutual Fund LH[2015]7002, Guizhou Province Department of Education Fund KY[2015]391,[2016]046, Guizhou Province Department of Education teaching reform project[2015]337, Guizhou Province Science and technology fund (qian ke he ji chu)[2016]1160,[2016]1161, Zunyi Science and technology talents[2016]15.

Conflict of Interest

All authors declare no conflicts of interest in this paper.

References

1. A.E. Bashirov and N.I. Mahmudov, *On concepts of controllability for deterministic and stochastic systems,* SIAM J. Control Optim, **37** (1999), 1808-1821.
2. H.F. Bohnenblust and S. Karlin, *On a Theorem of Ville, in: Contributions to the Theory of Games*, Princeton University Press, Princeton, NJ, 1950, 155-160.
3. P. Cannarsa, G. Fragnelli, and D. Rocchetti, *Controllability results for a class of one-dimensional degenerate parabolic problems in nondivergence form,* J. Evol. Equ., **8** (2008), 583-616.
4. S. Carl and D. Motreanu, *Extremal solutions of quasilinear parabolic inclusions with generalized Clarke's gradient,* J. Differ. Equa., **191** (2003), 206-233.
5. F.H. Clarke, *Optimization and Nonsmooth Analysis,* Wiley, New York, 1983.
6. Z. Denkowski, S. Migórski, and N.S. Papageorgiou, *An Introduction to Non-linear Analysis: Theory*, Kluwer Academic/Plenum Publishers, Boston, Dordrecht, London, New York, 2003.

7. S. Hu and N.S. Papageorgiou, *Handbook of multivalued Analysis (Theory),* Kluwer Academic Publishers, Dordrecht Boston, London, 1997.

8. A.A. Kilbas, H.M. Srivastava, and J.J. Trujillo, *Theory and Applications of Fractional Differential Equations,* North-Holland Math. Studies, Elservier Science B.V. Amsterdam, **204**, 2006.

9. M. Kisielewicz, *Differential Inclusions and Optimal Control,* Kluwer, Dordrecht, The Netherlands, 1991.

10. A. Kulig, *Nonlinear evolution inclusions and hemivariational inequalities for nonsmooth problems in contact mechanics. PhD thesis*, Jagiellonian University, Krakow, Poland, 2010.

11. S. Kumar and N. Sukavanam, *Approximate controllability of fractional order semilinear systems with bounded delay*, J. Differ. Equa., **252** (2012), 6163-6174.

12. Z.H. Liu, *A class of evolution hemivariational inequalities,* Nonlinear Anal. Theory Methods Appl., **36** (1999), 91-100.

13. Z.H. Liu, *Anti-periodic solutions to nonlinear evolution equations,* J. Funct. Anal., **258** (2010), 2026-2033.

14. Z.H. Liu and X.W. Li, *Existence and uniqueness of solutions for the nonlinear impulsive fractional differential equations,* Commun. Nonlinear Sci. Numer. Simulat., **18** (2013), 1362-1373.

15. Z.H. Liu and X.W. Li, *On the Controllability of Impulsive Fractional Evolution Inclusions in Banach Spaces*, J. Optim. Theory Appl., **156** (2013), 167-182.

16. Z.H. Liu and J.Y. Lv, R. Sakthivel, *Approximate controllability of fractional functional evolution inclusions with delay in Hilbert spaces*, IMA. J. Math. Control Info., **31** (2014), 363-383.

17. N.I. Mahmudov, *Approximate controllability of semilinear deterministic and stochastic evolution equations in abstract spaces*, SIMA. J. Control Optim., **42** (2003), 1604-1622.

18. S. Migórski and A. Ochal, *Existence of solutions for second order evolution inclusions with application to mechanical contact problems*, Optimization, **55** (2006), 101-120.

19. S. Migórski and A. Ochal, *Quasi-static hemivariational inequality via vanishing acceleration approach*, SIAM J. Math. Anal., **41** (2009), 1415-1435.

20. S. Migórski, A. Ochal, and M. Sofonea, *Nonlinear Inclusions and Hemivariational Inequalities. Models and Analysis of Contact Problems,* Advances in Mechanics and Mathematics, Springer, New York, **26**, 2013.

21. S. Migorski, *Existence of Solutions to Nonlinear Second Order Evolution Inclusions without and with Impulses,* Dynamics of Continuous, Discrete and Impulsive Systems, Series B, **18** (2011), 493-520.

22. S. Migorski and A. Ochal, *Nonlinear Impulsive Evolution Inclusions of Second Order,* Dynam. Syst. Appl., **16** (2007), 155-174.

23. P.D. Panagiotopoulos, *Nonconvex superpotentials in sense of F.H. Clarke and applications,* Mech. Res. Comm., **8** (1981), 335-340.

24. P.D. Panagiotopoulos, *Hemivariational inequalities*, Applications in Mechanics and Engineering, Springer, Berlin, 1993.

25. P.D. Panagiotopoulos, *Hemivariational inequality and Fan-variational inequality*, New Applications and Results, Atti. Sem. Mat. Fis. Univ. Modena XLIII, (1995), 159-191.

26. I. Podlubny, *Fractional Differential Equations*, Academic Press, San Diego, 1999.

27. K. Rykaczewski, *Approximate conrtollability of differential inclusions in Hilbert spaces*, Nonlinear Analysis, **75** (2012), 2701-2712.

28. Y. Zhou and F. Jiao, *Existence of mild solutions for fractional neutral evolution equations*, Compu. Math. Appl., **59** (2010), 1063-1077.

29. E. Zuazua, *Controllability of a system of linear thermoelasticity,* J. Math. Pures Appl.,**74** (1995), 291-315.

30. Jinrong Wang, M. Fe̋ckan, and Y. Zhou, *On the new concept of solutions and existence results for impulsive fractional evolution equations*, Dynamics of PDE, **8** (2011), 345-361.

8

Relations between the dynamics of network systems and their subnetworks

Yunjiao Wang[1,*], **Kiran Chilakamarri**[1], **Demetrios Kazakos**[1] **and Maria C. Leite** [2]

[1] Department of Mathematics, Texas Southern University, 3100 Cleburne, Houston, TX, 77004, USA
[2] Department of Mathematics, University of South Florida at St. Pete, 140 7th Avenue South St. Petersburg, Florida 33701, USA

[*] **Correspondence:** wangyx@tsu.edu

Abstract: Statistical analysis of the connectivity of real world networks have revealed interesting features such as community structure, network motif and as on. Such discoveries tempt us to understand the dynamics of a complex network system by studying those of its subnetworks. This approach is feasible only if the dynamics of the subnetwork systems can somehow be preserved or partially preserved in the whole system. Most works studied the connectivity structures of networks while very few considered the possibility of translating the dynamics of a subnetwork system to the whole system. In this paper, we address this issue by focusing on considering the relations between cycles and fixed points of a network system and those of its subnetworks based on Boolean framework. We proved that at a condition we called agreeable, if X_0 is a fixed point of the whole system, then X_0 restricted to the phase-space of one of the subnetwork systems must be a fixed point as well. An equivalent statement on cycles follows from this result. In addition, we discussed the relations between the product of the transition diagrams (a representation of trajectories) of subnetwork systems and the transition diagram of the whole system.

Keywords: Boolean network systems; cycle; fixed point; subnetwork systems; dynamics
Mathematics Subject Classification: 37N99

1. Introduction

Biological networks such as gene regulatory networks, neural networks, and metabolic networks are generally complex even from the network topology point of view [17, 18]. However, the understanding of the dynamics of such network systems is crucial to identify mechanisms behind many kinds of biological processes and diseases, and to decode the mysteries of life. Statistical studies on the topology of real world networks revealed some very intriguing features [17] including power-law degree distributions [3, 25, 35], local community structures [4, 11, 13] and network motifs [6, 14]. *A community* is defined to be a subnetwork within which the number of edges is much larger than the expected number in an equivalent network with edges placed at random [17]. On the other hand, a *network motif* is defined as a subnetwork that occurs more often in a complex network than in random networks. The discoveries of community structures and network motifs lead us to wonder about the possibility of using modular idea to study dynamics of network systems: the dynamics of a complex network can be understood by studying its subnetwork systems. In order for this idea to work, the dynamics of the subnetworks needs to be preserved or partially preserved in the original network. A simple example where this is true is when a subnetwork does not receive any input from the rest of the network. However, the situation becomes quite subtle when the subnetwork and its complementary subnetwork have mutual interactions.

There is a large body of work devoted to identifying communities or motifs in biological networks [6, 14, 17, 18, 22, 23, 34]. Interestingly, very few works focus on the relations between the dynamics of subnetworks and that of the whole system. In this work, we address the issue based on the Boolean network framework.

Mathematical models have proven to be indispensable tools for studying network systems. Among various mathematical modeling frameworks, coupled differential equations and Boolean networks are popular for modeling regulatory networks [1, 2, 5, 7, 10, 12, 15, 16, 20, 21, 26, 28–31, 33]. Network systems are often represented by directed graphs, wherein components are represented by nodes and interactions by arrows. An n-node Boolean network system is a discrete dynamical system with the form of

$$X(t+1) = F(X(t)) \tag{1.1}$$

where $X = (x_1, \cdots, x_n)$ and x_i represents the state variable of the ith node, $F = (f_1, \cdots, f_n)$ and f_i is the governing function of the ith node with its value being either 0 or 1. They can be set up in situations where information on the detailed kinetic interactions is not available and can provide many valuable insights [8, 9, 12, 19, 24, 27, 32].

In this work, we particularly consider networks formed by two subnetworks connected at a *cutting node*, which we will define next. A node is called a *cutting node* of a connected network if the removal of the node leads to two or more disjoint subnetworks. We introduce the notion of a network being *agreeable.* Let G be the network of the whole system formed by G_1 and G_2 connected at a cutting node c. Let $x_c(t, *)$ be the value of the cutting node in the system $*$ (here $*$ can be G_1, G, or G_2) at time t. We say that G is agreeable if $x_c(t, G) = z_0$ whenever $x_c(t, G_1) = x_c(t, G_2) = z_0$. We first show that if a network is agreeable and its subnetworks have only cycles, then the whole system has only cycles. We then prove that if X_0 is a fixed point of G, then X_0 restricted to the phase-space of one of the subnetwork systems must be a fixed point of that system. In addition, we discuss the relations between the product of the transition diagrams (a representation of trajectories) of the subnetwork systems and that of the whole system.

The paper is structured as follows: In Section 2, we introduce terminology related to Boolean network systems and prove a property of such network systems. Section 3 defines agreeable networks and gives an example of updating scheme for the cutting node that guarantees a network system to be agreeable. In Section 4, we prove results on the relations between cycles and fixed points of whole network system and its subnetworks. In Section 5, we discuss the relations between the transition diagram of a network system and the product diagram of its subnetworks. Finally, in Section 6, we introduce an algorithm to construct the transition diagram of the whole network from the transition diagram of the product system.

2. Boolean network systems

In this section, we will introduce several basic terminologies for Boolean network systems and give a general property (Proposition 2.1) for deterministic Boolean network systems. We use the standard gene regulatory network topological representation for a Boolean network - the species are represented by nodes and interactions between species by arrows. We also allow two types of arrows from the tail node to the head node: one representing activation ($\to$) and the other representing inhibition ($\dashv$). For example, in Fig. 1(a) the inhibition from node 2 to node 1 is represented by an arrow $\dashv$ from node 2 to node 1.

The dynamics of a Boolean network system can be represented by a *transition diagram*, which we represent by $X_0 \to X_1$ if $F(X_0) = X_1$, where F is a Boolean map. For example, suppose the Boolean map associated to the network in Fig. 1(a) is as shown in the table in Fig. 1. In that case, the state $(0, 0)$ transits to $(1, 0)$, (1,0) to (1,1) and so on. So, the transition diagram of the system can be set up as the one in Fig. 1(c). If $X_0 \to X_1$ occur in the transition diagram, we say that $X_0 \to X_1$ is a *transition* of the network system.

The *trajectory* of a given state X_0 is defined to be the sequence $\{F^n(X_0)\}_{n=0}^{\infty}$. X_0 is called a *fixed point* if $F(X_0) = X_0$ and a trajectory of X_0 is called a *cycle* with length $n > 0$ if $F^{k+n}(X_0) = F^k(X_0)$ for any nonnegative integer k and $F^{k+m}(X_0) \neq F^k(X_0)$ for some k if $m < n$. Finally, for a deterministic dynamical system, the trajectory of a given state is unique.

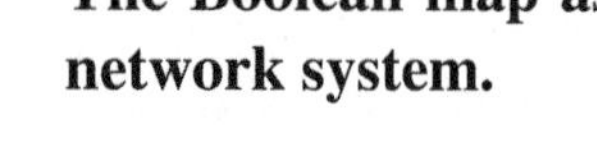

(a) Network

(x_1, x_2)	$F(x_1, x_2)$
$(0, 0)$	$(1, 0)$
$(1, 0)$	$(1, 1)$
$(1, 1)$	$(0, 1)$
$(0, 1)$	$(0, 0)$

(b) Boolean map

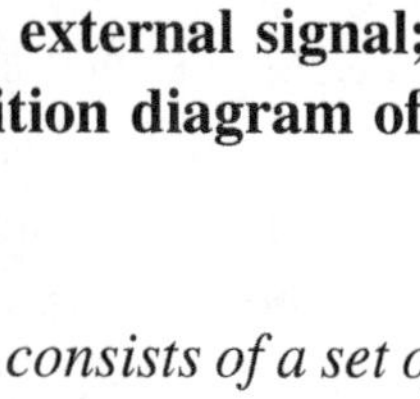

(c) Transition diagram

Figure 1. (a) A two-node network system, where S represents an external signal; (b) The Boolean map associated to the network in (a); (c) The transition diagram of the network system.

Proposition 2.1. *The transition diagram of a given Boolean network system consists of a set of disjoint connected sub-diagrams, in which the trajectory of any state in the same connected sub-diagram ends at the same steady-state: either a fixed point or a cycle.*

Proof. First note that there are only a finite number of states, the trajectory starting from any state will either end up in a cycle or a fixed point. Also note that for any two states, say X_1 and X_2, in the same connected sub-diagram, there must exist integers $n_1 \geq 0$ and $n_2 \geq 0$, so that $F^{n_1}(X_1) = F^{n_2}(X_2)$. The reason is as follows. Suppose the trajectories of X_1 and X_2 are in the same connected diagram. There then exists a state X^* and two integers $m_1 > 0$, $m_2 > 0$, such that $X^* = F^{m_1}(X_1) = F^{m_1}(X_2)$.

Let $F^{n_1}(X_1) = F^{n_2}(X_2) = Y_0$. Since there are only a finite number of states, the trajectory starting from Y_0 will either end up in a cycle or a fixed point.

◇

3. Agreeable networks

We first introduce the definition of an agreeable network system. Then we present an updating scheme for the cutting node that guarantees that the network $\mathcal{G}$ is agreeable. We would like to point out that there is no restriction on the updating schemes for the rest of the nodes in the network. Also note that the update scheme in this section is just an example. The results we present later on, except for the example in Section 6, are valid even if the scheme is not satisfied.

Let $\mathcal{G}$ be a network formed by two subnetworks $\mathcal{G}_1$ and $\mathcal{G}_2$, connected via a cutting node. For example, the network in Fig. 2 can be formed by two subnetworks in Fig. 3, which are connected at node 2. i.e. node 2 is a cutting node..

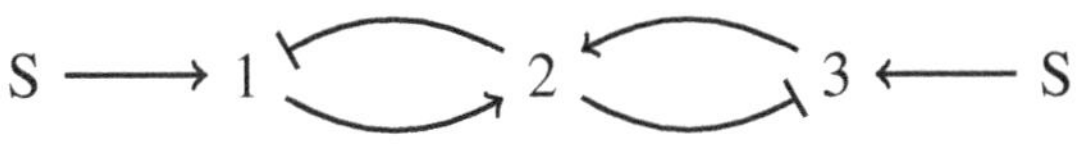

Figure 2. Network with one cutting node (node 2). The node 'S' represents an external signal.

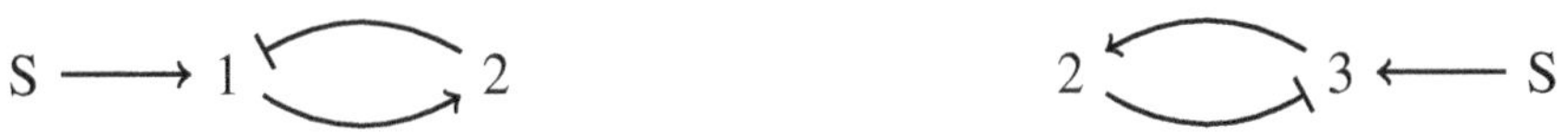

(a) **(b)**

Figure 3. Network in Fig. 2 can be formed by the two subnetworks that are connected at node 2.

Let c be the cutting node, C_1 be the set of the nodes in $\mathcal{G}_1\backslash\{c\}$ and C_2 be the set of nodes in $\mathcal{G}_2\backslash\{c\}$. Let x be the state variables of C_1, y be the state variables of C_2, z be the state variable of c in $\mathcal{G}_1$ and $\overline{z}$ be the state variable of c in $\mathcal{G}_2$. Suppose the governing equations of $\mathcal{G}_1$ are

$$\begin{cases} x(t+1) &= h(x(t), z(t)) \\ z(t+1) &= g_1(x(t), z(t)) \end{cases} \tag{3.1}$$

and those of $\mathcal{G}_2$ are

$$\begin{cases} \overline{z}(t+1) &= g_2(\overline{z}(t), y(t)) \\ y(t+1) &= f(\overline{z}(t), y(t)) \end{cases} \tag{3.2}$$

Since node c is the only node that connects the two subnetworks, the governing system of $\mathcal{G}$ can be written in the form of

$$\begin{cases} x(t+1) &= h(x(t), z(t)) \\ z(t+1) &= g(x(t), z(t), y(t)) \\ y(t+1) &= f(z(t), h(t)) \end{cases} \tag{3.3}$$

Definition 3.1. *The network system of $\mathcal{G}$ is* agreeable *if*

$$g_1(x, z) = g_2(z, y) \qquad \textit{implies} \qquad g(x, z, y) = g_1(x, z) = g_2(z, y) \tag{3.4}$$

We will show next an updating scheme of the cutting node with which the network $\mathcal{G}$ is agreeable. We will refer to the updating scheme as **Axioms**.

Updating schemes that guarantee network $\mathcal{G}$ to be agreeable

1. The effects of activators and inhibitors are never additive, but rather, inhibitors are dominant;
2. The activity of a node will be 'on' in the next time step if at least one of its activators is 'on' and all inhibitors are 'off';
3. The activity of a node will be 'off' in the next time step if none of its activators are 'on'.
4. If a node has an external/background activation, then we assume that the node has an activator that is permanently 'on'.

Let z be the state variable of the cutting node c. Define
$B(z)$ = {Inhibitors in $\mathcal{G}_1$ that are on}
$D(z)$ = {Inhibitors in $\mathcal{G}_2$ that are on},
$A(z)$ = {activators in $\mathcal{G}_1$ that are on},
$C(z)$ = {Activators in $\mathcal{G}_2$ that are on}.

Suppose the updating scheme for the cutting node satisfies the **Axioms**. Then the first three axioms can be rewritten as

1. If at time t, $B(z) \cup D(z) \neq \emptyset$, then $z = 0$ at time $t + 1$.
2. If at time t, $B(z) \cup D(z) = \emptyset$ and $A(z) \cup C(z) \neq \emptyset$, then $z = 1$ at time $t + 1$
3. If $A(z) \cup C(z) = \emptyset$, then $z = 0$ at time $t + 1$.

Theorem 3.2. *Suppose the updating scheme for the cutting node c follows* **Axioms**, *then $\mathcal{G}$ is agreeable.*

Proof. Suppose $g_1(z, x) = g_2(z, y) = z'$.

Case I $z' = 0$. From $\mathcal{G}_1$ system, $B(z) \neq \emptyset$ or $A(z) = \emptyset$; From $\mathcal{G}_2$ system $D(z) \neq \emptyset$ or $C(z) = \emptyset$. This implies that

$$B(z) \cup D(z) \neq \emptyset \text{ or } A(z) \cup C(z) = \emptyset$$

Following the system of $\mathcal{G}$, $g(x, z, y) = 0$. So $g_1(z, y) = g_2(z, y) = g(x, z, y)$.

Case I $z' = 1$. From $\mathcal{G}_1$ system, $B(z) = \emptyset$ and $A(z) \neq \emptyset$; From $\mathcal{G}_2$ system $D(z) = \emptyset$ and $C(z) \neq \emptyset$. This implies that

$$B(z) \cup D(z) = \emptyset \text{ and } A(z) \cup C(z) \neq \emptyset$$

Following the system of $\mathcal{G}$, $g(x, z, y) = 1$. So $g_1(z, x) = g_2(z, y) = g(x, z, y)$. Therefore $\mathcal{G}$ is agreeable.

◇

Remark 3.3. *The system of $\mathcal{G}$ is not agreeable any more if $g(x, z, y) = 1$ only when both $x = y = 1$. That is, when the activation of the cutting node requires inputs from both subnetworks, $\mathcal{G}$ is not agreeable. Similarly, when deactivation of the cutting node requires inputs from both subnetworks, $\mathcal{G}$ is also not agreeable. So the condition agreeable means the requirement on certain independency of the subnetworks.*

4. Relations between Dynamics of $\mathcal{G}$ and its subnetworks

In this section, we present our results on the relations between fixed points and cycles of the network system of $\mathcal{G}$ and those of its subnetwork systems $\mathcal{G}_1$ and $\mathcal{G}_2$. We prove that if the subnetwork systems have only cycles, then the whole system also has only cycles; on the other hand, if the whole network system has a fixed point, then the projection of the fixed point to the phase space of one of the subnetwork systems is a fixed point of that network systems. In addition, we show an example that the subnetwork systems have only fixed point(s) while the whole system has a cycle.

We assume throughout this section that the systems associated to $\mathcal{G}_1$, $\mathcal{G}_2$ and $\mathcal{G}$ are (3.1), (3.2) and (3.3) respectively.

Theorem 4.1. *Suppose the system of $\mathcal{G}$ is agreeable, and suppose both subnetwork systems, $\mathcal{G}_1$ and $\mathcal{G}_2$, have only cycles, then $\mathcal{G}$ system also has only cycles.*

Proof. Let the associated system of $\mathcal{G}_1$, $\mathcal{G}_2$ and $\mathcal{G}$ be of the form of equations (3.1), (3.2) and (3.3) respectively. Then we only need to prove that

$$(h(x,z), g(x,z,y), f(z,y)) \neq (x,z,y) \tag{4.1}$$

for any state (x, z, y) since any synchronous Boolean system has only a finite number of states, any state will repeat itself in a finite number of steps.

Let (x_0, z_0, y_0) be an initial state. First we show that

$$(h(x_0,z_0), g(x_0,z_0,y_0), f(z_0,y_0)) \neq (x_0,z_0,y_0) \tag{4.2}$$

Note that if $(h(x_0,z_0), g(x_0,z_0,y_0), f(z_0,y_0)) = (x_1, 1-z_0, y_1)$, then we are done. Otherwise, if $(h(x_0,z_0), g(x_0,z_0,y_0), f(z_0,y_0)) = (x_1, z_0, y_1)$, we claim that either $x_1 \neq x_0$ or $y_1 \neq y_0$. We can show this by using contradiction. Suppose $x_1 = x_0$ and $y_1 = y_0$. i.e. $h(x_0,z_0) = x_1 = x_0$ and $f(z_0,y_0) = y_1 = y_0$. Then, because the subnetwork systems $\mathcal{G}_1$ and $\mathcal{G}_2$ have only cycles, we have

$$(h(x_0,z_0), g_1(x_0,z_0)) = (x_0, 1-z_0)$$

and

$$(g_2(z_0,y_0), f(z_0,y_0) = (1-z_0, y_0)$$

By the condition $\mathcal{G}$ being agreeable, $(h(x_0,z_0), g(x_0,z_0,y_0), f(z_0,y_0)) = (x_0, 1-z_0, y_0)$ which contradicts with the assumption $G(x_0,z_0,y_0) = (x_1,z_0,y_1)$. Hence, either $x_1 \neq x_0$ or $y_1 \neq y_0$. Therefore, $(h(x_0,z_0), g(x_0,z_0,y_0), f(z_0,y_0)) \neq (x_0,z_0,y_0)$. It follows that the system of $\mathcal{G}$ has no fixed point. i.e. it only has cycles.

◇

Corollary 4.2. *Suppose the network system of $\mathcal{G}$ is agreeable. If the system of $\mathcal{G}$ has a fixed point, then $\mathcal{G}_1$ or $\mathcal{G}_2$ must have a fixed point.*

Proof. This Corollary follows directly from Theorem 4.1.

Corollary 4.3. *Suppose the network system of $\mathcal{G}$ is agreeable.*

1. *If the system of $\mathcal{G}_1$ and $\mathcal{G}_2$ have fixed points (x_0, z_0) and (z_0, y_0) respectively, then (x_0, z_0, y_0) is a fixed point of $\mathcal{G}$.*
2. *If (x_0, y_0, z_0) is a fixed point of the system of $\mathcal{G}$, then either (x_0, y_0) is a fixed point of the system of $\mathcal{G}_1$ or (y_0, z_0) is a fixed point of the system of $\mathcal{G}_2$.*

Proof. 1. By Theorem 3.2, $z_0 = g_1(x_0, z_0) = g_2(z_0, y_0)$ implies $g(x_0, z_0, y_0) = z_0$. Also because (x_0, z_0) and (z_0, y_0) are the fixed points of $\mathcal{G}_1$ and $\mathcal{G}_2$, $f(x_0, z_0) = x_0$ and $h(z_0, y_0) = y_0$. Therefore, (x, z, y) is a fixed point of $\mathcal{G}$.

2. If (x_0, y_0, z_0) is a fixed point of the system of $\mathcal{G}$, then $g(x_0, z_0, y_0) = z_0$. Note that $g_1(x_0, z_0) = z_0$ or $1 - z_0$ and $g_2(z_0, y_0) = z_0$ or $1 - z_0$. Because the system $\mathcal{G}$ is agreeable, either $g_1(x_0, z_0) = z_0$ or $g_2(z_0, y_0) = z_0$.

◇

Corollary 4.3 implies that when $\mathcal{G}$ is agreeable, the fixed points of whole network system can be obtained by first looking at the fixed points of the subnetwork systems.

Next, we show an example that both subnetwork systems have only fixed points while the whole network system has cycles. We consider the network in Fig. 4. Suppose the associate Boolean system satisfies the **Axioms**. Then a straightforward calculation shows that the system of network in Fig. 4 has a cycle $(010) \to (011) \to (111) \to (110) \to (010)$ while its two subnetwork systems shown in Fig. 5 have only fixed points.

Figure 4. A network consists of two feedback loops with S as an external signal. The system of the network has a cycle $(010) \to (011) \to (111) \to (110)$**.**

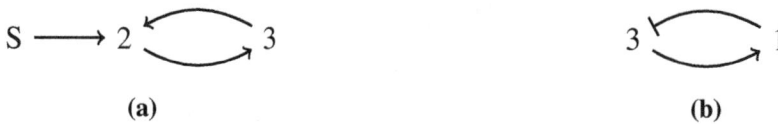

(a) **(b)**

Figure 5. The system of subnetwork (a) has only one steady-state $(x_2, x_3) = (1, 1)$ **and the system of subnetwork (b) has also only one steady-state** $(x_1, x_3) = (0, 0)$**.**

5. Relations between transition diagrams

In a Boolean network system, the transition diagram of the system represents the trajectory space of the discrete dynamical system. That is, the transition diagram represents the dynamics of the system. Note that if the dynamics of the subnetworks are all independent, then the dynamics of the whole network is just the product of the subnetworks. However, when they are not independent, the relation is not all that transparent. In this section, we explore the relations between the transition diagrams of a network system and its subnetwork systems.

Product network systems

Let the associated systems of $\mathcal{G}_1$ and $\mathcal{G}_2$ be of the form of (3.1) and (3.2) respectively. Then the associated system of the *product network* $\mathcal{G}_1 \times \mathcal{G}_2$ is defined to be

$$\begin{cases} x(t+1) &= h(x(t), z(t)) \\ z(t+1) &= g_1(x(t), z(t)) \\ \bar{z}(t+1) &= g_2(\bar{z}(t), y(t)) \\ y(t+1)) &= f(\bar{z}(t), y(t)) \end{cases} \tag{5.1}$$

That is, if $(x_0, z_0) \to (x_1, z_1)$ is a transition of the system of $\mathcal{G}_1$ and $(\bar{z}_0, y_0) \to (\bar{z}_1, y_1)$ is a transition of the system of $\mathcal{G}_2$, then $(x_0, z_0, \bar{z}_0, y_0) \to (x_1, z_1, \bar{z}_1, y_1)$ is a transition of the system of $\mathcal{G}_1 \times \mathcal{G}_2$. This can be represented by the diagram in Fig. 6, where peach-colored arrows represent transitions occurring in $\mathcal{G}_2$ system, green arrows represent transitions occurring in $\mathcal{G}_1$ system, and the blue arrow is the transition occurring in the product system.

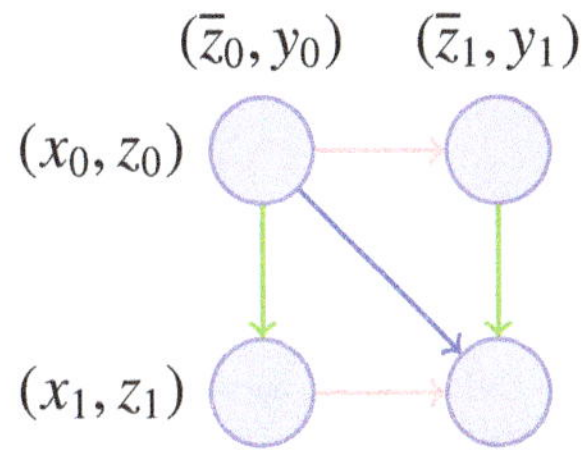

Figure 6. Peach-colored arrows represent transitions occurring in $\mathcal{G}_2$ system, green arrows represent transitions occurring in $\mathcal{G}_1$ system, and the blue arrow represents the transition occurring in the product system.

It is obvious that the phase space of the system of $\mathcal{G}$ can be embedded in the phase space of the system of $\mathcal{G}_1 \times \mathcal{G}_2$ by the map $J : \{0,1\}^3 \to \{0,1\}^4$ defined by $J(x,z,y) = (x,z,z,y)$. In order to simplify this notation, we identify point (x,z,y) in the phase space of $\mathcal{G}$ with the point (x,z,z,y) in the phase space of $\mathcal{G}_1 \times \mathcal{G}_2$. Now the question is: Does a transition such as $(x_0, z_0, z_0, y_0) \to (x_1, z_1, z_1, y_1)$ in the system of $\mathcal{G}_1 \times \mathcal{G}_2$ imply a transition of $(x_0, z_0, y_0) \to (x_1, z_1, y_1)$ in the system of $\mathcal{G}$? The answer is yes provided $\mathcal{G}$ is agreeable.

Proposition 5.1. *Suppose the network system of $\mathcal{G}$ is agreeable. Then, a transition in the system of the product network $\mathcal{G}_1 \times \mathcal{G}_2$ of the form $(x_0, z_0, z_0, y_0) \to (x_1, z_1, z_1, y_1)$ implies a transition $(x_0, z_0, y_0) \to (x_1, z_1, y_1)$ in the network system $\mathcal{G}$.*

Proof. By definition of product system, saying that $(x_0, z_0, z_0, y_0) \to (x_1, z_1, z_1, y_1)$ is a transition of the system of the product network $\mathcal{G}_1 \times \mathcal{G}_2$ means that $f(x_0, z_0) = x_1$, $g_1(x_0, z_0) = z_1$, $g_2(z_0, y_0) = z_1$ and $g(z_0, y_0) = y_1$. Because $\mathcal{G}$ is agreeable, $g(x_0, z_0, y_0) = z_1$. Hence, $(x_0, z_0, y_0) \to (x_1, z_1, y_1)$ is a transition of the network system $\mathcal{G}$ ◇

We would like to point out that the reverse of Proposition 5.1 does not hold. That is, $(x_0, z_0, y_0) \to (x_1, z_1, y_1)$ in the system of $\mathcal{G}$ does not imply $(x_0, z_0, z_0, y_0) \to (x_1, z_1, z_1, y_1)$ of the product system. More precisely, there exists a transition such as $(x_0, z_0, z_0, y_0) \to (x_1, z_1, \bar{z}_1, y_1)$ with $z_1 \neq \bar{z}_1$. We will see this in the example introduced in the next section.

Relation between transition diagrams

Next we will study relations between the transition diagram of $\mathcal{G}$ and that of the product system of its subnetworks. We achieve this goal by exploring the possibility of constructing the transition diagram of $\mathcal{G}$ from a product system.

Proposition 5.1 tells us that we can derive some transitions of $\mathcal{G}$ by simply translating $(x_0, z_0, z_0, y_0) \to (x_1, z_1, z_1, y_1)$ in the system of of $\mathcal{G}_1 \times \mathcal{G}_2$ to $(x_0, z_0, y_0) \to (x_1, z_1, y_1)$ in the system of $\mathcal{G}$. As pointed out earlier, there exists transition such as $(x_0, z_0, z_0, y_0) \to (x_1, z_1, \bar{z}_1, y_1)$ with $z_1 \neq \bar{z}_1$. In this case, we can not derive to which state (x_0, z_0, y_0) transits to based on the information of the product space. However, by the definition of the product system, it is certain that (x_0, z_0, y_0) transits to (x_1, z, y_1) where x_1 and y_1 can be read off from the transition $(x_0, z_0, z_0, y_0) \to (x_1, z_1, \bar{z}_1, y_1)$ while the value of z needs to be determined by g evaluated at (x_0, z_0, y_0).

Proposition 5.2. *If* $(x_0, z_0, z_0, y_0) \to (x_1, z_1, \bar{z}_1, y_1)$ *with* $z_1 \neq \bar{z}_1$ *is a transition of the product system* $\mathcal{G}_1 \times \mathcal{G}_2$*, then* $(x_0, z_0, y_0) \to (x_0, 0, y_0)$ *or* $(x_0, z_0, y_0) \to (x_0, 1, y_0)$ *is a transition of the system of* $\mathcal{G}$*.*

Proof. By the definition of the product system (5.1), $(x_0, z_0, z_0, y_0) \to (x_1, z_1, \bar{z}_1, y_1)$ means $h(x_0, z_0) = x_1$, $g_1(x_0, z_0) = z_1$, $g_2(z_0, y_0) = \bar{z}_1$ and $f(z_0, y_0) = y_1$. Following the definition of the system of $\mathcal{G}$ (system (3.3)), (x_0, z_0, y_0) transits to $(h(x_0, z_0), g(x_0, z_0, y_0), f(z_0, y_0)) = (x_1, g(x_0, z_0, y_0), y_1)$. Since g is a Boolean function, $g(x_0, z_0, y_0) = 0$ or $g(x_0, z_0, y_0) = 1$. i.e. $(x_0, z_0, y_0) \to (x_0, 0, y_0)$ or $(x_0, z_0, y_0) \to (x_0, 1, y_0)$. ◇

6. Construct transition diagram from that of sub-networks

In this section, by using an example, we discuss the construction of the transition diagram of a network $\mathcal{G}$ from the transition diagram of the product system of its subnetworks $\mathcal{G}_1$ and $\mathcal{G}_2$.

Example. Consider the network in Fig. 2 that is formed by the two subnetworks in Fig. 3. Suppose the updating scheme for all the nodes in the networks follow **Axioms**. Then, the transition diagrams of the subnetworks in Fig. 3(a),(b) are shown in the Fig. 7(a),(b), respectively.

(a) **(b)**

Figure 7. The transition diagrams from Fig. 3.

The product of the transition diagrams in Fig. 7 is shown in Fig. 8 (left). By the definition of product network systems, the state $(0, 0, 0, 0)$ transits to $(1, 0, 0, 1)$ since $h(0, 0) = 1$, $g_1(0, 0) = 0$, $g_2(0, 0) = 0$ and $f(0, 0) = 1$; the state $(1, 0, 0, 1)$ transit to $(1, 1, 1, 1)$ since $h(1, 0) = 1$, $g_1(1, 0) = 1$, $g_2(1, 0) = 1$ and $f(1, 0) = 1$ and so on. As a result, we can get the transition diagram of the product network system as represented by the diagram with blue arrows on the right of Fig. 8.

Next we show how we can construct the transition diagram of the original three-node network system based on the transition diagram of the product system. Since the phase space of a three-node

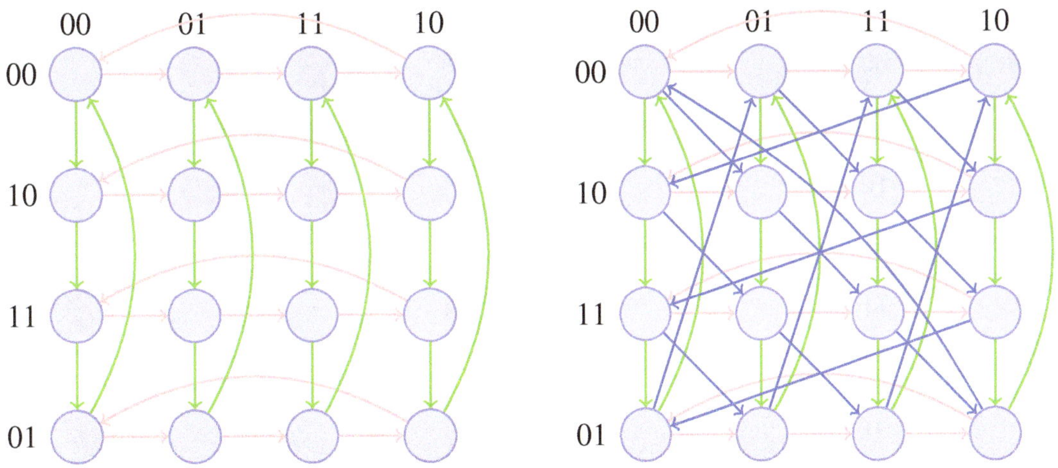

Figure 8. (Left) Product of transition diagrams in Figs. 7; (Right) The blue arrows represent the transitions of the product network systems.

network system can be embedded in the phase space of the product system by the map J, we colored those states of the form (x, z, z, y) in red and removed all arrows that are not from those nodes as shown in Fig. 10 (left). We then identify states (x, z, y) of the three-node system with states (x, z, z, y) of the product network system and consider where each state in red transits in the system of three-node network system. By Proposition 5.1, $(x_0, z_0, z_0, y_0) \to (x_1, z_1, z_1, y_1)$ implies a transition $(x_0, z_0, y_0) \to (x_1, z_1, y_1)$. That means the transitions in blue in Fig. 10 (right) are part of the transitions of the three-node network system. Next we need to determine transitions for remaining red states. The state (1000) transit to (1101) in the product network, see Fig. 10(left). However, (1101) is not a state of the three-node network system. On the other hand, the transition (1000) $\to$ (1101) in the product system implies (100) $\to$ $(1z1)$ in the three-node system. i.e. $h(1,0) = 1$, $g(1,0,0) = z$ and $f(0,0) = 1$ where z needs to be determined by g. Since $g(1,0,0) = 1$, (100) transit to (111) in the three-node system. We mark a transition: (1000) transit to (1111) in the product space. Similarly, we find (0111) $\to$ (0110), (1110) $\to$ (0110) and (0001) $\to$ (1111) as shown in Fig.10(right).

Now we have found the transitions for all the red states. The diagram that consists of only the red nodes and the arrows is the transition diagram of the three-node system – which is identical to the transition diagram we obtained directly using the rule for the three-node network system (Fig. 2) as shown in Fig.10(right).

Algorithm of constructing transition diagram. We summarize how to construct the transition diagram of the whole network from the product systems of its subnetworks as follows.

Suppose $\mathcal{G}$ is agreeable at the cutting node.

1. Let $\mathcal{T}_1$ be the set of all transitions of the system of $\mathcal{G}_1$ and expressed by

$$\mathcal{T}_1 = \left\{ (x_0, z_0) \to (x_1, z_1) \;\middle|\; \begin{array}{l} x_i \in \{0,1\}^k, z_i \in \{0,1\}, i = 0, 1 \\ h(x_0, z_0) = x_1, g_1(x_0, z_0) = z_1 \end{array} \right\}$$

and let $\mathcal{T}_2$ be the set of all transitions of the system of $\mathcal{G}_2$ and expressed by

$$\mathcal{T}_2 = \left\{ (\bar{z}_0, y_0) \to (\bar{z}_1, y_1) \;\middle|\; \begin{array}{l} y_i \in \{0,1\}^m, \bar{z}_i \in \{0,1\}, i = 0, 1 \\ g_2(\bar{z}_0, y_0) = \bar{z}_1, f(\bar{z}_0, y_0) = y_1 \end{array} \right\}$$

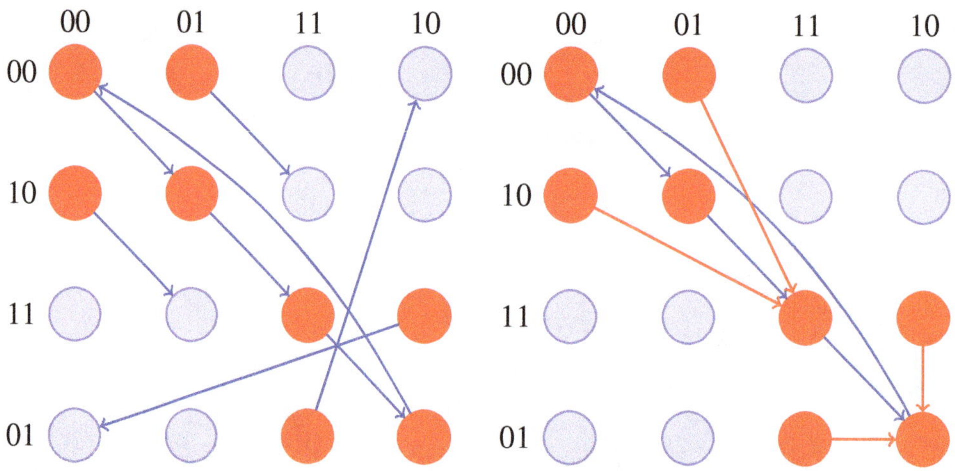

Figure 9. (Left) Part of transition diagram that involves states of the form (x, z, z, y) – colored in red; (Right) The transition diagram corresponds to three-node network system.

(x, z, y)	$G(x, z, y)$
$(0, 0, 0)$	$(1, 0, 1)$
$(1, 0, 1)$	$(1, 1, 1)$
$(1, 1, 1)$	$(0, 1, 0)$
$(0, 1, 0)$	$(0, 0, 0)$
$(0, 1, 1)$	$(0, 1, 0)$
$(1, 1, 0)$	$(0, 1, 0)$
$(0, 0, 1)$	$(1, 1, 1)$
$(1, 0, 0)$	$(1, 1, 1)$

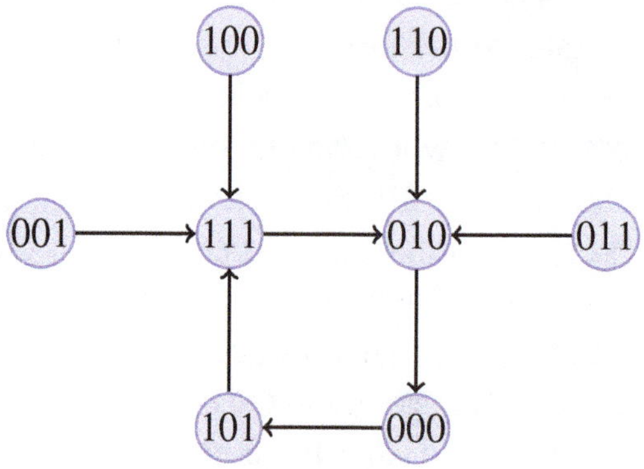

Figure 10. The truth table and transition diagram of the three-node network in Fig. 10. (Left) Truth table; (Right) Transition diagram

Then the set Q of all transitions of the product system is

$$Q = \left\{ (x_0, z_0, \bar{z}_0, y_0) \to (x_1, z_1, \bar{z}_1, y_1) \;\middle|\; \begin{array}{l} x_i \in \{0,1\}^k, z_i \in \{0,1\}, y_i \in \{0,1\}^m, \\ \bar{z}_i \in \{0,1\}, i = 0, 1 \\ h(x_0, z_0) = x_1, g_1(x_0, z_0) = z_1 \\ g_2(\bar{z}_0, y_0) = \bar{z}_1, f(\bar{z}_0, y_0) = y_1 \end{array} \right\}$$

Set the set of all transition of the system of $\mathcal{G}$ be $\mathcal{T}$.

2. Find all the states of the form of (x, z, z, y) and their transitions in the product system.
3. If $(x_0, z_0, z_0, y_0) \to (x_1, z_1, z_1, y_1) \in Q$, then add $(x_0, z_0, y_0) \to (x_1, z_1, y_1)$ to the set $\mathcal{T}$. If $(x_0, z_0, z_0, y_0) \to (x_1, z_1, \bar{z}_1, y_1) \in Q$ with $z_1 \neq \bar{z}_1$, then add $(x_0, z_0, y_0) \to (x_1, g(x_0, z_0, y_0), y_1)$ to $\mathcal{T}$.

Remark 6.1. The transitions on the diagonal of the product network are always transitions of the whole network.

Remark 6.2. The algorithm is rather straight forward. However, it can be very useful when the subnetworks are large and their transition diagrams are ready to use. Also note that even though we add the condition agreeable on the system of $\mathcal{G}$, we can easily modify it to the case that the condition fail. We just need to change the step 3 to *If* $(x_0, z_0, z_0, y_0) \to (x_1, z_1, \bar{z}_1, y_1) \in Q$, *then add* $(x_0, z_0, y_0) \to (x_1, g(x_0, z_0, y_0), y_1)$ *to* $\mathcal{T}$. On the other hand, the algorithm provide a rather clear view on the relations between the dynamics of the whole network and that of its subnetworks.

Acknowledgement

A few days before KC passed away, he discussed with YW about finishing this work in the summer of 2015. YW is grateful for KC's mentoring and his friendship, and will cherish the memory of him. YW was supported by DHS-14-ST-062-001.

Conflict of Interest

All authors declare no conflicts of interest in this paper.

References

1. W. Abuo-Jaoude, D. Ouattara, and M. Kaufman, *From structure to dynamics: frequency tuning in the p53-mdm2 network: I. logical approach*, Journal of Theoretical Biology, **258** (2009), 561-577.
2. R. Albert and H. Othmer, *The topology of the regulatory interactions perdicts the expression pattern of the segment polarity genes in drosophila melanogaster*, Journal of Theoretical Biology, **233** (2003) , 1-18.
3. Reka Albert, *Scale-free networks in cell biology*, Journal of Cell Science, **118** (2005), 4947-4957.
4. Sergio A. Alcalá-Corona, Tadeo E. Velázquez-Caldelas, Jesús Espinal-Enríquez, and Enrique Hernández-Lemus, *Community structure reveals biologically functional modules in mef2c transcriptional regulatory network*, Front Physiol, **7** (2016), 184.
5. B. B. Aldrige, J. M. Burke, D. A. Lauffenburge, and P.K. Sorger, *Physicochemical modeling of cell signaling pathways*, Nature Cell Biol, **8** (2006).
6. U. Alon, *Network motifs: Theory and experimental approaches*, Nature Reviews Genetics, **8** (2007), 450.
7. C. Campbell, J. Thakar, and R. Albert, *Network analysis reveals cross-links of the immune pathways activated by bacteria and allergen*, Physical Reveiw. E, Statistical physics, plasmas, fluids and related interdisciplinary topics, **84** (2011), 031929.
8. R. Edwards and L. Glass, *Combinatorial explosion in model gene networks*, Chaos: An inerdisciplinary Journal of Nonlinear Science, **11** (2000), 691-704.
9. R. Edwards, H. Siegelmann, K. Aziza, and L. Glass, *Symbolic dynamics and computation in model gene models*, Chaos: An inerdisciplinary Journal of Nonlinear Science, **11** (2001), 160-169.
10. C. Espionza-Soto, P. Padilla-Longoria, and E.R. Alvarez-Buylla, *A gene regulatory network model for cell-fate determination during arabidopsis thaliana flower development that is robus and recovers experiental gene expression profiles*, Plant Cell, **16** (2004), 2923-2939.

11.Newman M. E and Girvan M., *Community structure in social and biological networks*, Proc. Natl. Acad. Sci. U.S.A., **99** (2002), 7821-7826.

12.L. Glass and S. A. Kauffman, *The logical analysis of continuous, nonlinear biochemical control network*, J. Theor. Biol., **39** (1973), 103-139.

13.Cantini L., Medico E., Fortunato S., and Caselle M *Detection of gene communities in multi-networks reveals cancer drivers.*, Sci. Rep., **5** (2015), 17386.

14.R. Milo, S. Shen-Orr, S. Itzkovitz, N. Kashtan, D. Chklovskii, and U. Alon *Network motifs: Simple building blocks of complex networks*, Science, **298** (2002), 824-827.

15.A. Mogilner, R. Wollman, and W. F. Marshall *Quantitative modeling in cell biology*, Developmental Cell, **11** (2006), 279-287.

16.S. Li, S. M. AssMann, and R. Albert *Predicting essential components of signal transduction networks: A dynamic model of guard cell abscisic acid signaling*, PLoS Biol., **4** (2006), e312.

17.M. E. J. Newman, *Modularity and community structure in networks*, Proceedings of the National Academy of Sciences, **103** (2006), 8577-8582.

18.M. E. J. Newman, *Communities, modules and large scale structure*, Nature Physics, **8** (2012), 25-31.

19.A. Saadatpour, R. Albert, and T. C. Reluga, *A reduction method for boolean network models proven to conserve attractors*, SIAM J. Appl. Dyn. Sys., **12** (2013), 1997-2011.

20.J. Saez-Rodriguez, L. Simeoni, J. A. Lindquist, R. Hemenway, U. Bommhardt, U. U. Haus B. Arndt, R. Weismantel, E. D. Gilles, S. Klamt, and B. Schraven, *A logical model provides insights into t cell receptor signaling*, PLoS Computational Biology, **3** (2007), e163.

21.L. Sanchez and D. Thieffry, *A logical analysis of the drosophila gap-gene system*, J. Theor. Biol., **211** (2001), 115-141.

22.Adam J. Schwarz, Alessandro Gozzi, and Angelo Bifone, *Community structure and modularity in networks of correlated brain activity*, Magnetic Resonance Imaging., **27** (2008), 914-920.

23.S. S. Shen-Orr, R. Milo, S. Mangan, and U. Alon, *Network motifs in the transcriptional regulation network of escherichia coli*, Nature Genet **31** (2002), 64-68.

24.E. Snoussi, *Qualitative dynamics of piecewise differential equations: a discrete mapping approach.*, Dynamical and Stability of Systems, **4** (1989), 189-207.

25.R Tanaka, *Scale-rich metabolic networks.*, Phys. Rev. Lett., **94** (2005), 168101.

26.J. Thakar, A. K. Pathak, L. Murphy, R. Albert, I. Cattadori, and R. J. De Boer, *Network model of immune responses reveals key efectors to single and co-infection dynamics by a respiratory bacterium and a gatrointestinal helminth*, PLoS Computational Biology, **8** (2012), 1.

27.R. Thomas, *Biological feedback*, CRC, 1990.

28.D. Turner, P. Paszek, D.J. Woodcock, D. E. Nelson, C.A.Horton, Y. Wang, D.G. Spiller, D. A. Rand, M. R. H. White, and C. V. Harper, *Physiological levels of tnfalpha stimulation induce stochastic dynamics of nf-kappab responses in single living cells*, Journal of Cell Biology, **123** (2010), 2834-2843.

29. J. Tyson and B. Novak *Functional motifs in biochemical reaction networks*, Annu. Rev. Phys. Chem., **61** (2010), 219-240.

30. J. J. Tyson, K. C. Chen, and B. Novak, *Network dynamics and cell physiology*, Nature Rev. Mol. Cell Biol, **2** (2001), 908-916.

31. J. J. Tyson, K. C. Chen, and B. Novak, *Sniffers, buzzers, toggles and blinkers: dynamics of regulatory and signaling pathways in the cell*, Curr. Op. Cell Biol, **15** (2003), 221-231.

32. A. Veliz-Cuba, A. Kumar, and K. Josic, *Piecewise linear and boolean models of chemical reaction networks*, J. Math. Bio., **76** (2014), 2945-2984.

33. R. S. Wang and R. Albert, *Discrete dynamical modeling of cellular signaling networks*, Methods in Enzymology, **467** (2009) , 281-306.

34. Sebastian Wernicke, *A faster algorithm for detecting network motifs*, Proc. 5th WABI-05, **3692** (2005).

35. S. H. Yook, Z. N. Oltvai, and A. L. Barabási, *Functional and topological characterization of protein interaction networks*, Proteomics, **4** (2004), 928-942.

9

Localized Orthogonal Decomposition for two-scale Helmholtz-type problems

Mario Ohlberger and Barbara Verfürth*

Applied Mathematics, University of Münster, D-48149 Münster, Germany

* **Correspondence:** barbara.verfuerth@uni-muenster.de

Abstract: In this paper, we present a Localized Orthogonal Decomposition (LOD) in Petrov-Galerkin formulation for a two-scale Helmholtz-type problem. The two-scale problem is, for instance, motivated from the homogenization of the Helmholtz equation with high contrast, studied together with a corresponding multiscale method in a previous paper of the authors. There, an unavoidable resolution condition on the mesh sizes in terms of the wave number has been observed, which is known as "pollution effect" in the finite element literature. Following ideas of Gallistl and Peterseim, we use standard finite element functions for the trial space, whereas the test functions are enriched by solutions of subscsale problems (solved on a finer grid) on local patches. Provided that the oversampling parameter m, which indicates the size of the patches, is coupled logarithmically to the wave number, we obtain a quasi-optimal method under a reasonable resolution of a few degrees of freedom per wave length, thus overcoming the pollution effect. In the two-scale setting, the main challenges for the LOD lie in the coupling of the function spaces and in the periodic boundary conditions.

Keywords: Multiscale method; pollution effect; Helmholtz equation; finite elements; numerical homogenization
Mathematics Subject Classification: Primary: 65N12, 65N15, 65N30; Secondary: 35J05, 35B27, 78M40

1. Introduction

The numerical solution of high frequency Helmholtz problems with standard finite element methods is still a very challenging task due to the oscillatory nature of the solutions. This manifests itself in the so-called pollution effect [3]: A much smaller mesh size than needed for a meaningful approximation of the solution are required for the stability and convergence of the numerical scheme. Typically, this leads to a resolution condition $k^\alpha H = O(1)$ with $\alpha > 1$ instead of $kH = O(1)$, i.e. few degrees of freedom per wave length. The challenge becomes even greater when additionally studying the

Helmholtz problem in a heterogeneous medium, such as locally periodic structures.

(Locally) periodic media, such as photonic crystals, can exhibit astonishing properties such as band gaps, artificial magnetism, or negative refraction [11,29,33,39]. One popular and successful modelling setup are scatterers made up of two materials with high contrast in the permittivity [4–7, 10, 15, 28]. As the materials' fine-scale structures are much smaller than the wavelength, homogenization and corresponding numerical multiscale methods, such as the Heterogeneous Multiscale Method (HMM), are efficient tools to reduce the problem's complexity. Homogenization theory gives an effective Helmholtz problem, which describes the macroscopic behavior of the original problem, but involves the solution of some additional cell problems to determine the effective material parameters. This procedure can also be coupled in the so-called *two-scale equation*, which gives the macroscopic and the cell problems in one variational problem. The Heterogeneous Multiscale Method can be seen as a Galerkin discetization of this two-scale equation and it gives good approximations even of the solution in the heterogeneous medium, see [34]. However, as every standard Galerkin method, the HMM for Helmholtz-type problems also suffers from the pollution effect described in the beginning [14, 32, 40]. There are several attempts to reduce this pollution effect, e.g. high-order finite element methods [14, 32], (hybridizable) discontinuous Galerkin methods [9, 17], or (plane wave) Trefftz methods [23, 24, 36].

Recently, the works [8, 16, 37] suggested a multiscale Petrov-Galerkin method for the Helmholtz equation to reduce the pollution effect. The method is based on a so called Localized Orthogonal Decomposition (LOD), as first introduced in [30]. The LOD builds on ideas from the Variational Multiscale Method (VMM) [25, 26] by splitting the solution space into a coarse and a fine part. The coarse standard finite element functions are modified by adding a correction from the fine space, which is constructed as the kernel of an interpolation operator. The corrections are problem dependent and computed by solving PDEs on a finer grid. In most cases the corrections show exponential decay, which justifies to truncate them to patches of coarse elements. Since its development, the LOD has been successfully applied to elliptic boundary problems [22], eigenvalue problems [31], mixed problems [18], parabolic problems [31], the wave equation [1] or elasticity [21]. A review is given in [38]. As already discussed in [12] and analyzed in more detail in [38], Petrov-Galerkin formulations show the same stability and convergence behavior as the symmetric Galerkin methods while being less expensive with respect to communication. An extensive discussion on implementation aspects of the LOD, such as how to exploit a priori known structures to reduce the number of local subscale problems, is given in [13] and some remarks for acoustic scattering in [16].

In this article, we present how the LOD can be applied to two-scale Helmholtz-type problems. Following the ideas in [16], the test functions are modified by local subscale correction, which are solved on a fine grid fulfilling the resolution condition for standard Galerkin finite element methods. The resulting test functions have support over patches of size mH_c and mh_c; H_c being the mesh size of the grid for the macroscopic domain G, h_c being the mesh size of the grid for the unit square Y and m being the adjustable oversampling parameter. Under the condition that $m \approx \log(k)$ and $k(H_c + h_c) \lesssim 1$, the (two-scale) LOD is stable and quasi-optimal, i.e. the error between the LOD-approximation and the analytical solution of the two-scale equation is of the order of the best-approximation error. These are the results expected from the one-scale setting. The main contribution is the rigorous (theoretical) analysis for this LOD in the two-scale setting. The novelty here is that first, we have to deal with coupled functional spaces and sesquilinear forms. This coupling has to be taken care of in all new definitions of (again coupled) spaces and also in the estimates, where we have to jump back and forth between the two-scale norms and the properties of each individual space and interpolation operator. Second, our coupled two-scale functional spaces involve periodic boundary conditions on the unit cube, which also have to be paid attention at when defining the interpolation operators and the oversampling patches at the boundary.

Finally, let us summarize the connection of the LOD in two-scale setting to original Helmholtz problem with high contrast, but also emphasize the general nature of our (theoretical) findings. Combining the HMM of [34] and the LOD of the present paper, we have solved the Helmholtz problem with high contrast by a double approximation procedure: First removing the challenges from the fine-scale structures related to the periodicity δ and then reducing the pollution effect related to the wave number k, where we recall the three-scale nature $\delta \ll k^{-1} < 1$. Our numerical experiments in [34] have not experienced great restriction from the resolution condition, but more sophisticated examples or even higher wave numbers may make the application of the LOD favorable, which hereby has the necessary theoretical footing. Concerning possible generalizations of our work, let us note that we concentrate on the two-scale setting described in [34], but the analysis can also be adapted to the case without high contrast [27] and might be useful for future numerical studies of the three-dimensional cases [4,6,7,10,28]. Moreover, the definition of the patches and interpolation operators in the periodic case and also the coupling and decoupling of various function spaces may be of general interest.

The paper is structured as follows: Section 2 introduces the notation and problem setting, see also [34]. In Section 3, the LOD in Petrov-Galerkin formulation is defined in the two-scale setting, where we also give some general comments on implementation aspects. Stability and convergence of the error are discussed in Section 4. Section 5 is devoted to the detailed proof of the decay of the correctors.

2. The two-scale Helmholtz problem

In this section, we introduce what we call the two-scale Helmholtz problem and the necessary notation. For further details on the derivation of this two-scale model and its practical relevance we refer to [34].

For the remainder of this article, let $\Omega \subset\subset G \subset \mathbb{R}^2$ be two domains with (polygonal) Lipschitz boundary at least. Throughout this paper, we use standard notation for Lebesgue and Sobolev spaces: For a domain ω, $p \in [1, \infty)$, and $s \in \mathbb{R}_{\geq 0}$, $L^p(\omega)$ denotes the complex Lebesgue space and $H^s(\omega)$ denotes the complex (fractional) Sobolev space. The dot denotes a normal (real) scalar product, for a complex scalar product we explicitly conjugate the second component by using v^* as the conjugate complex of v. The complex L^2 scalar product on a domain ω is abbreviated by $(\cdot, \cdot)_\omega$ and the domain is omitted if no confusion can arise. For $v \in H^1(\omega)$, we frequently use the k-dependent norm

$$\|v\|_{1,k,\omega} := (\|\nabla v\|_\omega^2 + k^2 \|v\|_\omega^2)^{1/2},$$

which is obviously equivalent to the H^1 norm. We write $Y := [-\frac{1}{2}, \frac{1}{2})^2$ to denote the two-dimensional unit square. We indicate Y-periodic functions [2] with the subscript $\sharp$. For example, $H^1_{\sharp,0}(Y)$ is the space of Y-periodic functions from $H^1_{\text{loc}}(\mathbb{R}^2)$ with zero average over Y. For $Y^* \subset Y$, we denote by $H^1_{\sharp,0}(Y^*)$ the restriction of functions in $H^1_{\sharp,0}(Y)$ to Y^*. For $D \subset\subset Y$, $H^1_0(D)$ can be interpreted as subspace of $H^1_\sharp(Y)$ and we write $H^1_0(D)_\sharp$ to emphasize this periodic extension. By $L^p(\Omega; X)$ we denote Bochner-Lebesgue spaces over the Banach space X.

The two-scale Helmholtz problem is now formulated as follows: We seek $\mathbf{u} := (u, u_1, u_2) \in \mathcal{H}$ such that

$$\mathcal{B}(\mathbf{u}, \boldsymbol{\psi}) = \int_{\partial G} g\psi^* \, d\sigma \qquad \forall \boldsymbol{\psi} := (\psi, \psi_1, \psi_2) \in \mathcal{H} \tag{2.1}$$

with the two-scale function space $\mathcal{H} := H^1(G) \times L^2(\Omega; H^1_{\sharp,0}(Y^*)) \times L^2(\Omega; H^1_0(D)_\sharp)$ and the two-scale sesquilinear form $\mathcal{B}$ defined by

$$\begin{aligned}
&\mathcal{B}(\mathbf{v}, \boldsymbol{\psi}) \\
&:= \int_\Omega \int_{Y^*} \varepsilon_e^{-1}(\nabla v + \nabla_y v_1) \cdot (\nabla \psi^* + \nabla_y \psi_1^*) \, dy dx + \int_G \int_D \varepsilon_i^{-1} \nabla_y v_2 \cdot \nabla_y \psi_2^* \, dy dx \\
&\quad - k^2 \int_G \int_Y (v + \chi_D v_2)(\psi^* + \chi_D \psi_2^*) \, dy dx + \int_{G \setminus \overline{\Omega}} \nabla v \cdot \nabla \psi^* \, dx - ik \int_{\partial G} v\psi^* \, d\sigma.
\end{aligned}$$

Beside the natural norm on $\mathcal{H}$ it is convenient for error estimation to define the following two-scale energy norm

$$\|(v, v_1, v_2)\|^2_{e,\omega\times R} := \|\nabla v + \nabla_y v_1\|^2_{\omega\times R_1} + \|\nabla_y v_2\|^2_{\omega\times R_2} + k^2 \|v + \chi_D v_2\|^2_{\omega\times R} \tag{2.2}$$

for a subdomain $\omega \times R \subset G \times Y$ with $R_1 := R \cap Y^*$ and $R_2 := R \cap D$. Furthermore, we introduce a version of the H^1 semi-norm on $\mathcal{H}$ via

$$\|(v, v_1, v_2)\|^2_{1,e,\omega\times R} := \|\nabla v + \nabla_y v_1\|^2_{\omega\times R_1} + \|\nabla_y v_2\|^2_{\omega\times R_2}, \tag{2.3}$$

which is induced by the sesquilinear form

$$(\mathbf{v}, \boldsymbol{\psi})_{1,e,\omega\times R} := (\nabla v + \nabla_y v_1, \nabla \psi_1 + \nabla_y \psi_2)_{\omega\times R_1} + (\nabla_y v_2, \nabla_y \psi_2)_{\omega\times R_2} \qquad \forall \mathbf{v}, \boldsymbol{\psi} \in \mathcal{H}.$$

We omit the subscript for the subdomain if $\omega \times R = G \times Y$. In [34], we proved the following Lemma.

Lemma 2.1. *There exist constants $C_B > 0$ and $C_{\min} := \min\{1, \varepsilon_e^{-1}, \mathrm{Re}(\varepsilon_i^{-1})\} > 0$ depending only on the parameters and the geometry, such that $\mathcal{B}$ is continuous with constant C_B and fulfills a Gårding inequality with constant $C_{\min}$, i.e.*

$$|\mathcal{B}(\mathbf{v}, \boldsymbol{\psi})| \le C_B \|\mathbf{v}\|_e \|\boldsymbol{\psi}\|_e \quad \textit{and} \quad \mathrm{Re}\, \mathcal{B}(\mathbf{v}, \mathbf{v}) + 2k^2 \|v + \chi_D v_2\|^2_{G\times Y} \ge C_{\min} \|\mathbf{v}\|^2_e.$$

Note that $C_{\min}$ can also be used to estimate the gradient terms in $\mathcal{B}$ by $C_{\min}\|\cdot\|_{1,e}$ from below. Furthermore, the unique solution $\mathbf{u} \in \mathcal{H}$ to (2.1) (with additional volume term $f \in L^2(G)$ on the right-hand side) fulfills the following stability estimate

$$\|\mathbf{u}\|_e \le Ck^q (\|f\|_{L^2(G)} + \|g\|_{H^{1/2}(\partial G)}) \qquad \text{for some} \quad q \in \mathbb{N}_0, \tag{2.4}$$

see [34] for a proof with $q = 3$.

For the error analysis, we will compare the solution of the Localized Orthogonal Decomposition to a discrete reference solution, which is only needed for the theory and will never be computed in practical implementations. We introduce conforming and shape regular triangulations $\mathcal{T}_H$ and $\mathcal{T}_h$ of G and Y, respectively. Additionally, we assume that $\mathcal{T}_H$ resolves the partition into Ω and $G \setminus \overline{\Omega}$ and that $\mathcal{T}_h$ resolves the partition into D and Y^* and is periodic in the sense that it can be wrapped to a regular triangulation of the torus (without hanging nodes). We use the conforming subspace $\mathbf{V}_{H,h} \subset \mathcal{H}$ made up of linear Lagrange elements via $\mathbf{V}_{H,h} := V_H^1 \times L^2(\Omega; \widetilde{V}_h^1(Y^*)) \times L^2(\Omega; V_h^1(D))$, where the L^2 spaces with respect to x can be additionally approximated by piece-wise polynomials. The discrete reference solution $\mathbf{u}_{H,h} = (u_H, u_{h,1}, u_{h,2}) \in \mathbf{V}_{H,h}$ is the solution to

$$\mathcal{B}(\mathbf{u}_{H,h}, \boldsymbol{\psi}_{H,h}) = \int_{\partial G} g\psi_H^* \, d\sigma \qquad \forall \boldsymbol{\psi}_{H,h} := (\psi_H, \psi_{h,1}, \psi_{h,2}) \in \mathbf{V}_{H,h}. \tag{2.5}$$

We assume that this direct discretization is stable in the following sense: The (fine) mesh sizes H and h are small enough (in dependence on the wave number k) that there is a constant C_{HMM} such that

$$(C_{\text{HMM}}\, k^{q+1})^{-1} \leq \inf_{\mathbf{v}_{H,h} \in \mathbf{V}_{H,h}} \sup_{\boldsymbol{\psi}_{H,h} \in \mathbf{V}_{H,h}} \frac{\operatorname{Re} \mathcal{B}(\mathbf{v}_{H,h}, \boldsymbol{\psi}_{H,h})}{\|\mathbf{v}_{H,h}\|_e \|\boldsymbol{\psi}_{h,h}\|_e}. \tag{2.6}$$

In [34], we discussed that this direct discretization can be re-cast into the traditional formulation of a Heterogeneous Multiscale Method and we proved that the discrete inf-sup-condition (2.6) holds under the classical resolution condition $k^{q+2}(H + h) = O(1)$.

Remark 1. We demonstrate the LOD at the specific example at the two-scale Helmholtz problem obtained in [34]. However, the theory can easily be extended to more general two-scale Helmholtz problems, which fulfill the following assumptions

- the variational problem (2.1) involves a continuous sesquilinear form with Gårding inequality, i.e. an analogue of Lemma 2.1;
- the analytical solution fulfills a stability estimate (2.4);
- the (direct) Galerkin discretization (2.5) is stable (2.6).

3. The Localized Orthogonal Decomposition

In this section, we introduce the notation on meshes, finite element spaces, and (quasi)-interpolation operators and define the Localized Orthogonal decomposition in Petrov-Galerkin formulation for the two-scale setting. We close with some remarks regarding an implementation of the two-scale LOD.

3.1. Meshes and finite element spaces

Let the (fine) meshes $\mathcal{T}_H$ of G and $\mathcal{T}_h$ of Y be given as in the previous section, we assume that H and h are small enough that (2.6) is fulfilled. We consider a second, coarse discretization scale $H_c > H$ and $h_c > h$: Let $\mathcal{T}_{H_c}$ and $\mathcal{T}_{h_c}$ denote corresponding conforming, quasi-uniform, and shape regular triangulations of G and Y, respectively. As for the fine grids, we additionally assume that $\mathcal{T}_{h_c}$ is periodic and that $\mathcal{T}_{H_c}$ and $\mathcal{T}_{h_c}$ resolve the partition of G into Ω and its complement and of Y into D and Y^*, respectively. We denote by $\mathcal{T}_{h_c}(Y^*)$ and $\mathcal{T}_{h_c}(D)$ the parts of $\mathcal{T}_{h_c}$ triangulating Y^* and D, respectively. The global mesh sizes are defined as $H_c := \max\{\operatorname{diam}(T) | T \in \mathcal{T}_{H_c}\}$ and $h_c := \max\{\operatorname{diam}(S) | S \in \mathcal{T}_{h_c}\}$. For the sake of simplicity we assume that $\mathcal{T}_H$ and $\mathcal{T}_h$ are derived from $\mathcal{T}_{H_c}$ and $\mathcal{T}_{h_c}$, respectively, by some regular, possibly non-uniform, mesh refinement including at least one global refinement. We consider simplicial partitions, but the theory of this paper carries over to quadrilateral partitions [16] and even meshless methods would be possible [19].

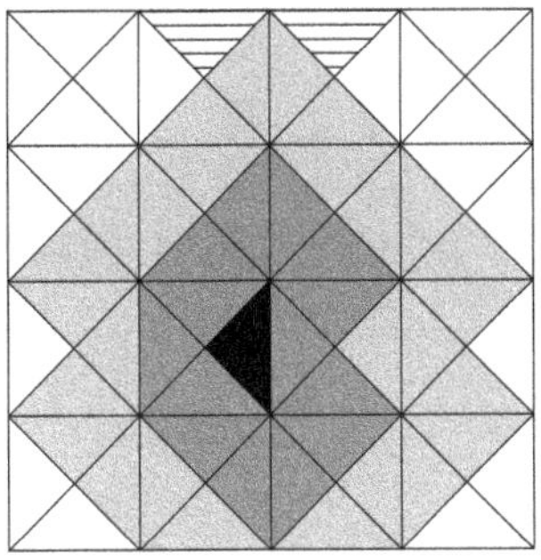

Figure 3.1. Triangle T (in black) and its first and second order patches (additional elements for $\mathrm{N}(T)$ in dark gray and additional elements for $\mathrm{N}^2(T)$ in light gray). Striped triangles belong to $\mathrm{N}^2(T)$ in the case of periodic boundary conditions.

Given any subdomain $\omega \subset \overline{G}$ define its neighborhood via

$$\mathrm{N}(\omega) := \operatorname{int}(\cup\{T \in \mathcal{T}_{H_c} | T \cap \overline{\omega} \neq \emptyset\})$$

and for any $m \geq 2$ the patches

$$\mathrm{N}^1(\omega) := \mathrm{N}(\omega) \qquad \text{and} \qquad \mathrm{N}^m(\omega) := \mathrm{N}(\mathrm{N}^{m-1}(\omega)),$$

see Figure 3.1 for an example. The shape regularity implies that there is a uniform bound $C_{\mathrm{ol},m,G}$ on the number of elements in the m-th order patch

$$\max_{T \in \mathcal{T}_{H_c}} \sharp\{K \in \mathcal{T}_{H_c} | K \subset \overline{\mathrm{N}^m(T)}\} \leq C_{\mathrm{ol},m,G}$$

and the quasi-uniformity implies that $C_{\mathrm{ol},m,G}$ depends polynomially on m. We abbreviate $C_{\mathrm{ol},G} := C_{\mathrm{ol},1,G}$. The patches can also be defined in a similar way for a subdomain $R \subset \overline{Y}$. Here, we split $R = R_1 \cup R_2$ with $R_1 = R \cap D$ and $R_2 = R \cap Y^*$, where R_1 or R_2 may be empty, and we write in short $\mathrm{N}^m(R) := \mathrm{N}^m(R_1) \cup \mathrm{N}^m(R_2)$. $\mathrm{N}^m(R_1)$ is defined in the same way as before, in particular, it ends at the boundary ∂D. For the patch $\mathrm{N}^m(R_2)$ we interpret $\overline{Y^*}$ as part of the torus. This implies that $\mathrm{N}^m(R_2)$ ends at the inner boundary ∂D, but is continued periodically over the outer boundary ∂Y. This means that also the striped triangles in Figure 3.1 belong to the second patch for the periodic setting. We denote the overlap constants by $C_{\mathrm{ol},m,Y}$ and $C_{\mathrm{ol},Y}$. By slight abuse of notation, we write $\mathrm{N}^m(\omega \times R) := \mathrm{N}^m(\omega) \times \mathrm{N}^m(R)$ for a subdomain $\omega \times R \subset \overline{G} \times \overline{Y}$.

We denote the conforming finite element triple space consisting of lowest order Lagrange elements with respect to the meshes $\mathcal{T}_{H_c}$ and $\mathcal{T}_{h_c}$ by $\mathbf{V}_{H_c,h_c}$ as in the previous section. Again, we have $\mathbf{V}_{H_c,h_c} := V^1_{H_c} \times L^2(\Omega; \widetilde{V}^1_{h_c}(Y^*)) \times L^2(\Omega; V^1_{h_c}(D))$ and we moreover note that $\mathbf{V}_{H_c,h_c} \subset \mathbf{V}_{H,h} \subset \mathcal{H}$.

3.2. Quasi-interpolation

A key tool in the definition and the analysis is a bounded linear surjective (quasi)-interpolation operator $\mathbf{I}_{H_c,h_c} : \mathbf{V}_{H,h} \to \mathbf{V}_{H_c,h_c}$ that acts as a stable quasi-local projection in the following sense: It is a projection, i.e. $\mathbf{I}_{H_c,h_c} \circ \mathbf{I}_{H_c,h_c} = \mathbf{I}_{H_c,h_c}$, and it is constructed as $\mathbf{I}_{H_c,h_c} := (I_{H_c}, I^{Y^*}_{h_c}, I^D_{h_c})$, where each (quasi)-interpolation operator fulfills the following. There exist constants $C_{I_{H_c}}$, $C_{I_{h_c}^{Y^*}}$, and $C_{I_{h_c}^D}$ such that for all $\mathbf{v}_{H,h} := (v_H, v_{h,1}, v_{h,2}) \in \mathbf{V}_{H,h}$ and for all $T \in \mathcal{T}_{H_c}$, $S_1 \in \mathcal{T}_{h_c}(Y^*)$ and $S_2 \in \mathcal{T}_{h_c}(D)$

$$\begin{aligned} H_c^{-1}\|v_H - I_{H_c}(v_H)\|_T + \|\nabla I_{H_c}(v_H)\|_T &\leq C_{I_{H_c}}\|\nabla v_H\|_{\mathrm{N}(T)}, \\ h_c^{-1}\|v_{h,1} - I_{h_c}^{Y^*}(v_{h,1})\|_{T\times S_1} + \|\nabla_y I_{h_c}^{Y^*}(v_{h,1})\|_{T\times S_1} &\leq C_{I_{H_c}^{Y^*}}\|\nabla_y v_{h,1}\|_{T\times\mathrm{N}(S_1)}, \\ h_c^{-1}\|v_{h,2} - I_{h_c}^{D}(v_{h,2})\|_{T\times S_2} + \|\nabla_y I_{h_c}^{D}(v_{h,2})\|_{T\times S_2} &\leq C_{I_{H_c}^{D}}\|\nabla_y v_{h,2}\|_{T\times\mathrm{N}(S_2)}. \end{aligned} \tag{3.1}$$

We abbreviate $C_{\mathbf{I}} := \max\{C_{I_{H_c}}, C_{I_{h_c}^{Y^*}}, C_{I_{H_c}^D}\}$. Under the mesh condition that $k(H_c + h_c) \lesssim 1$, this implies stability in the two-scale energy norm

$$\|\mathbf{I}_{H_c,h_c}\mathbf{v}_{H,h}\|_e \leq C_{\mathbf{I},e}\|\mathbf{v}_{H,h}\|_e \qquad \forall \mathbf{v}_{H,h} \in \mathbf{V}_{H,h}. \tag{3.2}$$

The quasi-interpolation operator $\mathbf{I}_{H_c,h_c}$ is not unique: A different choice might lead to a different Localized Orthogonal Decomposition and this can even affect the practical performance of the method [37]. One popular choice is the concatenation of the l^2 projection onto piece-wise polynomials and the Oswald interpolation operator. Other choices are discussed in [13, 38]. For the operators $I_{h_c}^{Y^*}$ and $I_{h_c}^D$ not that they only act with respect to the second variable y. For $I_{h_c}^{Y^*}$, one can preserve periodicity as follows: The averaging process of the Oswald interpolation operator has to be continued over the periodic boundary (as for the patches before).

3.3. *Definition of the LOD*

The method approximates the discrete two-scale solution $\mathbf{u}_{H,h} := (u_H, u_{h,1}, u_{h,2})$ to (2.5) for given (fine) mesh sizes H, h. It is determined by the choice of the coarse mesh sizes H_c and h_c and the oversampling parameter m explained in the following. We assign to any $(T, S_1, S_2) \in \mathcal{T}_{H_c} \times \mathcal{T}_{h_c}(Y^*) \times \mathcal{T}_{h_c}(D)$ its m-th order patch $G_T \times Y_S^* \times D_S := \mathrm{N}^m(T) \times \mathrm{N}^m(S_1) \times \mathrm{N}^m(S_2)$ and define for any $\mathbf{v}_{H,h} = (v_H, v_{h,1}, v_{h,2}), \boldsymbol{\psi}_{H,h} = (\psi_H, \psi_{h,1}, \psi_{h,2}) \in \mathbf{V}_{H,h}$ the localized sesquilinear form

$$\begin{aligned} &\mathcal{B}_{G_T\times Y_S}(\mathbf{v}_{H,h}, \boldsymbol{\psi}_{H,h}) \\ &\quad := (\varepsilon_e^{-1}(\nabla v_H + \nabla_y v_{h,1}), \nabla\psi_H + \nabla_y\psi_{h,1})_{(G_T\cap\Omega)\times Y_S^*} + (\varepsilon_i^{-1}\nabla_y v_{h,2}, \nabla_y\psi_{h,2})_{G_T\times D_S} \\ &\qquad + (\nabla v_H, \nabla\psi_H)_{G_T\cap(G\setminus\overline{\Omega})} - k^2(v_H + \chi_D v_{h,2}, \psi_H, +\chi_D\psi_{h,2})_{G_T\times Y_S} \\ &\qquad - ik(v_H, \psi_H)_{\partial G_T\cap\partial G} \end{aligned}$$

with $Y_S := D_S \cup Y_S^*$. For $m = 0$ (i.e. $\mathrm{N}^m(T) = T$), we write $\mathcal{B}_{T\times S}$ with $S = S_1 \cup S_2$. Note that the oversampling parameter does not have to be the same for G, Y^*, and D. We could as well introduce patches $\mathrm{N}^{m_0}(T) \times \mathrm{N}^{m_1}(S_1) \times \mathrm{N}^{m_2}(S_2)$, but we choose $m_0 = m_1 = m_2 =: m$ for simplicity of presentation and to improve readability.

We define the (truncated) finite element functions on the fine-scale meshes as

$$\begin{aligned} V_H(G_T) &:= \{v_H \in V_H^1 | v_H = 0 \text{ outside } G_T\}, \\ L^2(\Omega_T, \widetilde{V}_h^1(Y_S^*)) &:= \{v_{h,1} \in L^2(\Omega; \widetilde{V}_h^1(Y^*)) | v_{h,1} = 0 \text{ outside } (G_T \cap \Omega) \times (Y^*)_S\} \end{aligned}$$

and $L^2(\Omega_T; V_h^1(D_S))$ in a similar way. Define the null space

$$\mathbf{W}_{H,h}(G_T \times Y_S)$$

$$:= \{\mathbf{w}_{H,h} \in V_H(G_T) \times L^2(\Omega_T; \tilde{V}_h(Y_S^*)) \times L^2(\Omega_T; V_h(D_S)) | \mathbf{I}_{H_c,h_c}(\mathbf{w}_{H,h}) = 0\}$$

and note that $\mathbf{W}_{H,h}(G_T \times Y_S) := W_H(G_T) \times L^2(\Omega; W_h(Y_S^*)) \times L^2(\Omega; W_h(D_S))$, where W_H and W_h are defined as the kernels of the corresponding (single) interpolation operators I_{H_C} and $I_{h_c}^{Y^*}$ and $I_{h_c}^D$, respectively. For given $\mathbf{v}_{H_c,h_c} \in \mathbf{V}_{H_c,h_c}$ we define the localized correction $\mathbf{Q}_m(\mathbf{v}_{H_c,h_c}) := (Q_m(v_{H_c}), Q_{m,1}(v_{h_c,1}), Q_{m,2}(v_{h_c,2}))$ as

$$\mathbf{Q}_m(\mathbf{v}_{H_c,h_c}) := \sum_{(T,S_1,S_2)\in \mathcal{T}_{H_C} \times \mathcal{T}_{h_c}(Y^*) \times \mathcal{T}_{h_c}(D)} \mathbf{Q}_{T\times S,m}(\mathbf{v}_{H_c,h_c}|_{T\times S}),$$

where $\mathbf{Q}_{T\times S,m}(\mathbf{v}_{H_c,h_c}|_{T\times S}) \in \mathbf{W}_{H,h}(G_T \times Y_S)$ solves the following subscale corrector problem

$$\mathcal{B}_{G_T\times Y_S}(\mathbf{w}, \mathbf{Q}_{T\times S,m}(\mathbf{v}_{H_c,h_c}|_{T\times S})) = \mathcal{B}_{T\times S}(\mathbf{w}, \mathbf{v}_{H_c,h_c}) \quad \forall \mathbf{w} \in \mathbf{W}_{H,h}(G_T \times Y_S). \tag{3.3}$$

The space of test functions then reads

$$\overline{\mathbf{V}}_{H_c,h_c,m} := (1 - \mathbf{Q}_m)(\mathbf{V}_{H_c,h_c})$$

and can be written as triple

$$\overline{\mathbf{V}}_{H_c,h_c,m} = \overline{V}_{H_c,m} \times L^2(\Omega; \overline{V}_{h_c,m}(Y^*)) \times L^2(\Omega; \overline{V}_{h_c,m}(D)).$$

We emphasize that $\dim \overline{\mathbf{V}}_{H_c,h_c,m} = \dim \mathbf{V}_{H_c,h_c}$ is low-dimensional and the dimension does not depend on H, h, or m.

Definition 3.1. The two-scale Localized Orthogonal Decomposition in Petrov-Galerkin formulation seeks $\mathbf{u}_{H_c,h_c} \in \mathbf{V}_{H_c,h_c}$ such that

$$\mathcal{B}(\mathbf{u}_{H_c,h_c}, \overline{\boldsymbol{\psi}}_{H_c,h_c}) = (g, \overline{\psi}_{H_c})_{\partial G} \qquad \forall \overline{\boldsymbol{\psi}}_{H_c,h_c} := (\overline{\psi}_{H_c}, \overline{\psi}_{h_c,1}, \overline{\psi}_{h_c,2}) \in \overline{\mathbf{V}}_{H_c,h_c,m}. \tag{3.4}$$

The error analysis will show that the choice $k(H_c + h_c) \lesssim 1$ and $m \approx \log k$ suffices to guarantee stability and quasi-optimality of the method, provided that the direct discretization (2.5) (with mesh widths H, h) is stable.

As discussed in [37], further stable variants of the method are possible: The local subscale correction procedure can be applied to only the test functions, only the ansatz functions, or both ansatz and test functions.

3.4. Remarks on implementation aspects

The present approach of the LOD exploits the two-scale structure of the underlying problem. In practice, one cannot work with the space triples such as $\mathbf{V}_{H_c,h_c}$, but will look at each of the function spaces separately. The LOD consists of two main steps: First, the modified basis functions in $\overline{\mathbf{V}}_{H_c,h_c,m}$ have to be determined, which includes the solution of the localized subscale corrector problems (3.3). Second, the actual LOD-approximation is computed as solution to (3.4). In this section, we explain how the computations in the macroscopic domain G and on the unit square Y can be decoupled in both steps. For general considerations on how to implement an LOD, for example algebraic realizations of the problems, we refer to [13].

Computation of modified bases. We observe that due to the sesquilinearity of $\mathcal{B}$ the following linearity for the correction operators $\mathbf{Q}_m$ holds

$$\begin{aligned}\mathbf{Q}_m \mathbf{v}_{H_c,h_c} &= \mathbf{Q}_m(v_{H_c}, 0, 0) + \mathbf{Q}_m(0, v_{h_c,1}, 0) + \mathbf{Q}_m(0, 0, v_{h_c,2}) \\ &= (Q_m(v_{H_c}), 0, 0) + (0, Q_{m,1}(v_{h_c}), 0) + (0, 0, Q_{m,2}(v_{h_c,2})).\end{aligned}$$

This means that the corrections of the basis functions in $V^1_{H_c}$, $\widetilde{V}^1_{h_c}(Y^*)$ and $V^1_{h_c}(D)$ can be computed separately in the following way:

1. Choose a basis $\{\lambda_x\}$ of $V^1_{H_c}$, $\{\lambda_{y,1}\}$ of $\widetilde{V}^1_{h_c}(Y^*)$ and $\{\lambda_{y,2}\}$ of $V^1_{h_c}(D)$.
2. For each basis function λ_x, $\lambda_{y,1}$ and $\lambda_{y,2}$ do
   (a) Find the solutions $Q_{T\times S,m}(\lambda_x)$, $Q_{T\times S,m,1}(\lambda_{y,1})$ and $Q_{T\times S,m,2}(\lambda_{y,2})$ of the corrector problem (3.3) for each $T \in \mathcal{T}_{H_c}$, $S_1 \in \mathcal{T}_{h_c}(Y^*)$ and $S_2 \in \mathcal{T}_{h_c}(D)$. This needs the determination of $W_H(G_T)$, $W_h(Y^*_S)$ and $W_h(D_S)$.
   (b) Build up the modified bases $\overline{\lambda}_x$ of $\overline{V}_{H_c,m}$, $\overline{\lambda}_{y,1}$ of $\overline{V}_{h_c,m}(Y^*)$ and $\overline{\lambda}_{y,2}$ of $\overline{V}_{h_c,m}(D)$ via $\overline{\lambda}_x := \lambda_x - \sum_{(T,S_1,S_2)\in\mathcal{T}_{H_c}\times\mathcal{T}_{h_c}(Y^*)\times\mathcal{T}_{h_c}(D)} Q_{T\times S,m}(\lambda_x)$, etc.

Note that no communication between the basis functions on G, Y^*, and D is needed and therefore, the computation of the modified bases can be easily parallelized. Only if the parameters ε_i and ε_e are constant w.r.t. x as here, the corrections $Q_{T\times S,m,1}$ and $Q_{T\times S,m,2}$ are x-independent. Depending on the choice of the interpolation operator, Lagrange multipliers can be employed to decode that a function belongs to W_H or W_h, see [13].

We can further decrease the computational complexity of the localized corrector problems by decoupling the integrals over G and Y and by reducing the number of correction problems. The potential gain of course hinges on (additional) structure of the parameters and the meshes with the following general observations:

- The corrections $Q_{T\times S,m,1}$ and $Q_{T\times S,m,2}$ only have to be computed for $T \in \mathcal{T}_{H_c}$ with $T \cap \Omega \neq \emptyset$.
- It is sufficient to choose test functions of the form $\mathbf{w} = (w, 0, 0)$ for $Q_{T\times S,m}$, $\mathbf{w} = (0, w_1, 0)$ for $Q_{T\times S,m,1}$, and $\mathbf{w} = (0, 0, w_2)$ for $Q_{T\times S,m,2}$.
- In the case of constant parameters ε_e and ε_i, the corrector problems for $Q_{T\times S,m,1}$ and $Q_{T\times S,m,2}$ include information on T only in form of the weights $|T|$ and $|\Omega_T|$; and the problems for $Q_{T\times S,m}$ only depend on S in form of the weights $|S_1|$ and $|Y^*_S|$.
- In case of structured meshes $\mathcal{T}_{H_c}$ and $\mathcal{T}_{h_c}$ and constant parameters, we can exploit symmetries to reduce the number of corrector problems [16].

Computation of the LOD-approximation. The LOD-approximation is defined as the solution to (3.4). This problem is similar to the discrete two-scale equation (2.5), only the test functions have been modified. Therefore, the LOD-approximation can be re-interpreted as an HMM-approximation with modified test functions and corrector problems. To be more explicit, $\mathbf{u}_{H_c,h_c} \in \mathbf{V}_{H_c,h_c}$ from Definition 3.1 can be characterized as $\mathbf{u}_{H_c,h_c} = (u_{H_c}, K_{h_c,1}(u_{H_c}), K_{h_c,2}(u_{H_c}))$, where $u_{H_c} \in V^1_{H_c}$ is the solution to a HMM with modified test functions and the corrections $K_{h_c,1}(u_{H_c})$ and $K_{h_c,2}(u_{H_c})$ are computed from u_{H_c} and its reconstructions as described in [34]; see also [20, 35] for similar reformulations in different settings. The HMM with modified test functions involves the following two steps:

1. Solve the cell problems for the reconstructions R_1 and R_2 around each quadrature point of the macroscopic triangulation $\mathcal{T}_{H_c}$ using test functions in $\overline{V}_{h_c,m}(Y^*)$ and $\overline{V}_{h_c,m}(D)$.

2. Assemble the macroscopic sesquilinear form B_H with the computed reconstructions and the test functions in $\overline{V}_{H_c,m}$.

Note that the reconstructions R_1 and R_2 as well as the fine-scale correctors $K_{h,1}$ and $K_{h,2}$ are different from those in [34] because of the modified test functions. This reformulation of the LOD-approximation as solution to a (modified) HMM decouples the computations on Y and G and no function triple spaces have to be considered. This is one great advantage of the present Petrov-Galerkin ansatz for the LOD in comparison to a Galerkin ansatz: We only need to compute reconstructions of standard Lagrange basis functions in $V^1_{H_c}$, but not of the basis functions in $\overline{V}_{H_c,m}$.

4. Error analysis

The error analysis is based on the observation that the localized subscale corrector problems (3.3) can be seen as perturbation of idealized subscale problems posed on the whole domain $G \times Y$. So let us introduce idealized counterparts of the correction operators $\mathbf{Q}_{T\times S,m}$ and $\mathbf{Q}_m$ where the patch $G_T \times Y_S$ equals $G \times Y$, roughly speaking "$m = \infty$". Define the null space

$$\mathbf{W}_{H,h} := W_H \times L^2(\Omega; W_h(Y^*)) \times L^2(\Omega; W_h(D)) := \{\mathbf{v}_{H,h} \in \mathbf{V}_{H,h} | \mathbf{I}_{H_c,h_c}(\mathbf{v}_{H,h}) = 0\}.$$

For any $\mathbf{v}_{H,h} \in \mathbf{V}_{H,h}$, the idealized element corrector problem seeks $\mathbf{Q}_{T\times S,\infty}\mathbf{v}_{H,h} \in \mathbf{W}_{H,h}$ such that

$$\mathcal{B}(\mathbf{w}, \mathbf{Q}_{T\times S,\infty}\mathbf{v}_{H,h}) = \mathcal{B}_{T\times S}(\mathbf{w}, \mathbf{v}_{H,h}) \qquad \forall \mathbf{w} \in \mathbf{W}_{H,h}, \tag{4.1}$$

and we define

$$\mathbf{Q}_\infty(\mathbf{v}_{H,h}) := \sum_{(T,S_1,S_2)\in\mathcal{T}_{H_c}\times\mathcal{T}_{h_c}(Y^*)\times\mathcal{T}_{h_c}(D)} \mathbf{Q}_{T\times S,\infty}(\mathbf{v}_{H,h}). \tag{4.2}$$

The following result implies the well-posedness of the idealized corrector problems.

Lemma 4.1. *Under the assumption*

$$k(C_{I_{H_c}} \sqrt{C_{ol,G}}\, H_c + C_{I^D_{h_c}} \sqrt{C_{ol,Y}}\, h_c) \le \sqrt{C_{\min}/2}, \tag{4.3}$$

we have for all $\mathbf{w}_{h,h} := (w_H, w_{h,1}, w_{h,2}) \in \mathbf{W}_{H,h}$ *the following equivalence of norms*

$$\|(w_H, w_{h,1}, w_{h,2})\|_{1,e} \le \|(w_H, w_{h,2}, w_{h,2})\|_e \le \sqrt{1 + C_{\min}/2}\, \|(w_H, w_{h,1}, w_{h,2})\|_{1,e},$$

and coercivity

$$C_{\min}/2\, \|(w_H, w_{h,1}, w_{h,2})\|^2_{1,e} \le \operatorname{Re} \mathcal{B}(\mathbf{w}_{H,h}, \mathbf{w}_{H,h}),$$

where the H^1*-semi norm* $\|\cdot\|_{1,e}$ *is defined in* (2.3).

Proof. The essential observation is that for any $(w_h, w_{h,1}, w_{h,2}) \in \mathbf{W}_{H,h}$ the property of the quasi-interpolation operators (3.1) implies that

$$\begin{aligned}
&k^2\|w_H + \chi_D w_{h,2}\|^2_{G\times Y}\\
&\quad\le k^2(\|w_H\|_G + \|w_{h,2}\|_{G\times D})^2\\
&\quad= k^2(\|w_H - I_{H_c}(w_H)\|_G + \|w_{h,2} - I^D_{h_c}(w_{h,2})\|_{G\times D})^2
\end{aligned}$$

$$\leq k^2(H_c\, C_{I_{H_c}}\, \sqrt{C_{ol,G}}\, \|\nabla w_H\|_G + h_c\, C_{I^D_{h_c}}\, \sqrt{C_{ol,Y}} \|\nabla_y w_{h,2}\|_{G\times D})^2.$$

This directly yields the equivalence of norms on $\mathbf{W}_{H,h}$ under the resolution condition (4.3). For the coercivity we observe that

$$\operatorname{Re} \mathcal{B}(\mathbf{w}_{H,h}, \mathbf{w}_{H,h}) \geq C_{\min}\|\mathbf{w}_{H,h}\|^2_{1,e} - k^2\|w_H + \chi_D w_{h,2}\|^2_{G\times Y}.$$

□

As the sesquilinear form $\mathcal{B}$ is also continuous (see Lemma 2.1), Lemma 4.1 implies that the idealized corrector problem (4.1) is well-posed and that the idealized correctors $\mathbf{Q}_\infty$ defined by (4.2) are continuous w.r.t. the two-scale energy norm

$$\|\mathbf{Q}_\infty(\mathbf{v}_{H,h})\|_e \leq C_{\mathbf{Q}}\, \|\mathbf{v}_{H,h}\|_e \qquad \text{for all} \quad \mathbf{v}_{H,h} \in \mathbf{V}_{H,h}.$$

Since the inclusion $\mathbf{W}_{H,h}(G_T \times Y_S) \subset \mathbf{W}_{H,h}$ holds, the well-posedness result carries over to the localized corrector problems (3.3) with the same constant.

The proof of the well-posedness of the two-scale LOD in Petrov-Galerkin formulation (3.4) relies on the fact that $(\mathbf{Q}_\infty - \mathbf{Q}_m)(\mathbf{v})$ decays exponentially with the distance from supp($\mathbf{v}$). The difference between idealized and localized correctors is quantified in the next theorem. The proof is given in Section 5 and is based on the observation that $\mathbf{Q}_\infty(\mathbf{v}|_{T\times S})$ decays exponentially with distance from $T \times S$.

Theorem 4.2. *Under the resolution condition* (4.3) *there exist constants* C_1, C_2, *and* $0 < \beta < 1$, *independent of* H_c, h_c, H, *and* h, *such that for any* $\mathbf{v}_{H_c,h_c} \in \mathbf{V}_{H_c,h_c}$, *any* $(T, S_1, S_2) \in \mathcal{T}_{H_c}{\times}\mathcal{T}_{h_c}(Y^*){\times}\mathcal{T}_{h_c}(D)$ *and any* $m \in \mathbb{N}$ *it holds*

$$\|(\mathbf{Q}_{T\times S,\infty} - \mathbf{Q}_{T\times S,m})(\mathbf{v}_{H_c,h_c})\|_{1,e} \leq C_1\beta^m\|\mathbf{v}_{H_c,h_c}\|_{1,e,T\times S}, \tag{4.4}$$

$$\|(\mathbf{Q}_\infty - \mathbf{Q}_m)(\mathbf{v}_{H_c,h_c})\|_{1,e} \leq C_2(\sqrt{C_{ol,m,G}} + \sqrt{C_{ol,m,Y}})\beta^m\|\mathbf{v}_{H_c,h_c}\|_{1,e}. \tag{4.5}$$

The stability of the LOD requires the coupling of the oversampling parameter m to the stability-/inf-sup-constant of the HMM. Therefore, we assume that H and h are small enough that (2.6) holds.

Theorem 4.3 (Well-posedness of the LOD). *Under the resolution conditions* (4.3) *and* (2.6) *and the following oversampling condition*

$$m \geq \frac{(q+1)\log(k)+\log(2C_2C_{\mathbf{I}}C_{\mathbf{I},e}C_{\mathrm{HMM}}C_{\mathcal{B}}\sqrt{1 + C_{\min}/2}(\sqrt{C_{\mathrm{ol},m,G}}+\sqrt{C_{\mathrm{ol},m,Y}}))}{|\log(\beta)|}, \tag{4.6}$$

the two-scale LOD (3.4) *is well-posed and with the constant* $C_{\mathrm{LOD}} := 2C_{\mathrm{HMM}}C^2_{\mathbf{I},e}(1 + C_{\mathbf{Q}})$ *it holds*

$$(C_{\mathrm{LOD}}\, k^{q+1})^{-1} \leq \inf_{\mathbf{v}_{H_c,h_c}\in\mathbf{V}_{H_c,h_c}} \sup_{\overline{\boldsymbol{\psi}}_{H_c,h_c}\in\overline{\mathbf{V}}_{H_c,h_c,m}} \frac{\operatorname{Re}\mathcal{B}(\mathbf{v}_{H_c,h_c}, \overline{\boldsymbol{\psi}}_{H_c,h_c})}{\|\mathbf{v}_{H_c,h_c}\|_e\|\overline{\boldsymbol{\psi}}_{H_c,h_c}\|_e}.$$

As $C_{\mathrm{ol},m,G}$ and $C_{\mathrm{ol},m,Y}$ grow at most polynomially with m because of the quasi-uniformity of $\mathcal{T}_{H_c}$ and $\mathcal{T}_{h_c}$, condition (4.6) is indeed satisfiable and the choice of the oversampling parameter m will be dominated by the logarithm of the wave number, i.e. $m \approx \log k$ is a good choice. This condition for the oversampling parameter is standard for the LOD of Helmholtz problems, see for instance [37, equation (5.12)]. In the numerical examples, moderate choices of m were sufficient, see Remark 3.

Proof. Let $\mathbf{v}_{H_c,h_c} \in \mathbf{V}_{H_c,h_c}$ be given. From (2.6) we infer that there is $\psi \in \mathbf{V}_{H,h}$ such that

$$\operatorname{Re}\mathcal{B}(\mathbf{v}_{H_c,h_c} - (\mathbf{Q}_\infty(\mathbf{v}^*_{H_c,h_c}))^*, \psi) \geq (C_{\mathrm{HMM}}^{-1} k^{-(q+1)}) \|\mathbf{v}_{H_c,h_c} - (\mathbf{Q}_\infty(\mathbf{v}^*_{H_c,h_c}))^*\|_e \|\psi\|_e.$$

It follows from the structure of the sesquilinear form $\mathcal{B}$ that $(\mathbf{Q}_\infty(\mathbf{v}^*_{H_c,h_c}))^*$ solves the following adjoint corrector problem

$$\mathcal{B}((\mathbf{Q}_\infty(\mathbf{v}^*_{H_c,h_c}))^*, \mathbf{w}) = \mathcal{B}(\mathbf{v}_{H_c,h_c}, \mathbf{w}) \qquad \forall \mathbf{w} \in \mathbf{W}_{H,h}.$$

Let $\overline{\psi}_{H_c,h_c} := (1 - \mathbf{Q}_m)\mathbf{I}_{H_c,h_c}\psi \in \overline{\mathbf{V}}_{H_c,h_c,m}$. It obviously holds that

$$\mathcal{B}(\mathbf{v}_{H_c,h_c}, \overline{\psi}_{H_c,h_c}) = \mathcal{B}(\mathbf{v}_{H_c,h_c}, (1 - \mathbf{Q}_\infty)\mathbf{I}_{H_c,h_c}\psi) + \mathcal{B}(\mathbf{v}_{H_c,h_c}, (\mathbf{Q}_\infty - \mathbf{Q}_m)\mathbf{I}_{H_c,h_c}\psi). \tag{4.7}$$

Since $\mathbf{Q}_\infty$ is a projection onto $\mathbf{W}_{H,h}$ and $(1 - \mathbf{I}_{H_c,h_c})\psi \in \mathbf{W}_{H,h}$, we have $(1 - \mathbf{Q}_\infty)(1 - \mathbf{I}_{H_c,h_c})\psi = 0$ and thus, $(1 - \mathbf{Q}_\infty)\mathbf{I}_{H_c,h_c}\psi = (1 - \mathbf{Q}_\infty)\psi$. The solution property of $(\mathbf{Q}_\infty(\mathbf{v}^*_{H_c,h_c}))^*$ and the definition of $\mathbf{Q}_\infty$ in (4.1)–(4.2) gives

$$\begin{aligned}\mathcal{B}((\mathbf{Q}_\infty(\mathbf{v}^*_{H_c,h_c}))^*, \psi) &= \mathcal{B}((\mathbf{Q}_\infty(\mathbf{v}^*_{H_c,h_c}))^*, \mathbf{Q}_\infty\psi) + \mathcal{B}((\mathbf{Q}_\infty(\mathbf{v}^*_{H_c,h_c}))^*, (1 - \mathbf{Q}_\infty)\psi)\\ &= \mathcal{B}(\mathbf{v}_{H_c,h_c}, \mathbf{Q}_\infty\psi).\end{aligned}$$

Hence, we obtain

$$\begin{aligned}\operatorname{Re}\mathcal{B}(\mathbf{v}_{H_c,h_c}, (1 - \mathbf{Q}_\infty)\mathbf{I}_{H_c,h_c}\psi) &= \operatorname{Re}\mathcal{B}(\mathbf{v}_{H_c,h_c} - (\mathbf{Q}_\infty(\mathbf{v}^*_{H_c,h_c}))^*, \psi)\\ &\geq (C_{\mathrm{HMM}}\, k^{q+1})^{-1} \|\mathbf{v}_{H_c,h_c} - (\mathbf{Q}_\infty(\mathbf{v}^*_{H_c,h_c}))^*\|_e \|\psi\|_e.\end{aligned}$$

Furthermore, the estimate (3.2) implies

$$\|\mathbf{v}_{H_c,h_c}\|_e = \|\mathbf{I}_{H_c,h_c}(\mathbf{v}_{H_c,h_c} - (\mathbf{Q}_\infty(\mathbf{v}^*_{H_c,h_c}))^*)\|_e \leq C_{\mathbf{I},e}\|\mathbf{v}_{H_c,h_c} - (\mathbf{Q}_\infty(\mathbf{v}^*_{H_c,h_c}))^*\|_e$$

and

$$\|\overline{\psi}_{H_c,h_c}\|_e \leq C_{\mathbf{I},e}(1 + C_{\mathbf{Q}}) \|\psi\|_e.$$

The second term on the right-hand side of (4.7) satisfies with Lemma 4.1 and Theorem 4.2 that

$$\begin{aligned}&|\mathcal{B}(\mathbf{v}_{H_c,h_c}, (\mathbf{Q}_\infty - \mathbf{Q}_m)\mathbf{I}_{H_c,h_c}\psi)|\\ &\quad\leq \sqrt{1 + C_{\min}/2}\; C_B \|(\mathbf{Q}_\infty - \mathbf{Q}_m)\mathbf{I}_{H_c,h_c}\psi\|_{1,e} \|\mathbf{v}_{H_c,h_c}\|_e\\ &\quad\leq \sqrt{1 + C_{\min}/2}\; C_B C_2(\sqrt{C_{\mathrm{ol},m,G}} + \sqrt{C_{\mathrm{ol},m,Y}})\beta^m C_{\mathbf{I}} \|\psi\|_e \|\mathbf{v}_{H_c,h_c}\|_e.\end{aligned}$$

Altogether, this yields

$$\begin{aligned}&\operatorname{Re}\mathcal{B}(\mathbf{v}_{H_c,h_c}, \overline{\psi}_{H_c,h_c})\\ &\quad\geq \Big(\frac{1}{C_{\mathbf{I},e}C_{\mathrm{HMM}}k^4} - \sqrt{1 + C_{\min}/2}\; C_B C_2(\sqrt{C_{ol,m,G}} + \sqrt{C_{ol,m,Y}})\beta^m C_{\mathbf{I}}\Big)\\ &\qquad \cdot \|\mathbf{v}_{H_c,h_c}\|_e \|\psi\|_e\\ &\quad\geq \Big(\frac{1}{C_{\mathbf{I},e}C_{\mathrm{HMM}}k^4} - \sqrt{1 + C_{\min}/2}\; C_B C_2(\sqrt{C_{ol,m,G}} + \sqrt{C_{ol,m,Y}})\beta^m C_{\mathbf{I}}\Big)\\ &\qquad \cdot \frac{1}{C_{\mathbf{I},e}(1 + C_{\mathbf{Q}})} \|\mathbf{v}_{H_c,h_c}\|_e \|\overline{\psi}_{H_c,h_c}\|_e.\end{aligned}$$

Hence, the condition (4.6) implies the assertion. □

Remark 2 (Adjoint problem). Under the assumption of Theorem 4.3, problem (3.4) is well-posed. Thus, it follows from a dimension argument that also the adjoint problem to (3.4) is well-posed with the same stability constant as in Theorem 4.3, cf. [16, Remark 1].

Theorem 4.4 (Quasi-optimality). *Under the resolution conditions* (4.3) *and* (2.6) *and the oversampling conditions* (4.6) *and*

$$m \geq ((q+1)\log(k) + \log(2\sqrt{1 + C_{\min}/2}\; C_B^2 C_2 C_{\mathrm{LOD}}))/|\log(\beta)|, \tag{4.8}$$

the LOD-approximation $\mathbf{u}_{H_c,h_c}$*, solution to* (3.4)*, and the solution* $\mathbf{u}_{H,h}$ *of the direct discretization* (2.5) *satisfy*

$$\|\mathbf{u}_{H,h} - \mathbf{u}_{H_c,h_c}\|_e \leq C \min_{\mathbf{v}_{H_c,h_c} \in \mathbf{V}_{H_c,h_c}} \|\mathbf{u}_{H,h} - \mathbf{v}_{H_c,h_c}\|_e$$

with a generic constant C depending only on $C_{\mathbf{I},e}$.

Proof. Let $\mathbf{e} := \mathbf{u}_{H,h} - \mathbf{u}_{H_c,h_c}$. We prove that $\|\mathbf{e}\|_e \leq 2\|(1 - \mathbf{I}_{H_c,h_c})\mathbf{u}_{H,h}\|_e$, which gives the assertion because $\mathbf{I}_{H_c,h_c}$ is a projection. By the triangle inequality and the fact that $\mathbf{I}_{H_c,h_c}$ is a projection onto $\mathbf{V}_{H_c,h_c}$, we obtain

$$\|\mathbf{e}\|_e \leq \|(1 - \mathbf{I}_{H_c,h_c})\mathbf{u}_{H,h}\|_e + \|\mathbf{I}_{H_c,h_c}\mathbf{e}\|_e,$$

so that it only remains to bound the second term on the right-hand side. The proof employs a standard duality argument, the stability of the idealized method and the fact that the actual two-scale LOD can be seen as a perturbation of the idealized method. Let $\mathbf{z}_{H_c,h_c} \in \mathbf{V}_{H_c,h_c}$ be the solution to the dual problem

$$\mathcal{B}(\boldsymbol{\psi}_{H_c,h_c}, (1 - \mathbf{Q}_\infty)\mathbf{z}_{H_c,h_c}) = (\boldsymbol{\psi}_{H_c,h_c}, \mathbf{I}_{H_c,h_c}\mathbf{e})_e \qquad \forall \boldsymbol{\psi}_{H_c,h_c} \in \mathbf{V}_{H_c,h_c},$$

where $(\cdot,\cdot)_e$ denotes the scalar product which induces the two-scale energy norm (2.2). This adjoint problem is uniquely solvable as explained in Remark 2. Choosing the test function $\boldsymbol{\psi}_{H_c,h_c} = \mathbf{I}_{H_c,h_c}\mathbf{e}$ implies

$$\begin{aligned} \|\mathbf{I}_{H_c,h_c}\mathbf{e}\|_e^2 &= \mathcal{B}(\mathbf{I}_{H_c,h_c}\mathbf{e}, (1 - \mathbf{Q}_\infty)\mathbf{z}_{H_c,h_c}) \\ &= \mathcal{B}(\mathbf{I}_{H_c,h_c}\mathbf{e}, (\mathbf{Q}_m - \mathbf{Q}_\infty)\mathbf{z}_{H_c,h_c}) + \mathcal{B}(\mathbf{I}_{H_c,h_c}\mathbf{e}, (1 - \mathbf{Q}_m)\mathbf{z}_{H_c,h_c}). \end{aligned} \tag{4.9}$$

Since $(1 - \mathbf{Q}_m)\mathbf{z}_{H_c,h_c} \in \overline{\mathbf{V}}_{H_c,h_c,m}$ by definition, we have the Galerkin orthogonality

$$\mathcal{B}(\mathbf{u}_{H,h} - \mathbf{u}_{H_c,h_c}, (1 - \mathbf{Q}_m)\mathbf{z}_{H_c,h_c}) = 0.$$

Using this orthogonality and the fact that $\mathbf{I}_{H_c,h_c}\mathbf{u}_{H,h} - \mathbf{u}_{H,h} \in \mathbf{W}_{H,h}$ together with the definition of $\mathbf{Q}_\infty$ (4.1)–(4.2) implies for the second term

$$\begin{aligned} \mathcal{B}(\mathbf{I}_{H_c,h_c}\mathbf{e}, (1 - \mathbf{Q}_m)\mathbf{z}_{H_c,h_c}) &= \mathcal{B}(\mathbf{I}_{H_c,h_c}\mathbf{u}_{H,h} - \mathbf{u}_{H,h}, (1 - \mathbf{Q}_m)\mathbf{z}_{H_c,h_c}) \\ &= \mathcal{B}(\mathbf{I}_{H_c,h_c}\mathbf{u}_{H,h} - \mathbf{u}_{H,h}, (\mathbf{Q}_\infty - \mathbf{Q}_m)\mathbf{z}_{H_c,h_c}). \end{aligned}$$

Now the first and the (modified) second term of (4.9) are similar and can be treated with the same procedure. First, we note that $(\mathbf{Q}_\infty - \mathbf{Q}_m)\mathbf{z}_{H_c,h_c} \in \mathbf{W}_{H,h}$. Applying Lemma 4.1 and then the decay estimate (4.5) from Theorem 4.2, we obtain (for the second term)

$$\begin{aligned} &|\mathcal{B}(\mathbf{I}_{H_c,h_c}\mathbf{u}_{H,h} - \mathbf{u}_{H,h}, (\mathbf{Q}_\infty - \mathbf{Q}_m)\mathbf{z}_{H_c,h_c})| \\ &\quad \leq \sqrt{1 + C_{\min}/2}\; C_B \|(1 - \mathbf{I}_{H_c,h_c})\mathbf{u}_{H,h}\|_e \, \|(\mathbf{Q}_\infty - \mathbf{Q}_m)\mathbf{z}_{H_c,h_c}\|_{1,e} \\ &\quad \leq \sqrt{1 + C_{\min}/2}\; C_B C_2 (\sqrt{C_{ol,m,G}} + \sqrt{C_{ol,m,Y}})\beta^m \, \|(1 - \mathbf{I}_{H_c,h_c})\mathbf{u}_{H,h}\|_e \, \|\mathbf{z}_{H_c,h_c}\|_{1,e}. \end{aligned}$$

The stability of the adjoint problem from Remark 2 implies

$$\|\mathbf{z}_{H_c,h_c}\|_{1,e} \le C_{\mathrm{LOD}} k^{q+1} C_B \|\mathbf{I}_{H_c,h_c}\mathbf{e}\|_e.$$

Thus, we obtain for (4.9) after division by $\|\mathbf{I}_{H_c,h_c}\mathbf{e}\|_e$ that

$$\begin{aligned}\|\mathbf{I}_{H_c,h_c}\mathbf{e}\|_e \le \sqrt{1+C_{\min}/2}\; C_B^2 C_2(\sqrt{C_{ol,m,G}} + \sqrt{C_{ol,m,Y}})\beta^m C_{\mathrm{LOD}} k^{q+1} \\ \cdot (\|(1-\mathbf{I}_{H_c,h_c})\mathbf{u}_{H,h}\|_e + \|\mathbf{I}_{H_c,h_c}\mathbf{e}\|_e).\end{aligned}$$

The oversampling condition (4.8) implies that the constants can be bounded by $1/2$ and hence, the term $\|\mathbf{I}_{H_c,h_c}\mathbf{e}\|_e$ can be absorbed on the left-hand side. □

Corollary 1 (Full error). *Let* $\mathbf{u} := (u, u_1, u_2) \in \mathcal{H}$ *be the two-scale solution to* (2.1). *Under the assumptions of Theorem 4.4, the two-scale LOD-approximation* $\mathbf{u}_{H_c,h_c}$*, solution to* (3.4)*, satisfies with some generic constant* C

$$\|\mathbf{u}-\mathbf{u}_{H_c,h_c}\|_e \le \|\mathbf{u}-\mathbf{u}_{H,h}\|_e + C \min_{\mathbf{v}_{H_c,h_c}\in \mathbf{V}_{H_c,h_c}} \|\mathbf{u}_{H,h}-\mathbf{u}_{H_c,h_c}\|_e.$$

The full error is dominated by the best approximation error of $\mathbf{V}_{H_c,h_c}$, which can be quantified using standard interpolation operators and regularity results. The error of the HMM-approximation $\mathbf{u}-\mathbf{u}_{H,h}$ is estimated in detail in [34, Theorem 5.1, Corollary 5.2]: Assuming sufficient regularity of $\mathbf{u}$, this error converges like $k^{q+1}(H+h)$ if the resolution condition $k^{q+2}(H+h) \lesssim 1$ is fulfilled. As we consider the HMM-approximation as an (overkill) reference solution only needed for the error estimates and *not in practical computations*, H and h are much smaller than the actual mesh sizes H_c and h_c, so that the error $\mathbf{u}-\mathbf{u}_{H,h}$ indeed becomes negligible.

Remark 3 (Choice of LOD parameters in practice). Although this is a purely theoretical paper, we want to give a few comments on the practical choice of H_c, h_c and m. Our numerical experiment in [34] reveals a resolution condition of $k^2(H+h) \lesssim 1$, so that for a wavenumber $k = 2^6$, grids with mesh width of about $H = h = 2^{-9}$ up to $H = h = 2^{-12}$ are required for a reliable HMM-approximation. Numerical experiments by other authors in [8, 16, 37] for the standard Helmholtz problem showed that the LOD only needs meshes of about $H_c = h_c = 2^{-5}$ to give a good approximation for a wavenumber $k = 2^6$. Furthermore, an oversampling parameter of $m = 2$ already was sufficient for various wavenumbers and no problems with the well-posedness of the discrete scheme have been reported. This is a significant reduction of the mesh size, which might outweigh the overhead of the additional corrector computations, in particular when taking the decoupling of Section 3.4 into account.

5. Proof of the decay property for the correctors

In this section, we give a proof of the exponential decay result of Theorem 4.2, which is central for this method. The idea of the proof is the same as in the previous proofs for the Helmholtz equation [8, 16, 37] or in the context of diffusion problems [19, 30]. As in the previous sections, we have to take into account the two-scale nature of the problem and the spaces.

Let $\mathcal{I}_{H,h} := (\mathcal{I}_H, \mathcal{I}_h^{Y^*}, \mathcal{I}_h^D)$ with $\mathcal{I}_H : C^0(G) \to V_H^1$, $\mathcal{I}_h^{Y^*} : L^2(\Omega; C^0(Y^*)) \to L^2(\Omega; \widetilde{V}_h(Y^*))$, and $\mathcal{I}_h^D : L^2(\Omega; C^0(D)) \to L^2(\Omega; V_h^1(D))$ denote the nodal Lagrange interpolation operators, where $\mathcal{I}_h^{Y^*}$ and

$\mathcal{I}_H^D$ only act on the second variable. We note that periodicity is preserved when identifying degrees of freedom on the periodic boundary. Recall that the nodal Lagrange interpolation operator $\mathcal{I}$ is L^2- and H^1-stable on piece-wise polynomials on shape regular meshes due to inverse inequalities. Hence, for any $(T, S_1, S_2) \in \mathcal{T}_{H_c} \times \mathcal{T}_{h_c}(Y^*) \times \mathcal{T}_{h_c}(D)$ and all $\mathbf{q} \in \mathbb{P}_2(T) \times L^2(T; \mathbb{P}_2(S_1)) \times L^2(T; \mathbb{P}_2(S_2))$ we have the stability estimate

$$\|\mathcal{I}_{H,h}\mathbf{q}\|_{1,e,T\times S} \leq C_{\mathcal{I}} \|\mathbf{q}\|_{1,e,T\times S}. \tag{5.1}$$

In this section, we do not explicitly give the constants in the estimates. Instead we use a generic constant C, which is independent of the mesh sizes and the oversampling parameter, but may depend on the (quasi)-interpolation operators' norms, the overlap constants $C_{\mathrm{ol},G}$ and $C_{\mathrm{ol},Y}$ (not on $C_{\mathrm{ol},m,G}$ and $C_{\mathrm{ol},m,Y}$!), the constant for the cut-off functions (see below), and $C_{\min}$.

In the proofs, we will frequently use cut-off functions. We collect some basic properties in the following lemma, cf. also [16, Appendix A, Lemma 2].

Lemma 5.1. *Let $\boldsymbol{\eta} := (\eta_0, \eta_1, \eta_2) \in \mathbf{V}_{H_c,h_c}$ be a function triple with η_i, $i = 1, 2, 3$, having values in the interval $[0, 1]$ and satisfying the bounds*

$$\|\nabla \eta_0\|_{L^\infty(G)} \leq C_\eta H_c^{-1} \qquad \text{and} \qquad \|\nabla_y \eta_i\|_{L^2(\Omega, L^\infty(Y))} \leq C_\eta h_c^{-1} \quad i = 1, 2. \tag{5.2}$$

By writing $\nabla\boldsymbol{\eta}$, we mean the triple $(\nabla\eta_0, \nabla_y\eta_1, \nabla_y\eta_2)$. Let $\mathbf{w} := (w, w_1, w_2) \in \mathbf{W}_{H,h}$ be arbitrary and define $\boldsymbol{\eta}\mathbf{w} := (\eta_0 w, \eta_1 w_1, \eta_2 w_2)$. Given any subset $(\mathcal{K}_0, \mathcal{K}_1, \mathcal{K}_2) \subset \mathcal{T}_{H_c} \times \mathcal{T}_{h_c}(Y^*) \times \mathcal{T}_{h_c}(D)$, $\mathbf{w}$ fulfills for $\mathcal{S}_i = \cup\mathcal{K}_i$ and $\mathcal{S} := \mathcal{S}_0 \times (\mathcal{S}_1 \cup \mathcal{S}_2)$ that

$$\|\mathbf{w}\|_{\mathcal{S}} \leq C(H_c\|\nabla w\|_{\mathrm{N}(\mathcal{S}_0)} + \sum_{i=1}^{2} h_c\|\nabla_y w_i\|_{\mathcal{S}_0\times\mathrm{N}(\mathcal{S}_i)}), \tag{5.3}$$

$$\|\boldsymbol{\eta}\mathbf{w}\|_{1,e,\mathcal{S}} \leq C(\|\mathbf{w}\|_{1,e,\mathcal{S}\cap\mathrm{supp}(\boldsymbol{\eta})} + \|\mathbf{w}\|_{1,e,\mathrm{N}(\mathcal{S}\cap\mathrm{supp}(\nabla\boldsymbol{\eta}))}), \tag{5.4}$$

$$\|(1 - \mathbf{I}_{H_c,h_c})\mathcal{I}_{H,h}(\boldsymbol{\eta}\mathbf{w})\|_{\mathcal{S}} \leq C(H_c + h_c)\|\boldsymbol{\eta}\mathbf{w}\|_{1,e,\mathrm{N}(\mathcal{S})}. \tag{5.5}$$

Proof. The properties (3.1) directly imply (5.3). For the proof of (5.4) the product rule and (5.2) yield

$$\begin{aligned}\|\boldsymbol{\eta}\mathbf{w}\|_{1,e,\mathcal{S}} \leq\ & \|\mathbf{w}\|_{1,e,\mathcal{S}\cap\mathrm{supp}(\boldsymbol{\eta})} + C_\eta H_c^{-1}\|w\|_{\mathcal{S}_0\cap\mathrm{supp}(\nabla\eta_0)} \\ & + \sum_{i=1}^{2} C_\eta h_c^{-1}\|w_i\|_{\mathcal{S}_0\times(\mathcal{S}_i\cap\mathrm{supp}(\nabla\eta_i))}.\end{aligned}$$

The combination with (5.3) gives the assertion. For a proof of (5.5), apply (3.1). The estimate then follows from the H^1-stability of $\mathcal{I}_{H,h}$ (5.1) on the piece-wise polynomial function $\boldsymbol{\eta}\mathbf{w}$. □

Proposition 1. *Under the resolution condition (4.3), there exists $0 < \beta < 1$ such that, for any $\mathbf{v}_{H_c,h_c} \in \mathbf{V}_{H_c,h_c}$ and all $(T, S_1, S_2) \in \mathcal{T}_{H_c} \times \mathcal{T}_{h_c}(Y^*) \times \mathcal{T}_{h_c}(D)$ and $m \in \mathbb{N}$*

$$\|\mathbf{Q}_{T\times S,\infty}\mathbf{v}_{H_c,h_c}\|_{1,e,(G\times Y)\setminus\mathrm{N}^m(T\times S)} \leq C\beta^m\|\mathbf{v}_{H_c,h_c}\|_{1,e,T\times S}.$$

Proof. For $m \geq 5$, we define the cut-off functions $\eta_0 \in V^1_{H_c}$, $\eta_1 \in L^2(\Omega; \widetilde{V}^1_{h_c}(Y^*))$, $\eta_2 \in L^2(\Omega; V^1_{h_c}(D))$ via

$$\eta_0 = 0 \quad \text{in } \mathrm{N}^{m-3}(T) \qquad \text{and} \qquad \eta_0 = 1 \quad \text{in } G \setminus \mathrm{N}^{m-2}(T),$$

$$\eta_1 = 0 \quad \text{in } \mathrm{N}^{m-3}(T \times S_1) \qquad \text{and} \qquad \eta_1 = 1 \quad \text{in } (\Omega \times Y^*) \setminus \mathrm{N}^{m-2}(T \times S_1),$$
$$\eta_2 = 0 \quad \text{in } \mathrm{N}^{m-3}(T \times S_2) \qquad \text{and} \qquad \eta_2 = 1 \quad \text{in } (\Omega \times D) \setminus \mathrm{N}^{m-2}(T \times S_2),$$

where η_1 and η_2 w.l.o.g. are chosen piece-wise x-constant. The shape regularity implies that η_i satisfies (5.2). Denote $\boldsymbol{\eta} := (\eta_0, \eta_1, \eta_2)$ and $\mathcal{R} := \operatorname{supp}(\nabla \boldsymbol{\eta})$. Let $\mathbf{v}_{H_c,h_c} \in \mathbf{V}_{H_c,h_c}$ and denote $\boldsymbol{\phi} := \mathbf{Q}_{T\times S,\infty}\mathbf{v}_{H_c,h_c} \in \mathbf{W}_{H,h}$. Elementary estimates yield

$$\begin{aligned}
&\|\boldsymbol{\phi}\|^2_{1,e,(G\times Y)\setminus \mathrm{N}^m(T\times S)} \\
&\quad = |\operatorname{Re}(\boldsymbol{\phi}, \boldsymbol{\phi})_{1,e,(G\times Y)\setminus \mathrm{N}^m(T\times S)}| \\
&\quad \le |\operatorname{Re}((\nabla\phi + \nabla_y\phi_1, \eta_0\nabla\phi + \eta_1\nabla_y\phi_1)_{G\times Y^*} + (\nabla_y\phi_2, \eta_2\nabla_y\phi_2)_{G\times D})| \\
&\quad \le |\operatorname{Re}(\boldsymbol{\phi}, \boldsymbol{\eta}\boldsymbol{\phi})_{1,e}| + |\operatorname{Re}((\nabla\phi+\nabla_y\phi_1, \phi\nabla\eta_0 + \phi_1\nabla_y\eta_1)_{G\times Y^*} + (\nabla_y\phi_2, \phi_2\nabla_y\eta_2)_{G\times D})| \\
&\quad \le M_1 + M_2 + M_3 + M_4
\end{aligned}$$

for

$$\begin{aligned}
M_1 &:= |\operatorname{Re}(\boldsymbol{\phi}, (1 - \mathcal{I}_{H,h})(\boldsymbol{\eta}\boldsymbol{\phi}))_{1,e}|, \\
M_2 &:= |\operatorname{Re}(\boldsymbol{\phi}, (1 - \mathbf{I}_{H_c,h_c})\mathcal{I}_{H,h}(\boldsymbol{\eta}\boldsymbol{\phi}))_{1,e}|, \\
M_3 &:= |\operatorname{Re}(\boldsymbol{\phi}, \mathbf{I}_{H_c,h_c}\mathcal{I}_{H,h}(\boldsymbol{\eta}\boldsymbol{\phi}))_{1,e}|, \\
M_4 &:= |\operatorname{Re}((\nabla\phi + \nabla_y\phi_1, \phi\nabla\eta_0 + \phi_1\nabla_y\eta_1)_{G\times Y^*} + (\nabla_y\phi_2, \phi_2\nabla_y\eta_2)_{G\times D})|.
\end{aligned}$$

With the stability of $\mathcal{I}_{H,h}$ on polynomials (5.1) and estimate (5.4) we obtain

$$M_1 \le C\|\boldsymbol{\phi}\|_{1,e,\mathcal{R}}\, \|\boldsymbol{\eta}\boldsymbol{\phi} - \mathcal{I}_{H,h}(\boldsymbol{\eta}\boldsymbol{\phi})\|_{1,e,\mathcal{R}} \le C\|\boldsymbol{\phi}\|_{1,e,\mathcal{R}}\, \|\boldsymbol{\phi}\|_{1,e,\mathrm{N}(\mathcal{R})}.$$

Since $\mathbf{w} := (1 - \mathbf{I}_{H_c,h_c})\mathcal{I}_{H,h}(\boldsymbol{\eta}\boldsymbol{\phi}) \in \mathbf{W}_{H,h}$, the idealized corrector problem (4.1) and the fact that $\mathbf{w}$ has support only outside $T \times S$ imply $\mathcal{B}(\mathbf{w}, \boldsymbol{\phi}) = \mathcal{B}_{T\times S}(\mathbf{w}, \mathbf{v}_{H_c,h_c}) = 0$. Therefore, we obtain

$$\begin{aligned}
M_2 := |\operatorname{Re}(\boldsymbol{\phi}, \mathbf{w})_{1,e}| &\le C_{\min}^{-1}\, |\operatorname{Re}(\mathcal{B}(\mathbf{w}, \boldsymbol{\phi}) + k^2(w + \chi_D w_2, \phi + \chi_D\phi_2)_{G\times Y})| \\
&= C_{\min}^{-1} |\operatorname{Re} k^2(w + \chi_D w_2, \phi + \chi_D\phi_2)_{G\times Y}|.
\end{aligned}$$

Hence, estimates (5.5) and (5.4) give with the resolution condition (4.3)

$$\begin{aligned}
M_2 &\le C_{\min}^{-1}k^2(H_c^2 C_{I_{H_c}}^2 C_{\mathrm{ol},G} + h_c^2 C_{I_{h_c}^D}^2 C_{\mathrm{ol},Y})\|\boldsymbol{\phi}\|^2_{1,e,(G\times Y)\setminus \mathrm{N}^m(T\times S)} \\
&\quad + C_{\min}^{-1}k^2(H_c^2 C_{I_{H_c}}^2 C_{\mathrm{ol},G} + h_c^2 C_{I_{h_c}^D}^2 C_{\mathrm{ol},Y})C_{\mathbf{I}}C_{\mathcal{I}}C_\eta\|\boldsymbol{\phi}\|^2_{1,e,\mathrm{N}^2(\mathcal{R})} \\
&\le \frac{1}{2}\|\boldsymbol{\phi}\|^2_{1,e,(G\times Y)\setminus \mathrm{N}^m(T\times S)} + C\|\boldsymbol{\phi}\|^2_{1,e,\mathrm{N}^2(\mathcal{R})},
\end{aligned}$$

so that the first term can be absorbed. Because of $\operatorname{supp}(\mathbf{I}_{H_c,h_c}\mathcal{I}_{H,h}(\boldsymbol{\eta}\boldsymbol{\phi})) \subset \mathrm{N}(\mathcal{R})$, the properties (3.1) of $\mathbf{I}_{H_c,h_c}$, (5.1) of $\mathcal{I}_{H,h}$, and estimate (5.4) lead to

$$M_3 \le \|\boldsymbol{\phi}\|_{1,e,\mathrm{N}(\mathcal{R})}\, \|\mathbf{I}_{H_c,h_c}\mathcal{I}_{H,h}(\boldsymbol{\eta}\boldsymbol{\phi})\|_{1,e,\mathrm{N}(\mathcal{R})} \le C\|\boldsymbol{\phi}\|^2_{1,e,\mathrm{N}^2(\mathcal{R})}.$$

For the last term, the Lipschitz bound (5.2) on the cut-off functions and estimate (5.3) show

$$M_4 \le C_\eta(H_c^{-1}\|\phi\|_{\operatorname{supp}(\nabla\eta_0)} + \sum_{i=1}^{2} h_c^{-1}\|\phi_i\|_{\operatorname{supp}(\nabla\eta_i)})\|\boldsymbol{\phi}\|_{1,e,\mathcal{R}} \le C\|\boldsymbol{\phi}\|^2_{1,e,\mathrm{N}(\mathcal{R})}.$$

All in all, it follows for some $\tilde{C} > 0$ that

$$\frac{1}{2}\|\boldsymbol{\phi}\|^2_{1,e,(G\times Y)\setminus \mathrm{N}^m(T\times S)} \leq \tilde{C}\|\boldsymbol{\phi}\|^2_{1,e,\mathrm{N}^2(\mathcal{R})},$$

where we recall that $\mathrm{N}^2(\mathcal{R}) = \mathrm{N}^m(T\times S)\setminus \mathrm{N}^{m-5}(T\times S)$. Because of

$$\|\boldsymbol{\phi}\|^2_{1,e,(G\times Y)\setminus \mathrm{N}^m(T\times S)} + \|\boldsymbol{\phi}\|^2_{1,e,\mathrm{N}^m(T\times S)\setminus \mathrm{N}^{m-5}(T\times S)} = \|\boldsymbol{\phi}\|^2_{1,e,(G\times Y)\setminus \mathrm{N}^{m-5}(T\times S)},$$

we obtain

$$(1 + (2\tilde{C})^{-1})\|\boldsymbol{\phi}\|^2_{1,e,(G\times Y)\setminus \mathrm{N}^m(T\times S)} \leq \|\boldsymbol{\phi}\|^2_{1,e,(G\times Y)\setminus \mathrm{N}^{m-5}(T\times S)}.$$

Repeated application of this argument gives for $\tilde{\beta} := 2\tilde{C}/(2\tilde{C}+1) < 1$ together with the stability of $\mathbf{Q}_{T\times S,\infty}$ that

$$\|\boldsymbol{\phi}\|^2_{1,e,(G\times Y)\setminus \mathrm{N}^m(T\times S)} \leq \tilde{\beta}^{\lfloor m/5\rfloor}\|\boldsymbol{\phi}\|^2_{1,e} \leq C_Q\tilde{\beta}^{\lfloor m/5\rfloor}\|\mathbf{v}_{H_c,h_c}\|^2_{1,e,T\times S},$$

which gives the assertion after some algebraic manipulations. □

As the localized correctors $\mathbf{Q}_m$ are the Galerkin approximations of the idealized correctors $\mathbf{Q}_\infty$, the decay property carries over to $\mathbf{Q}_m$. This is the main observation for the proof of Theorem 4.2.

Proof of Theorem 4.2. For $m \geq 3$, we define the cut-off functions $\eta_0 \in V^1_{H_c}$, $\eta_1 \in L^2(\Omega; \widetilde{V}^1_{h_c}(Y^*))$ and $\eta_2 \in L^2(\Omega; V^1_{h_c}(D))$ via

$$\begin{aligned}
\eta_0 &= 0 \quad \text{in } G\setminus \mathrm{N}^{m-1}(T) && \text{and} && \eta_0 = 1 \quad \text{in } \mathrm{N}^{m-2}(T),\\
\eta_1 &= 0 \quad \text{in } (\Omega\times Y^*)\setminus \mathrm{N}^{m-1}(T\times S_1) && \text{and} && \eta_1 = 1 \quad \text{in } \mathrm{N}^{m-2}(T\times S_1),\\
\eta_2 &= 0 \quad \text{in } (\Omega\times D)\setminus \mathrm{N}^{m-1}(T\times S_2) && \text{and} && \eta_2 = 1 \quad \text{in } \mathrm{N}^{m-2}(T\times S_2),
\end{aligned}$$

where again η_1 and η_2 are w.l.o.g. piece-wise x-constant. The cut-off functions η_i satisfy the bounds (5.2). Set again $\boldsymbol{\eta} := (\eta_0, \eta_1, \eta_2)$. As already discussed, $\mathbf{Q}_{T\times S,m}$ can be interpreted as Galerkin approximation of $\mathbf{Q}_{T\times S,\infty}$ in the discrete subspace $\mathbf{W}_{H,h}(G_T\times Y_S) \subset \mathbf{W}_{H,h}$. Hence, Céa's Lemma gives for any $\mathbf{w}_{H,h} \in \mathbf{W}_{H,h}(G_T\times Y_S)$

$$\|(\mathbf{Q}_{T\times S,\infty} - \mathbf{Q}_{T\times S,m})\mathbf{v}\|^2_{1,e} \leq C\|\mathbf{Q}_{T\times S,\infty}\mathbf{v} - \mathbf{w}_{H,h}\|^2_e.$$

We choose $\mathbf{w}_{H,h} := (1 - \mathbf{I}_{H_c,h_c})\mathcal{I}_{H,h}(\boldsymbol{\eta}\mathbf{Q}_{T\times S,\infty}\mathbf{v}) \in \mathbf{W}_{H,h}(G_T\times Y_S)$ and obtain with the identity $\mathbf{I}_{H_c,h_c}\mathbf{Q}_{T\times S,\infty}\mathbf{v} = 0$, the estimate (5.5), the approximation and stability estimates (3.1) and (5.1), the resolution condition (4.3) and estimate (5.4) that

$$\begin{aligned}
&\|(\mathbf{Q}_{T\times S,\infty} - \mathbf{Q}_{T\times S,m})\mathbf{v}\|^2_{1,e}\\
&\leq C\|\mathbf{Q}_{T\times S,\infty}\mathbf{v} - (1 - \mathbf{I}_{H_c,h_c})\mathcal{I}_{H,h}(\boldsymbol{\eta}\mathbf{Q}_{T\times S,\infty}\mathbf{v})\|^2_e\\
&= C\|(1 - \mathbf{I}_{H_c,h_c})\mathcal{I}_{H,h}(\mathbf{Q}_{T\times S,\infty}\mathbf{v} - \boldsymbol{\eta}\mathbf{Q}_{T\times S,\infty}\mathbf{v})\|^2_{e,(G\times Y)\setminus\{\boldsymbol{\eta}=1\}}\\
&\leq C\|(1 - \boldsymbol{\eta})\mathbf{Q}_{T\times S,\infty}\mathbf{v}\|^2_{1,e,\mathrm{N}((G\times Y)\setminus\{\boldsymbol{\eta}=1\})}\\
&\leq C\|\mathbf{Q}_{T\times S,\infty}\mathbf{v}\|^2_{1,e,\mathrm{N}((G\times Y)\setminus\{\boldsymbol{\eta}=1\})}.
\end{aligned}$$

Note that $\mathrm{N}((G\times Y)\setminus\{\boldsymbol{\eta}=1\}) = (G\times Y)\setminus \mathrm{N}^{m-3}(T\times S)$. Together with Proposition 1, this proves (4.4).

Define $\mathbf{z} := (\mathbf{Q}_\infty - \mathbf{Q}_m)\mathbf{v}$ and $\mathbf{z}_{T\times S} := (\mathbf{Q}_{T\times S,\infty} - \mathbf{Q}_{T\times S,m})\mathbf{v}$. The ellipticity from Lemma 4.1 yields

$$\|\mathbf{z}\|_{1,e}^2 \leq C\Big|\sum_{(T,S_1,S_2)\in\mathcal{T}_{H_c}\times\mathcal{T}_{h_c}(Y^*)\times\mathcal{T}_{h_c}(D)} \mathcal{B}(\mathbf{z}, \mathbf{z}_{T\times S})\Big|.$$

We define the cut-off functions $\eta_0 \in V^1_{H_c}$, $\eta_1 \in L^2(\Omega; \widetilde{V}^1_{h_c}(Y^*))$, and $\eta_2 \in L^2(\Omega; V^1_{h_c}(D))$ via

$$\begin{aligned}
\eta_0 &= 1 \quad \text{in } G \setminus \mathrm{N}^{m+2}(T) && \text{and} && \eta_0 = 0 \quad \text{in } \mathrm{N}^{m+1}(T),\\
\eta_1 &= 1 \quad \text{in } (\Omega\times Y^*) \setminus \mathrm{N}^{m+2}(T\times S_1) && \text{and} && \eta_1 = 0 \quad \text{in } \mathrm{N}^{m+1}(T\times S_1),\\
\eta_2 &= 1 \quad \text{in } (\Omega\times D) \setminus \mathrm{N}^{m+2}(T\times S_2) && \text{and} && \eta_2 = 0 \quad \text{in } \mathrm{N}^{m+1}(T\times S_2),
\end{aligned}$$

where again η_1 and η_2 are w.l.o.g. piece-wise x-constant. The cut-off functions fulfill (5.2) and we set $\boldsymbol{\eta} := (\eta_0, \eta_1, \eta_2)$. For any $(T, S_1, S_2) \in \mathcal{T}_{H_c} \times \mathcal{T}_{h_c}(Y^*) \times \mathcal{T}_{h_c}(D)$ we have $(1 - \mathbf{I}_{H_c,h_c})\mathcal{I}_{H,h}(\boldsymbol{\eta}\mathbf{z}) \in \mathbf{W}_{H,h}$ with support outside $G_T \times Y_S$. Hence, we deduce with $\mathbf{z} = \mathcal{I}_{H,h}\mathbf{z}$ that

$$\mathcal{B}(\mathbf{z}, \mathbf{z}_{T\times S}) = \mathcal{B}(\mathcal{I}_{H,h}(\mathbf{z} - \boldsymbol{\eta}\mathbf{z}), \mathbf{z}_{T\times S}) + \mathcal{B}(\mathbf{I}_{H_c,h_c}\mathcal{I}_{H,h}(\boldsymbol{\eta}\mathbf{z}), \mathbf{z}_{T\times S}).$$

The function $\mathbf{z} - \mathcal{I}_{H,h}(\boldsymbol{\eta}\mathbf{z})$ vanishes on $\{\boldsymbol{\eta} = 1\}$. Thus, the first term on the right-hand satisfies

$$|\mathcal{B}(\mathcal{I}_{H,h}(\mathbf{z} - \boldsymbol{\eta}\mathbf{z}), \mathbf{z}_{T\times S})| \leq C_B \|\mathcal{I}_{H,h}(\mathbf{z} - \boldsymbol{\eta}\mathbf{z})\|_{e,\mathrm{supp}(1-\boldsymbol{\eta})} \|\mathbf{z}_{T\times S}\|_e.$$

The L^2- and H^1-stability of $\mathcal{I}_{H,h}$ on piece-wise polynomials gives together with the estimate (5.4) applied to the cut-off function $1 - \boldsymbol{\eta}$

$$\|\mathcal{I}_{H,h}(\mathbf{z} - \boldsymbol{\eta}\mathbf{z})\|_{e,\mathrm{supp}(1-\boldsymbol{\eta})} \leq C\|\mathbf{z}\|_{e,\mathrm{N}(\mathrm{supp}(1-\boldsymbol{\eta}))}.$$

Furthermore, $\mathbf{I}_{H_c,h_c}\mathcal{I}_{H,h}(\boldsymbol{\eta}\mathbf{z})$ vanishes on $(G\times Y)\setminus \mathrm{N}(\mathrm{supp}(1-\boldsymbol{\eta}))$. Therefore, we infer from the stability (3.2) of $\mathbf{I}_{H_c,h_c}$, the stability of $\mathcal{I}_{H,h}$ as before, and estimate (5.4) that

$$|\mathcal{B}(\mathbf{I}_{H_c,h_c}\mathcal{I}_{H,h}(\boldsymbol{\eta}\mathbf{z}), \mathbf{z}_{T\times S})| \leq C\|\mathbf{z}\|_{e,\mathrm{N}^2(\mathrm{supp}(1-\boldsymbol{\eta}))} \|\mathbf{z}_{T\times S}\|_e.$$

Because of the resolution condition (4.3) and Lemma 4.1, it holds $\|\mathbf{z}\|_e \leq C\|\mathbf{z}\|_{1,e}$. Summing up over $(T, S_1, S_2) \in \mathcal{T}_{H_c} \times \mathcal{T}_{h_c}(Y^*) \times \mathcal{T}_{h_c}(D)$ yields with the Cauchy-Schwarz inequality and the finite overlap of patches that

$$\begin{aligned}
\|\mathbf{z}\|_{1,e}^2 &\leq C \sum_{(T,S_1,S_2)\in\mathcal{T}_{H_c}\times\mathcal{T}_{h_c}(Y^*)\times\mathcal{T}_{h_c}(D)} \|\mathbf{z}\|_{1,e,\mathrm{N}^2(\mathrm{supp}(1-\boldsymbol{\eta}))} \|\mathbf{z}_{T\times S}\|_e\\
&\leq C(\sqrt{C_{\mathrm{ol},m,G}} + \sqrt{C_{\mathrm{ol},m,Y}})\|\mathbf{z}\|_{1,e} \sqrt{\sum_{(T,S_1,S_2)\in\mathcal{T}_{H_c}\times\mathcal{T}_{h_c}(Y^*)\times\mathcal{T}_{h_c}(D)} \|\mathbf{z}_{T\times S}\|_e^2}.
\end{aligned}$$

Combining the last estimate with (4.4) concludes the proof. □

Conclusion

In this paper, we presented a Localized Orthogonal Decomposition in Petrov-Galerkin formulation for the two-scale Helmholtz-type problems coming from homogenization, see for instance [34]. Under the natural assumption of a few degrees of freedom per wave length, $k(H_c + h_c) \lesssim 1$, and suitably chosen oversampling patches, $m \approx \log(k)$, the method is stable and quasi-optimal without any further restrictions on the mesh width. Thereby, this work gives the theoretical foundation and justification for an application of the LOD to the two-scale setting in case the resolution condition poses a practical restriction for the direct discretization. We have demonstrated how the periodicity in the function spaces and the coupling of different spaces can be tackled in the LOD framework. Furthermore, this paper underlines the significance of viewing the Heterogeneous Multiscale Method as a direct discretization with numerical quadrature since only this viewpoint makes the additional application of the LOD possible.

Acknowledgments

Financial support by the German research Foundation (DFG) under the project "OH 98/6-1: Wave propagation in periodic structures and negative refraction mechanisms" is gratefully acknowledged. The authors would like to thank P. Henning, A. Lamacz, and B. Schweizer for fruitful discussions on the subject. We also thank the anonymous referee for his helpful remarks.

Conflict of interest

All authors declare no conflicts of interest in this paper.

References

1. A. Abdulle and P. Henning, *Localized orthogonal decomposition method for the wave equation with a continuum of scales,* Math. Comp., **86** (2017), 549-587.
2. G. Allaire, *Homogenization and two-scale convergence,* SIAM J. Math. Anal., **23** (1992), 1482-1518.
3. I. M. Babuška and S. A. Sauter, *Is the pollution effect of the FEM avoidable for the Helmholtz equation considering high wave numbers?,* SIAM Rev., **42** (2000), 451-484.
4. G. Bouchitté, C. Bourel, and D. Felbacq, *Homogenization of the 3D Maxwell system near resonances and artificial magnetism,* C. R. Math. Acad. Sci. Paris, **347** (2009), 571-576.
5. G. Bouchitté and D. Felbacq, *Homogenization near resonances and artificial magnetism from dielectrics,* C. R. Math. Acad. Sci. Paris, **339** (2004), 377-382.
6. G. Bouchitté and B. Schweizer, *Homogenization of Maxwell's equations in a split ring geometry,* Multiscale Model. Simul., **8** (2010), 717-750.
7. G. Bouchitté and B. Schweizer, *Plasmonic waves allow perfect transmission through subwavelength metallic gratings,* Netw. Heterog. Media, **8** (2013), 857-878.

8. D. L. Brown, D. Gallistl, and D. Peterseim, *Multiscale Petrov-Galerkin method for high-frequency heterogeneous Helmholtz equations,* In M. Griebel and M. A. Schweitzer, editors, Meshfree Methods for Partial Differential Equations VII, Lecture Notes in Computational Science and Engineering, 2016.

9. H. Chen, P. Lu, and X. Xu, *A hybridizable discontinuous Galerkin method for the Hlmholtz equation with high wave number,* SIAM J. Numer. Anal., **51** (2013), 2166-2188.

10. K. Cherednichenko and S. Cooper, *Homogenization of the system of high-contrast Maxwell equations,* Mathematika, **61** (2015), 475-500.

11. A. Efros and A. Pokrovsky, *Dielectric photonic crystal as medium with negative electric permittivity and magnetic permeability,* Solid State Communications, **129** (2004), 643-647.

12. D. Elfverson, V. Ginting, and P. Henning, *On multiscale methods in Petrov-Galerkin formulation,* Numer. Math., **131** (2015), 643-682.

13. C. Engwer, P. Henning, A.Målqvist, and D. Peterseim, *Efficient implementation of the localized orthogonal decomposition method,* arXiv.

14. S. Esterhazy and J. M. Melenk, *On stability of discretizations of the Helmholtz equation,* Numerical analysis of multiscale problems of Lect. Notes Comput. Sci. Eng., Springer, Heidelberg, **83** (2012), 285-324,

15. D. Felbacq and G. Bouchitté, *Homogenization of a set of parallel fibres,* Waves Random Media, **7** (1997), 245-256.

16. D. Gallistl and D. Peterseim, *Stable multiscale Petrov-Galerkin finite element method for high frequency acoustic scattering,* Comput. Methods Appl. Mech. Engrg., **295** (2015), 1-17.

17. R. Griesmaier and P. Monk, *Error analysis for a hybridizable discontinuous Galerkin method for the Helmholtz equation,* J. Sci. Comput., **49** (2011), 291-310.

18. F. Hellman, P. Henning, and A. Målqvist, *Multiscale mixed finite elements,* Discrete Contin. Dyn. Syst. Ser. S, **9** (2016), 12691298.

19. P. Henning, P. Morgenstern, and D. Peterseim, *Multiscale partition of unity of Lecture Notes in Computational Science and Engineering, chapter Meshfree Methods for Partial Differential Equations VII,* Springer International Publishing, **100** (2014), 185-204.

20. P. Henning , M.Ohlberger, and B.Verfürth, *A new Heterogeneous Multiscale Method for time-harmonic Maxwell's equations,* SIAM J. Numer. Anal., **54** (2016), 3493-3522.

21. P. Henning and A. Persson, *A multiscale method for linear elasticity reducing Poisson locking,* Comput. Methods Appl. Mech. Engrg., **310** (2016),156171.

22. P. Henning and A. Målqvist, *Localized orthogonal decomposition techniques for boundary value problems,* SIAM J. Sci. Comput., **36** (2014), A1609-A1634.

23. R. Hiptmair, A. Moiola, and I. Perugia, *A Survey of Trefftz methods for the Helmholtz Equation, Building Bridges: Connections and Challenges in Modern Approaches to Numerical Partial Differential Equations (eds. G. R. Barrenechea, A. Cangiani and E. H. Geogoulis), Lecture Notes in Computational Science and Engineering (LNCSE),* Springer, Accepted for publication; Preprint arXiv:1505.04521.

24. R. Hiptmair, A. Moiola, and I. Perugia, *Plane Wave Discontinuous Galerkin Methods: Exponential Convergence of the hp-version,* Found. Comp. Math., **16** (2016), 637675.

25. T. J. R. Hughes and G. Sangalli, *Variational multiscale analysis: the fine-scale Green's function, projection, optimization, localization, and stabilized methods,* SIAM J. Numer. Anal., **45** (2007), 539-557.

26. T. J. R. Hughes, G. R. Feijóo, L. Mazzei, and J. B. Quincy, *The variational multiscale method—a paradigm for computational mechanics,* Comput. Methods Appl. Mech. Engrg., **166** (1998), 3-24.

27. P. C. Jr. and C. Stohrer, *Finite Element Heterogeneous Multiscale Method for the classical Helmholtz Equation,* C. R. Acad. Sci., **352** (2014), 755-760.

28. A. Lamacz and B. Schweizer, *A negative index meta-material for Maxwell's equations,* SIAM J. Math. Anal., **48** (2016), 41554174.

29. C. Luo, S. G. Johnson, J. Joannopolous, and J. Pendry, *All-angle negative refraction without negative effective index,* Phys. Rev. B, **65** (2011).

30. A. Målqvist and D. Peterseim, *Localization of elliptic multiscale problems,* Math. Comp., **83** (2014), 2583-2603.

31. A. Målqvist and D. Peterseim, *Computation of eigenvalues by numerical upscaling,* Numer. Math., **130** (2015), 337-361.

32. J. M. Melenk and S. Sauter, *Wavenumber explicit convergence analysis for Galerkin discretizations of the Helmholtz equation,* SIAM J. Numer. Anal., **49** (2011), 1210-1243.

33. S. O'Brien and J. B. Pendry, *Photonic band-gap effects and magnetic activity in dielectric composites,* Journal of Physics: Condensed Matter, **14** (2002), 4035.

34. M. Ohlberger and B. Verfürth, *A new Heterogenous Multiscale Method for the Helmholtz equation with high contrast,* arXiv, 2016.

35. M. Ohlberger, *A posteriori error estimates for the heterogeneous multiscale finite element method for elliptic homogenization problems,* Multiscale Model. Simul., **4** (2005), 88-114.

36. I. Perugia, P. Pietra, and A. Russo, *A Plane Wave Virtual Element Method for the Helmholtz Problem,* ESAIM Math. Model. Numer. Anal., **50** (2016), 783-808.

37. D. Peterseim, *Eliminating the pollution effect in Helmholtz problems by local subscale correction,* Math. Comp., **86** (2017), 1005-1036.

38. D. Peterseim, *Variational multiscale stabilization and the exponential decay of fine-scale correctors,* Springer, **114** (2016).

39. A. Pokrovsky and A. Efros, *Diffraction theory and focusing of light by a slab of left-handed material,* Physica B: Condensed Matter, **338** (2003), 333-337,

40. S. A. Sauter, *A refined finite element convergence theory for highly indefinite Helmholtz problems,* Computing, **78** (2006), 101-115.

10

The exact traveling wave solutions of a class of generalized Black-Scholes equation

Weiping Gao and Yanxia Hu*

School of Mathematics and Physics, North China Electric Power University, Beijing, 102206, China

* **Correspondence:** yxiahu@163.com

Abstract: In this paper, the traveling wave solutions of a class of generalized Black-Scholes equation are considered. By using the first integral method and the G'/G-expansion method, several exact traveling wave solutions of the equation are obtained.

Keywords: The Black-Scholes equation; first integral method; the G'/G-expansion method; traveling wave solutions
Mathematics Subject Classification: 34A05,34A34

1. Introduction

In financial mathematics and financial engineering, the classical Black-Scholes equation is a practical partial differential equation. In 1973, Black and Scholes derived the famous Black-Scholes Option Pricing Model [1]. In [2], Sunday O. Edeki etc successfully calculated the European option valuation using the Projected Differential Transformation Method. The results obtained converge faster to their associated exact solutions. In [3], the author studied the Black-Scholes equation in stochastic volatility model. In [4], the author considered to deal with the Black-Scholes equation in financial mathematics by the volatility of a variable and the abstract boundary conditions.

In this paper, we consider to obtain the traveling wave solutions of a class of generalized Black-Scholes equation by using the first integral method and the G'/G-expansion method. In 2002, Feng first proposed the first integral method [5]. This method has been widely used to solve the exact solutions of some partial differential equations. In [6], authors applied first integral method and functional variable method to obtain optical solitons from the governing nonlinear Schrödinger equation with spatio-temporal dispersion. In [7], the first integral method is applied for solving the system of nonlinear partial differential equations which are $(2 + 1)$-dimensional Broer-Kaup-Kupershmidt system and $(3 + 1)$-dimensional Burgers equations exactly. In [8], authors applied first integral method to construct travelling wave solutions of modified Zakharov-Kuznetsov equation and ZK-MEW equation. This method can also be applied to other systems of nonlinear differential equations [9–12]. The advantage of the first integral method is that the calculation is concise. A more accurate traveling wave solution can be obtained by the first integral method. In 2008, Mingliang Wang et al introduced the G'/G-expansion method in detail [13]. In [14], the G'/G-expansion method is applied to address the resonant nonlinear Schrödinger equation with dual-power law nonlinearity. In [15], the author constructed the traveling wave solutions involving parameters for some nonlinear evolution equations in mathematical physics via the (2 + 1)-dimensional Painlevé integrable Burgers equations, the (2 + 1)-dimensional Nizhnik-Novikov-Vesselov equations, the (2 + 1)-dimensional Boiti-Leon-Pempinelli equations and the (2 + 1)-dimensional dispersive long wave equations by using the G'/G-expansion method. In [16], a generalized G'/G-expansion method is proposed to seek exact solutions of the mKdV equation with variable coefficients. The G'/G-expansion method has been proposed to construct more explicit travelling wave solutions to many nonlinear evolution equations [17–19]. The performance of this method is effective, simple, convenient and gives many new solutions.

The classical celebrated Black-Scholes option pricing model is as follows:

$$\frac{\partial f}{\partial \tau} + \frac{1}{2}S^2\delta^2\frac{\partial^2 f}{\partial S^2} + rS\frac{\partial f}{\partial S} - rf = 0.$$

We consider the class of generalized Black-Scholes equation is as follows:

$$v_t + \frac{1}{2}A^2x^2v_{xx} + Bxv_x - Cv + Dv^3 = 0, \tag{1}$$

where $v = v(t, x), A, B, C, D \neq 0$ are arbitrary constants, when $D = 0$, (1) changes to the classical Black-Scholes equation.

Using the wave transformation $v(t, x) = v(\xi)$, $\xi = \ln x - at$, and a is wave velocity, we have the following ordinary differential equation,

$$-\frac{1}{2}A^2v'' + (\frac{1}{2}A^2 - B + a)v' + Cv - Dv^3 = 0. \tag{2}$$

Letting $w = v'$, (2) is equivalent to the autonomous system,

$$\begin{cases} v' = w, \\ w' = \dfrac{A^2 - 2B + 2a}{A^2}w + \dfrac{2C}{A^2}v - \dfrac{2D}{A^2}v^3. \end{cases} \tag{3}$$

2. Traveling wave solutions of (1) by using the first integral method

In this section, we apply the first integral method to obtain the traveling wave solution to (1). We assume that $p(v, w) = \sum_{i=0}^{N} \alpha_i(v)w^i$ is an irreducible polynomial in $C[v, w]$, and $p(v, w) = \sum_{i=0}^{N} \alpha_i(v)w^i = 0$ is a first integral of (3), such that,

$$\left.\frac{dP}{d\xi}\right|_{(3)} = 0.$$

Owing to Division Theorem, there exists a polynomial $g(v) + h(v)w$ in the complex domain $C(v, w)$, such that,

$$\frac{dP}{d\xi} = \frac{\partial P}{\partial v}\frac{dv}{d\xi} + \frac{\partial P}{\partial w}\frac{dw}{d\xi} = [g(v) + h(v)w]\sum_{i=0}^{N} \alpha_i(v)w^i = 0. \tag{4}$$

Here, we mainly consider (4) in two cases: $N = 1$ and $N = 2$.

2.1. N=1

At present,

$$P(v, w) = \alpha_0(v) + \alpha_1(v)w = 0.$$

From (4), we have

$$\begin{aligned} \frac{dP}{d\xi} &= \frac{d\alpha_0}{dv}w + \frac{d\alpha_1}{dv}w^2 + \alpha_1(v)(\frac{A^2 - 2B + 2a}{A^2}w + \frac{2C}{A^2}v - \frac{2D}{A^2}v^3) \\ &= g(v)\alpha_0(v) + [g(v)\alpha_1(v) + h(v)\alpha_0(v)]w + h(v)\alpha_1(v)w^2. \end{aligned} \tag{5}$$

By observing the coefficients of $w^i (i = 0, 1)$ of the two sides of (5), obviously, we have

$$\begin{cases} \dfrac{d\alpha_1}{dv} = h(v)\alpha_1(v), \\ \dfrac{d\alpha_0}{dv} = g(v)\alpha_1(v) + h(v)\alpha_0(v) - \dfrac{A^2 - 2B + 2a}{A^2}\alpha_1(v), \\ \alpha_0(v)g(v) = \alpha_1(v)(\dfrac{2C}{A^2}v - \dfrac{2D}{A^2}v^3). \end{cases} \tag{6}$$

Since $\alpha_i(v)(i = 0, 1)$ are polynomials, then from the first equation of (6), we deduce that $\alpha_1(v)$ is constant and $h(v) = 0$. For simplification, taking $\alpha_1(v) = 1$. In order to keep balancing the degree of $g(v)$, $\alpha_1(v)$ and $\alpha_0(v)$ in the second and the third equations of (6), one can conclude that $\deg g(v) = 1$. Suppose that $g(v) = g_0 v + d_0$, where g_0 and d_0 are arbitrary constants. By solving the above equations, one can obtain

$$\begin{cases} \alpha_1(v) = 1, \\ \alpha_0(v) = \dfrac{1}{2}g_0 v^2 + (d_0 - \dfrac{A^2 - 2B + 2a}{A^2})v + d_1, \end{cases}$$

where d_1 is an integration constant. Substituting $g(v)$, $\alpha_1(v)$ and $\alpha_0(v)$ into the third equation of (6) and setting all the coefficients of $v^i (i = 0, 1, 2, 3)$ to be zeros, then one can get

$$\begin{cases} \dfrac{1}{2}g_0^2 = -\dfrac{2D}{A^2}, \\ \dfrac{3}{2}g_0 d_0 - g_0\dfrac{A^2 - 2B + 2a}{A^2} = 0, \\ g_0 d_1 + d_0^2 - d_0\dfrac{A^2 - 2B + 2a}{A^2} = \dfrac{2C}{A^2}, \\ d_0 d_1 = 0. \end{cases} \tag{7}$$

From the last equation of (7), we consider to solve (7) in two cases $d_0 = 0$ or $d_0 \neq 0$.

Case 1: $d_0 = 0$

By solving (7), we have

$$\begin{cases} g_0 = \pm 2\sqrt{\frac{-D}{A^2}}, \\ d_0 = 0, \\ d_1 = \pm \frac{C}{A^2}\sqrt{\frac{A^2}{-D}}, \end{cases}$$

and $D < 0, A^2 - 2B + 2a = 0$, such that,

$$\begin{cases} \alpha_0(v) = \pm \sqrt{\frac{-D}{A^2}}v^2 \pm \frac{C}{A^2}\sqrt{\frac{A^2}{-D}}, \\ \alpha_1(v) = 1. \end{cases}$$

So,

$$P(v, w) = \pm \sqrt{\frac{-D}{A^2}}v^2 \pm \frac{C}{A^2}\sqrt{\frac{A^2}{-D}} + w = 0.$$

One can obtain the following one order ordinary differential equation,

$$w = \frac{dv}{d\xi} = \mp \sqrt{\frac{-D}{A^2}}v^2 \mp \frac{C}{A^2}\sqrt{\frac{A^2}{-D}}. \tag{8}$$

Integrating (8) once with respect to ξ, we can get the following results.

I: $\frac{C}{D} > 0$, we have

$$v(\xi) = \frac{\sqrt{\frac{C}{D}}(1 + \xi_0 e^{\mp 2\sqrt{\frac{-D}{A^2}}\xi})}{1 - \xi_0 e^{\mp 2\sqrt{\frac{-D}{A^2}}\xi}},$$

where ξ_0 is an integration constant.

The traveling wave solution to (1) can be got as follows:

$$v(t, x) = \frac{\sqrt{\frac{C}{D}}(1 + \xi_0 e^{\mp 2\sqrt{\frac{-D}{A^2}}(\ln x - at)})}{1 - \xi_0 e^{\mp 2\sqrt{\frac{-D}{A^2}}(\ln x - at)}}.$$

II: $\frac{C}{D} < 0$, we have

$$v(\xi) = \sqrt{\frac{C}{-D}}\tan[\mp \sqrt{\frac{C}{A^2}}\xi + \xi_1],$$

where ξ_1 is an arbitrary integration constant.

The traveling wave solution to (1) can be got as follows:

$$v(t,x) = \sqrt{\frac{C}{-D}}\tan[\mp\sqrt{\frac{C}{A^2}}(\ln x - at) + \xi_1].$$

Case 2: $d_0 \neq 0$

From (7), we have

$$\begin{cases} g_0 = \pm 2\sqrt{\frac{-D}{A^2}}, \\ d_0 = \frac{2(A^2 - 2B + 2a)}{3A^2}, \\ d_1 = 0, \end{cases}$$

and $D < 0$, $6C + (A^2 - 2B + 2a) = 0$, such that,

$$\begin{cases} \alpha_0(v) = \pm\sqrt{\frac{-D}{A^2}}v^2 - \frac{A^2 - 2B + 2a}{3A^2}v, \\ \alpha_1(v) = 1. \end{cases}$$

So,

$$P(v,w) = \pm\sqrt{\frac{-D}{A^2}}v^2 - \frac{A^2 - 2B + 2a}{3A^2}v + w = 0.$$

One can obtain the following one order ordinary differential equation,

$$w = \frac{dv}{d\xi} = \mp\sqrt{\frac{-D}{A^2}}v^2 + \frac{A^2 - 2B + 2a}{3A^2}v. \tag{9}$$

Integrating equation (9) once with respect to ξ, we can get the following results,

$$v(\xi) = \frac{\frac{A^2-2B+2a}{3A^2}\xi_2 e^{\frac{A^2-2B+2a}{3A^2}\xi}}{1 \pm \sqrt{\frac{-D}{A^2}}\xi_2 e^{\frac{A^2-2B+2a}{3A^2}\xi}},$$

where ξ_2 is an integration constant.

The exact traveling wave solution to (1) can be got as follows:

$$v(t,x) = \frac{\frac{A^2-2B+2a}{3A^2}\xi_2 e^{\frac{A^2-2B+2a}{3A^2}(\ln x - at)}}{1 \pm \sqrt{\frac{-D}{A^2}}\xi_2 e^{\frac{A^2-2B+2a}{3A^2}(\ln x - at)}}.$$

2.2. $N = 2$

At present,

$$P(v, w) = \alpha_0(v) + \alpha_1(v)w + \alpha_2(v)w^2 = 0. \tag{10}$$

From (4), we have

$$\frac{dP}{d\xi} = \frac{d\alpha_0}{dv}w + \frac{d\alpha_1}{dv}w^2 + \alpha_1(v)(\frac{A^2 - 2B + 2a}{A^2}w + \frac{2C}{A^2}v - \frac{2D}{A^2}v^3)$$

$$+ \frac{d\alpha_2}{dv}w^3 + 2\alpha_2(v)w(\frac{A^2 - 2B + 2a}{A^2}w + \frac{2C}{A^2}v - \frac{2D}{A^2}v^3)$$

$$= g(v)\alpha_0(v) + [g(v)\alpha_1(v) + h(v)\alpha_0(v)]w + [g(v)\alpha_2(v) + h(v)\alpha_1(v)]w^2 + h(v)\alpha_2(v)w^3. \tag{11}$$

By observing the coefficients of $w^i (i = 0, 1, 2, 3)$ of the two sides of (11), obviously, we have

$$\begin{cases} \dfrac{d\alpha_2}{dv} = h(v)\alpha_2(v), \\ \dfrac{d\alpha_1}{dv} = g(v)\alpha_2(v) + h(v)\alpha_1(v) - 2\dfrac{A^2 - 2B + 2a}{A^2}\alpha_2(v), \\ \dfrac{d\alpha_0}{dv} = g(v)\alpha_1(v) + h(v)\alpha_0(v) - \dfrac{A^2 - 2B + 2a}{A^2}\alpha_1(v) - 2\alpha_2(v)(\dfrac{2C}{A^2}v - \dfrac{2D}{A^2}v^3), \\ \alpha_0(v)g(v) = \alpha_1(v)(\dfrac{2C}{A^2}v - \dfrac{2D}{A^2}v^3). \end{cases} \tag{12}$$

Similarly, from the first equation of (12), we deduce that $\alpha_2(v)$ is constant and $h(v) = 0$. From the second and the third equations of (12), we assume that deg $g(v) = k$, then deg $\alpha_1(v) = k + 1$ and deg $\alpha_0(v) = 2k + 2$. By the last equation of (12), we have the degree of the polynomial $\alpha_1(v)(\frac{2C}{A^2}v - \frac{2D}{A^2}v^3)$ is $k + 4$, and the degree of the polynomial $\alpha_0(v)g(v)$ is $3k + 2$. By balancing the degree of the last equation of (12), we have $k + 4 = 3k + 2$, obviously, $k = 1$. Specially, we conclude that deg $g(v) = 0$, then deg $\alpha_1(v) = 1$. The degree on both sides of the last equation of (12) is still true.

Case 1: deg $g(v) = 0$

Assume $g(v) = g_1$. From (12), we have

$$\begin{cases} \alpha_0(v) = \dfrac{D}{A^2}v^4 + (\dfrac{1}{2}g_1^2 - \dfrac{3}{2}g_1\dfrac{A^2 - 2B + 2a}{A^2} + \dfrac{(A^2 - 2B + 2a)^2}{A^4} - \dfrac{2C}{A^2})v^2 \\ +d_2(g_1 - \dfrac{A^2 - 2B + 2a}{A^2})v + d_3, \\ \alpha_1(v) = (g_1 - 2\dfrac{A^2 - 2B + 2a}{A^2})v + d_2, \\ \alpha_2(v) = 1, \end{cases} \tag{13}$$

where d_2, d_3 are integration constants. From the last one of equation (12), we have

$$\begin{cases} g_1\dfrac{D}{A^2} = -\dfrac{2D}{A^2}(g_1 - 2\dfrac{A^2-2B+2a}{A^2}), \\ -d_2\dfrac{2D}{A^2} = 0, \\ g_1(\dfrac{1}{2}g_1^2 - \dfrac{3}{2}g_1\dfrac{A^2-2B+2a}{A^2} + \dfrac{(A^2-2B+2a)^2}{A^4} - \dfrac{2C}{A^2}) = \dfrac{2C}{A^2}(g_1 - 2\dfrac{A^2-2B+2a}{A^2}), \\ g_1d_2(g_1 - \dfrac{A^2-2B+2a}{A^2}) = d_2\dfrac{2C}{A^2}, \\ g_1d_3 = 0. \end{cases} \tag{14}$$

Solving (14), we have

$$\begin{cases} g_1 = \dfrac{4(A^2-2B+2a)}{3A^2}, \\ 9CA^2 + (A^2-2B+2a)^2 = 0, \\ d_2 = 0, \\ d_3 = 0. \end{cases} \tag{15}$$

Then from (15), (13) and (10), one can obtain the following equation,

$$P(v,w) = \frac{D}{A^2}v^4 + \frac{(A^2-2B+2a)^2}{9A^4}v^2 - \frac{2(A^2-2B+2a)}{3A^2}vw + w^2 = 0.$$

Solving the above algebraic equation with respect to the variable w, we have

$$w = \frac{dv}{d\xi} = \pm\sqrt{\frac{-D}{A^2}}v^2 + \frac{A^2-2B+2a}{3A^2}v. \tag{16}$$

Integrating (16) once with respect to ξ, then we have

$$v(\xi) = \frac{\frac{A^2-2B+2a}{3A^2}\xi_3 e^{\frac{A^2-2B+2a}{3A^2}\xi}}{1 \mp \sqrt{\frac{-D}{A^2}}\xi_3 e^{\frac{A^2-2B+2a}{3A^2}\xi}},$$

where ξ_3 is an integration constant.

The exact traveling wave solution to (1) can be got as follows:

$$v(t,x) = \frac{\frac{A^2-2B+2a}{3A^2}\xi_3 e^{\frac{A^2-2B+2a}{3A^2}(\ln x - at)}}{1 \mp \sqrt{\frac{-D}{A^2}}\xi_3 e^{\frac{A^2-2B+2a}{3A^2}(\ln x - at)}}.$$

Case 2: deg $g(v) = 1$

Assume $g(v) = g_2 v + d_4$. From (12), we have

$$\begin{cases} \alpha_0(v) = \frac{1}{4}(\frac{1}{2}g_2^2 + \frac{4D}{A^2})v^4 + \frac{1}{3}(\frac{3}{2}g_2 d_4 - \frac{5}{2}g_2\frac{A^2-2B+2a}{A^2})v^3 + (d_4 d_5 - d_5\frac{A^2-2B+2a}{A^2})v \\ +\frac{1}{2}(g_2 d_5 + d_4^2 - 3d_4\frac{A^2-2B+2a}{A^2} + 2\frac{(A^2-2B+2a)^2}{A^4} - \frac{4C}{A^2})v^2 + d_6, \\ \alpha_1(v) = \frac{1}{2}g_2 v^2 + (d_4 - 2\frac{A^2-2B+2a}{A^2})v + d_5, \\ \alpha_2(v) = 1, \end{cases}$$

where d_5, d_6 are integration constants.

From the last equation of (12), we have

$$\begin{cases} -g_2\frac{D}{A^2} = \frac{1}{8}g_2^3 + g_2\frac{D}{A^2}, \\ -\frac{2D}{A^2}(d_4 - 2\frac{A^2-2B+2a}{A^2}) = \frac{5}{8}g_2^2 d_4 - \frac{5}{6}g_2^2\frac{A^2-2B+2a}{A^2} + d_4\frac{D}{A^2}, \\ g_2\frac{C}{A^2} - d_5\frac{2D}{A^2} = g_2 d_4^2 - \frac{7}{3}g_2 d_4\frac{A^2-2B+2a}{A^2} + \frac{1}{2}g_2^2 d_5 + g_2\frac{(A^2-2B+2a)^2}{A^4} - g_2\frac{2C}{A^2}, \\ \frac{-4C(A^2-2B+2a)}{3A^4} = \frac{3}{2}g_2 d_4 d_5 + \frac{1}{2}d_4^3 - \frac{3(A^2-2B+2a)}{2A^2}d_4^2 \\ +d_4(\frac{(A^2-2B+2a)^2}{A^4} - \frac{2C}{A^2}) - g_2 d_5\frac{A^2-2B+2a}{A^2}, \\ d_5\frac{2C}{A^2} = g_2 d_6 + d_4^2 d_5 - d_4 d_5\frac{A^2-2B+2a}{A^2}, \\ d_4 d_6 = 0. \end{cases} \tag{17}$$

From the last equation of (17), we need to discuss (17) in two cases $d_4 = 0$ or $d_4 \neq 0$.

Case I: $d_4 = 0$

Solving (17), we have

$$\begin{cases} g_2 = \pm 4\sqrt{\frac{-D}{A^2}}, \\ d_4 = 0, \\ d_5 = \mp 2\sqrt{\frac{-D}{A^2}}\frac{C}{D}, \\ d_6 = -\frac{C^2}{A^2 D}, \end{cases}$$

and $D < 0, A^2 - 2B + 2a = 0$, such that,

$$\begin{cases} \alpha_0(v) = \frac{-D}{A^2}v^4 + \frac{2C}{A^2}v^2 - \frac{C^2}{A^2 D}, \\ \alpha_1(v) = \pm 2\sqrt{\frac{-D}{A^2}}v^2 \mp 2\sqrt{\frac{-D}{A^2}\frac{C}{D}}, \\ \alpha_2(v) = 1. \end{cases}$$

So,

$$P(v,w) = \frac{-D}{A^2}v^4 + \frac{2C}{A^2}v^2 - \frac{C^2}{A^2 D} + (\pm 2\sqrt{\frac{-D}{A^2}}v^2 \mp 2\sqrt{\frac{-D}{A^2}\frac{C}{D}})w + w^2 = 0.$$

Solving the above algebraic equation with respect to the variable w, we have

$$w = \frac{dv}{d\xi} = \mp\sqrt{\frac{-D}{A^2}}v^2 \pm \sqrt{\frac{-D}{A^2}\frac{C}{D}}. \tag{18}$$

Integrating (18) once with respect to ξ, we can get the following results.

I: $\frac{C}{D} > 0$.

$$v(\xi) = \frac{\sqrt{\frac{C}{D}}(1 + \xi_4 e^{\mp 2\sqrt{\frac{-C}{A^2}}\xi})}{1 - \xi_4 e^{\mp 2\sqrt{\frac{-C}{A^2}}\xi}},$$

where ξ_4 is an integration constant.

The traveling wave solution to (1) can be got as follows:

$$v(t,x) = \frac{\sqrt{\frac{C}{D}}(1 + \xi_4 e^{\mp 2\sqrt{\frac{-C}{A^2}}(\ln x - at)})}{1 - \xi_4 e^{\mp 2\sqrt{\frac{-C}{A^2}}(\ln x - at)}}.$$

II: $\frac{C}{D} < 0$.

$$v(\xi) = \sqrt{\frac{C}{-D}}\tan[\mp\sqrt{\frac{C}{A^2}}\xi + \xi_5],$$

where ξ_5 is an integration constant.

The traveling wave solution to (1) can be got as follows:

$$v(t,x) = \sqrt{\frac{C}{-D}}\tan[\mp\sqrt{\frac{C}{A^2}}(\ln x - at) + \xi_5].$$

Case II: $d_4 \neq 0$

By solving (17), we have

$$\begin{cases} g_2 = \pm 4\sqrt{\dfrac{-D}{A^2}}, \\ d_4 = \dfrac{4(A^2 - 2B + 2a)}{3A^2}, \\ C = -\dfrac{(A^2 - 2B + 2a)^2}{9A^2}, \\ d_5 = 0, \\ d_6 = 0. \end{cases} \tag{19}$$

Substituting (19) into (13), we have

$$\begin{cases} \alpha_0(v) = -\dfrac{D}{A^2}v^4 \mp \dfrac{2}{3}\sqrt{\dfrac{-D}{A^2}}\dfrac{A^2 - 2B + 2a}{A^2}v^3 + \dfrac{(A^2 - 2B + 2a)^2}{9A^4}v^2, \\ \alpha_1(v) = \pm 2\sqrt{\dfrac{-D}{A^2}}v^2 - \dfrac{2(A^2 - 2B + 2a)}{3A^2}v, \\ \alpha_2(v) = 1. \end{cases}$$

So,

$$P(v, w) = -\frac{D}{A^2}v^4 \mp \frac{2}{3}\sqrt{\frac{-D}{A^2}}\frac{A^2 - 2B + 2a}{A^2}v^3 + \frac{(A^2 - 2B + 2a)^2}{9A^4}v^2 + (\pm 2\sqrt{\frac{-D}{A^2}}v^2 - \frac{2(A^2 - 2B + 2a)}{3A^2}v)w + w^2 = 0.$$

Solving the above algebraic equation with respect to the variable w, we have

$$w = \frac{dv}{d\xi} = \mp\sqrt{\frac{-D}{A^2}}v^2 + \frac{A^2 - 2B + 2a}{3A^2}v. \tag{20}$$

Integrating (20) once with respect to ξ, then we have

$$v(\xi) = \frac{\frac{A^2-2B+2a}{3A^2}\xi_6 e^{\frac{A^2-2B+2a}{3A^2}\xi}}{1 \pm \sqrt{\frac{-D}{A^2}}\xi_6 e^{\frac{A^2-2B+2a}{3A^2}\xi}},$$

where ξ_6 is an integration constant.

The traveling wave solution to (1) can be got as follows:

$$v(t, x) = \frac{\frac{A^2-2B+2a}{3A^2}\xi_6 e^{\frac{A^2-2B+2a}{3A^2}(\ln x - at)}}{1 \pm \sqrt{\frac{-D}{A^2}}\xi_6 e^{\frac{A^2-2B+2a}{3A^2}(\ln x - at)}}.$$

3. Traveling wave solutions of (1) by the G'/G-expansion method

In this section, we get the traveling wave solutions to (1) using the G'/G-expansion method. Introducing the solution $v(\xi)$ of equation (2) in a form of finite series:

$$v(\xi) = \sum_{l=0}^{N} a_l \left(\frac{G'(\xi)}{G(\xi)}\right)^l, \tag{21}$$

where a_l are real constants with $a_N \neq 0$ and N is a positive integer that needs to be determined. The function $G(\xi)$ is the solution of the auxiliary linear ODE

$$G''(\xi) + \lambda G'(\xi) + \mu G(\xi) = 0, \tag{22}$$

where λ and μ are real constants to be determined. By balancing v'' with v^3 in the equation (2), we have $N + 2 = 3N$, obviously, $N = 1$. Therefore, (21) is written as

$$v(\xi) = a_0 + a_1(\frac{G'(\xi)}{G(\xi)}). \tag{23}$$

Correspondingly,

$$v'(\xi) = -a_1\lambda\frac{G'}{G} - a_1(\frac{G'}{G})^2 - a_1\mu, \tag{24}$$

$$v''(\xi) = (a_1\lambda^2 + 2a_1\mu)\frac{G'}{G} + 3a_1\lambda(\frac{G'}{G})^2 + 2a_1(\frac{G'}{G})^3 + a_\lambda\mu. \tag{25}$$

Substituting (23), (24), (25) and v^3 into (2), we have

$$\begin{aligned}[-\frac{1}{2}A^2(a_1\lambda^2 + 2a_1\mu) - (\frac{1}{2}A^2 - B + a)a_1\lambda + Ca_1 - 3Da_0^2a_1]\frac{G'}{G}\\ - [\frac{3}{2}A^2a_1\lambda + (\frac{1}{2}A^2 - B + a)a_1 + 3Da_0a_1^2](\frac{G'}{G})^2\\ - (A^2a_1 + Da_1^3)(\frac{G'}{G})^3 - [\frac{1}{2}A^2a_1\lambda\mu + (\frac{1}{2}A^2 - B + a)a_1\mu - Ca_0 + Da_0^3] = 0.\end{aligned}$$

Setting all the coefficients of $(\frac{G'}{G})^i (i = 0, 1, 2, 3)$ to be zeros, we yield a set of algebraic equations for a_0, a_1, λ and μ as follows:

$$\begin{cases} -\frac{1}{2}A^2(a_1\lambda^2 + 2a_1\mu) - (\frac{1}{2}A^2 - B + a)a_1\lambda + Ca_1 - 3Da_0^2a_1 = 0, \\ \frac{3}{2}A^2a_1\lambda + (\frac{1}{2}A^2 - B + a)a_1 + 3Da_0a_1^2 = 0, \\ A^2a_1 + Da_1^3 = 0, \\ \frac{1}{2}A^2a_1\lambda\mu + (\frac{1}{2}A^2 - B + a)a_1\mu - Ca_0 + Da_0^3. \end{cases}$$

Solving the above algebraic equations, we have

$$\begin{cases} a_0 = \pm\left(\dfrac{\sqrt{\frac{A^2}{-D}}(A^2-2B+2a)}{3A^2} + \dfrac{\sqrt{\frac{A^2}{-D}}(A^2-2B+2a)^3}{27A^4C}\right), \\ a_1 = \pm\sqrt{\dfrac{A^2}{-D}}, \\ \lambda = -\left(\dfrac{16(\frac{1}{2}A^2-B+a)^3}{27A^4C} + \dfrac{A^2-2B+2a}{A^2}\right), \\ \mu = \dfrac{(A^2-2B+2a)^6}{729A^8C^2} + \dfrac{(A^2-2B+2a)^4}{27A^6C} + \dfrac{5(A^2-2B+2a)^2}{12A^4} + \dfrac{C}{A^2}. \end{cases} \tag{26}$$

From equations (23) and (26), the solution of (2) can be written as the following equation,

$$V(\xi) = \mp\left(\frac{\sqrt{\frac{A^2}{-D}}(A^2-2B+2a)}{3A^2} + \frac{\sqrt{\frac{A^2}{-D}}(A^2-2B+2a)^3}{27A^4C}\right) \pm \sqrt{\frac{A^2}{-D}}\left(\frac{G'}{G}\right). \tag{27}$$

By solving the second order linear ODE (22), then (27) gives three types traveling wave solutions.

Case 1: $\Delta = \lambda^2 - 4\mu = -\frac{2(A^2-2B+2a)^2}{3A^4} - \frac{4C}{A^2} > 0$, we obtain

$$\frac{G'}{G} = -\frac{\lambda}{2} + \frac{\sqrt{\lambda^2-4u}}{2}\frac{C_1e^{\frac{\sqrt{\lambda^2-4u}}{2}\xi} - C_2e^{-\frac{\sqrt{\lambda^2-4u}}{2}\xi}}{C_1e^{\frac{\sqrt{\lambda^2-4u}}{2}\xi} + C_2e^{-\frac{\sqrt{\lambda^2-4u}}{2}\xi}},$$

where C_1, C_2 are integration constants. The traveling wave solution to (1) can be got as follows:

$$v(t,x) = \mp\left(\frac{\sqrt{\frac{A^2}{-D}}(A^2-2B+2a)}{3A^2} + \frac{\sqrt{\frac{A^2}{-D}}(A^2-2B+2a)^3}{27A^4C}\right) \pm \sqrt{\frac{A^2}{-D}}\left\{\frac{(A^2-2B+2a)^3}{27A^4C} + \frac{A^2-2B+2a}{2A^2} + \right.$$

$$\left.\frac{1}{2}\sqrt{-\frac{2(A^2-2B+2a)^2}{3A^4} - \frac{4C}{A^2}}\left[\frac{C_1e^{H_1(\ln x-at)} - C_2e^{-H_1(\ln x-at)}}{C_1e^{H_1(\ln x-at)} + C_2e^{-H_1(\ln x-at)}}\right]\right\}.$$

where $H_1 = \frac{1}{2}\sqrt{-\frac{2(A^2-2B+2a)^2}{3A^4} - \frac{4C}{A^2}}$. Specially, when $C_1 = C_2$, we obtain traveling wave solution of (1) as follows:

$$v(t,x) = \mp\left(\frac{\sqrt{\frac{A^2}{-D}}(A^2-2B+2a)}{3A^2} + \frac{\sqrt{\frac{A^2}{-D}}(A^2-2B+2a)^3}{27A^4C}\right) \pm \sqrt{\frac{A^2}{-D}}\left\{\frac{(A^2-2B+2a)^3}{27A^4C} + \frac{A^2-2B+2a}{2A^2} + \right.$$

$$\left.\frac{1}{2}\sqrt{-\frac{2(A^2-2B+2a)^2}{3A^4} - \frac{4C}{A^2}}\tanh\left\{\frac{1}{2}\sqrt{-\frac{2(A^2-2B+2a)^2}{3A^4} - \frac{4C}{A^2}}(\ln x - at)\right\}\right\}.$$

Case 2: $\Delta = \lambda^2 - 4\mu = -\frac{2(A^2-2B+2a)^2}{3A^4} - \frac{4C}{A^2} < 0$, we obtain

$$\frac{G'}{G} = -\frac{\lambda}{2} + \frac{\sqrt{\lambda^2-4u}}{2}\frac{C_2\cos\frac{\sqrt{\lambda^2-4u}}{2}\xi - C_1\sin\frac{\sqrt{\lambda^2-4u}}{2}\xi}{C_1\cos\frac{\sqrt{\lambda^2-4u}}{2}\xi + C_2\sin\frac{\sqrt{\lambda^2-4u}}{2}\xi},$$

where C_1, C_2 are integration constants. The traveling wave solution to (1) can be got as follows:

$$v(t,x) = \mp\left(\frac{\sqrt{\frac{A^2}{-D}}(A^2-2B+2a)}{3A^2} + \frac{\sqrt{\frac{A^2}{-D}}(A^2-2B+2a)^3}{27A^4C}\right) \pm \sqrt{\frac{A^2}{-D}}\left\{\frac{(A^2-2B+2a)^3}{27A^4C}\right.$$

$$\left. + \frac{A^2-2B+2a}{2A^2} + \frac{1}{2}\sqrt{-\frac{2(A^2-2B+2a)^2}{3A^4} - \frac{4C}{A^2}}\left[\frac{C_2\cos H_2(\ln x - at) - C_1\sin H_2(\ln x - at)}{C_1\cos H_2(\ln x - at) + C_2\sin H_2(\ln x - at)}\right]\right\}.$$

where $H_2 = \frac{1}{2}\sqrt{-\frac{2(A^2-2B+2a)^2}{3A^4} - \frac{4C}{A^2}}$.

Case 3: $\Delta = \lambda^2 - 4\mu = -\frac{2(A^2-2B+2a)^2}{3A^4} - \frac{4C}{A^2} = 0$, we obtain

$$\frac{G'}{G} = -\frac{\lambda}{2} + \frac{C_2}{C_1 + C_2\xi},$$

where C_1, C_2 are integration constants. The traveling wave solution to (1) can be got as follows:

$$\begin{aligned} v(t,x) = & \mp \left(\frac{\sqrt{\frac{A^2}{-D}}(A^2-2B+2a)}{3A^2} + \frac{\sqrt{\frac{A^2}{-D}}(A^2-2B+2a)^3}{27A^4C}\right) \\ & \pm \sqrt{\frac{A^2}{-D}}\left\{\frac{(A^2-2B+2a)^3}{27A^4C} + \frac{A^2-2B+2a}{2A^2} + \frac{C_2}{C_1 + C_2(\ln x - at)}\right\}. \end{aligned}$$

4. Conclusion

We have obtained several traveling wave solutions of a class of generalized Black- Scholes equation under certain parametric conditions by using the first integral method and the G'/G-expansion method. From the above investigation, we see that the two methods are effective for obtaining the exact traveling wave solutions of the class of generalized Black-Scholes equation. The first integral method is based on the ring theory of commutative algebra, and supposes that (2) has rational first integrals of the variables v and v'. The G'/G-expansion method is to obtain a special class of solutions of (2) based on the balance principle. When $D = 0$, the class of generalized Black-Scholes equation is changed to the classical Black-Scholes equation, that is, the classical Black-Scholes equation is a special case of (2). The results obtained in this paper can provide some useful reference and help for the relevant financial research.

Conflict of Interest

All authors declare no conflicts of interest in this paper.

References

1. Fischer Black, Myron Scholes, *The Pricing of Options and Corporate Liabilities,* The Journal of Political Economy, **81** (1973), 637-654.
2. Sunday O. Edeki, Olabisi O. Ugbebor and Enahoro A Owoloko, *Analytical solutions of the Black-Scholes pricing model for European option valuation via a projected differential transformation method,* Entropy, **17** (2015), 7510-7521.
3. Erik Ekström, Johan Tysk, *The Black-Scholes equation in stochastic volatility models,* J. Math. Anal. Appl., **368** (2010), 498-507.
4. Rubén Figueroa and Maria do Rosário Grossinho, *On some nonlinear boundary value problems related to a Black-Scholes model with transaction costs,* Boundary value problems, 2015.
5. Zhaosheng Feng, *The first-integral method to study the Burgers-Korteweg-de Vries equation,*Journal of Physics A: Mathematical and General, **35** (2002), 343-349.
6. M. Mirzazadeha, Anjan Biswasb, *Optical solitons with spatio-temporal dispersion by first integral approach and functional variable method,* Optik, **125** (2014) 5467-5475.
7. Bin Lu, Hongqing Zhang and Fuding Xie, *Travelling wave solutions of nonlinear partial equations by using the first integral method,* Applied Mathematics and Computation, **216** (2010), 1329-1336.
8. Filiz Tascan, Ahmet Bekir and Murat Koparan, *Travelling wave solutions of nonlinear evolution equations by using the first integral method,* Commun Nonlinear Sci Numer Simulat, **14** (2009), 1810-1815.
9. Zhaosheng Feng, Xiaohui Wang, *The first method to the two-dimensional Burgers-Korteweg-de Vries equation,* Physics Letters A, **308** (2003), 173-178.
10. Zhaosheng Feng, *On explicit exact solutions to the compound Burgers-KdV equation,* Physics Letters A, **293** (2002), 57-66.
11. Zhaosheng Feng, *Exact solution to an approximate sine-Gordon equation in (n+1)-dimensional space,* Physics Letters A, **302** (2002), 64-76.
12. Yanxia Hu, Weiping Gao, *Exact and explicit solutions of a coupled nonlinear Schrödinger type equation,* Asian Journal of Mathematics and Computer Research, **9** (2016), 240-252.
13. Mingliang Wang, Xiangzheng Li and Jinliang Zhang, *The G'/G-expansion method and travling wave solutions of nonlinear evolution equations in mathematics physics,* Phys.Lett.A, **372** (2008), 417-423.
14. M.Mirzazadeha, M.Eslamib, Daniela Milovicc and Anjan Biswasd, *Topological solitons of resonant nonlinear Schödinger's equation with dual-power law nonlinearity by G'/G-expansion technique,* Optik, **125** (2014), 5480-5489.
15. E.M.E. Zayed, *The G'/G-expansion method and its applications to some nonlinear evolution equations in the mathematical physics,* J. Appl. Math. Comput., **30** (2009), 89-103.
16. Sheng Zhang, Jinglin Tong and Wei Wang, *A generalized G'/G-expansion method for the mKdV equation with variable coefficients,* Physics Letters A, **372** (2008), 2254-2257.

17. Turgut Öziş, İsmail Aslan, *Application of the G'/G-expansion method to Kawahara type equations using symbolic computation* Applied Mathematics and Computation, **216** (2010), 2360-2365.

18. Lingxiao Li, Mingliang Wang, *The G'/G-expansion method and travelling wave solutions for a higher-order nonlinear schrödinger equation* Applied Mathematics and Computation, **208** (2009), 440-445.

19. Hua Gao, Rongxia Zhao , *New application of the G'/G-expansion method to higher-order nonlinear equations,* Elsevier Science Inc., 2009.

11

A High-Order Symmetric Interior Penalty Discontinuous Galerkin Scheme to Simulate Vortex Dominated Incompressible Fluid Flow

Lunji Song[1,2,*] **and Charles O'Neill**[3]

[1] School of Mathematics and Statistics, Lanzhou University, Lanzhou, 730000, China

[2] Department of Mathematics, The University of Alabama, Tuscaloosa, AL 35487, USA

[3] Department of Aerospace Engineering and Mechanics, The University of Alabama, Tuscaloosa, AL 35487, USA

[*] **Correspondence:** Email:song@lzu.edu.cn

Abstract: A high-order Symmetric Interior Penalty discontinuous Galerkin (SIPG) method has been used for solving the incompressible Navier-Stokes equation. We apply the temporal splitting scheme in time and the SIPG discretization in space with the local Lax-Friedrichs flux for the discretization of nonlinear terms. A fully discrete semi-implicit splitting scheme has been presented and high-order discontinuous Galerkin (DG) finite elements are available. Under a constraint of the CFL condition, two benchmark problems in 2D are investigated: one is a lid-driven cavity flow to verify the high-order discontinuous Galerkin method is accurate and robust; the other is a flow past a circular cylinder, for which we mainly check the Strouhal numbers with the von Kármán vortex street, and also simulate the boundary layers with walls and corresponding dynamical behavior with Neumann conditions on the top and bottom boundaries, respectively. We predict the Strouhal number for the range of Reynolds number $50 \leq Re \leq 400$, making a comparison between the predicted values by our numerical method and the referenced values from physical experiments.

Keywords: Navier-Stokes equations; von Kármán vortex street; discontinuous Galerkin method; interior penalty

1. Introduction

As a fundamental equation of fluid dynamics, the Navier-Stokes equations have been investigated by many scientists conducting research on numerical schemes for their numerical solutions [7, 8, 10, 17, 18, 19, 22, 23]. Analytical solutions of real flow problems including complex geometries are not available, nor likely in the foreseeable future. There are two ways to provide reference data for such problems: One consists in the measurement of quantities of interest in physical experiments and the other is to perform careful numerical studies with highly accurate discretizations. With the prevalence and development of high-performance computers, advanced numerical algorithms are able to be tested for the validation of approaches and codes and for high-order convergence behavior of delicate discretizations. For example, Symmetric Interior Penalty Galerkin (SIPG) and Non-symmetric Interior Penalty Galerkin (NIPG) methods were first introduced originally for elliptic problems by Wheeler [24] and Rivière et al. [16]. Recently, some work based on the SIPG and NIPG methods has been successfully applied to the steady-state and transient Navier-Stokes equations [8, 9, 15], for which optimal error estimates for the velocity have been derived.

The Navier-Stokes equations are a concise physics model of low Knudsen number (i.e. non-rarefied) fluid dynamics. Phenomena described with the Navier-Stokes equations include boundary layers, shocks, flow separation, turbulence, and vortices, as well as integrated effects such as lift and drag. The physics of Navier-Stokes flows are non-dimensionalized by Mach number M and Reynolds number Re,

$$M = \frac{\mathbf{u}_\infty}{\mathbf{a}}, \tag{1}$$

$$Re = \frac{\rho \mathbf{u}_\infty D}{\mu}, \tag{2}$$

where ρ is the density of the fluid, and μ is the dynamic viscosity. The kinematic viscosity ν is the ratio of μ to ρ. At low Knudsen numbers, Navier-Stokes surface boundary conditions are effectively no-slip (i.e. zero velocity). Diffusion of momentum from freestream to surface no-slip velocities forms boundary layers decreasing in thickness as Reynolds number increases. Thus, the range of characteristic solution scale increases as the Reynolds number increases. Nonlinear convective terms couped with the strong velocity gradients in the Navier Stokes equations drive fluid flow at even moderate Reynolds numbers to inherently unsteady behavior. Rotational flow is measured in terms of the vorticity ω, defined as the curl of a velocity vector $\mathbf{v}$,

$$\omega = \nabla \times \mathbf{v} \tag{3}$$

The related concept of circulation Γ is defined as a contour integral of vorticity by

$$\Gamma = \oint_{\partial S} \mathbf{v} \cdot \mathbf{ds} = -\iint_{\mathbf{S}} \omega \cdot \hat{\mathbf{n}}\, \mathbf{dS} \tag{4}$$

The concept of a vortex is that of vorticity concentrated along a path [3].

Lid driven cavity flows are geometrically simple boundary conditions testing the convective and viscous portions of the Navier Stokes equation in a enclosed unsteady environment. The cavity flow is characterized by a quiescent flow with the driven upper lid providing energy transfer into the cavity through viscous stresses. Boundary layers along the side and lower surfaces develop as the Reynolds number increases, which tends to shift the vorticity center of rotation towards the center. A presence of the sharp corner at the downstream upper corner increasingly generates small scale flow features as the Reynolds number increases. Full cavity flows remain a strong research topic for acoustics and sensor deployment technologies.

For non-streamlined blunt bodies with a cross-flow, an adverse pressure gradient in the aft body tends to promote flow separation and an unsteady flow field. The velocity field develops into an oscillating separation line on the upper and lower surfaces. This manifests as a series of shed vortices forming and then convecting downstream with the mean flow. The von Kármán vortex street is named after the engineer and fluid dynamicist Theodore Kármán (1963; 1994). Vortex streets are ubiquitous in nature and are visibly seen in river currents downstream of obstacles, atmospheric phenomena, and the clouds of Jupiter (e.g. The Great Red Spot). Shed vortices are also the primary driver for the the zig-zag motion of bubbles in carbonated drinks. The bubble rising through the drink creates a wake of shed vorticity which impacts the integrated pressures causing side forces and thus side accelerations. The physics of sound generation with an Aeolean's harp operates by alternating vortices creating harmonic surfaces pressure variations leading to radiated acoustic tones. Tones generated by vortex shedding are the so-called Strouhal friction tones. If the diameter of the string, or cylinder immersed in the flow is D and the free stream velocity of the flow is $\mathbf{u}_\infty$ then the shedding frequency f of the sound is given by the Strouhal formula

$$St = \frac{fD}{\mathbf{u}_\infty}, \tag{5}$$

where $f = T^{-1}$, and St is the Strouhal number named after Vincent Strouhal, a Czech physicist who experimented in 1878 with wires experiencing vortex shedding and singing in the wind (Strouhal, 1878; White, 1999). The Strouhal formula provides a experimentally derived shedding frequency for fluid flow. Therefore, we are interested in an investigation of Stouhal numbers of incompressible flow at different Reynolds number.

Let Ω be a bounded polygonal domain in $\mathbb{R}^2$. The dynamics of an incompressible fluid flow in 2D is described by the Navier-Stokes equations, which include the equations of continuity and momentum, written in dimensionless form [23] as follows:

$$\frac{\partial \mathbf{u}}{\partial t} - \nu\Delta\mathbf{u} + (\mathbf{u}\cdot\nabla)\mathbf{u} + \nabla p = \mathbf{f}, \quad \text{in} \quad \Omega\times(0,T) \tag{6}$$

$$\nabla\cdot\mathbf{u} = 0, \quad \text{in} \quad \Omega\times(0,T) \tag{7}$$

$$\mathbf{u}|_{t=0} = \mathbf{u}_0, \tag{8}$$

subject to the boundary conditions on $\partial\Omega$:

$$\alpha\mathbf{u} + (1-\alpha)\frac{\partial \mathbf{u}}{\partial \mathbf{n}} = \mathbf{u}_\infty.$$

Here the parameter α has the limit values of 0 for the free-slip (no stress) condition (Neumann) and 1 for the no-slip condition (Dirichlet); $\mathbf{u} = (u, v)$ is the velocity; t is the time; and p is the pressure. In general, the external force $\mathbf{f}$ is not taken into account in Eq. (6).

Using the divergence free constraint, problem (6)-(8) can be rewritten in the following conservative flux form:

$$\frac{\partial \mathbf{u}}{\partial t} - \nu\Delta\mathbf{u} + \nabla\cdot F + \nabla p = \mathbf{f}, \quad \text{in} \quad \Omega\times(0,T) \tag{9}$$

$$\nabla\cdot\mathbf{u} = 0, \quad \text{in} \quad \Omega\times(0,T) \tag{10}$$

$$\mathbf{u}|_{t=0} = \mathbf{u}_0, \tag{11}$$

with the flux F being defined as

$$F(\mathbf{u}) = \mathbf{u} \otimes \mathbf{u} = \begin{bmatrix} u^2 & uv \\ uv & v^2 \end{bmatrix}. \tag{12}$$

and $\mathbf{u} \otimes \mathbf{v} = u_i v_j$, $i,\ j = 1, 2$. Indeed, it holds

$$\mathcal{N}(\mathbf{u}) = \begin{pmatrix} \frac{\partial(u^2)}{\partial x} + \frac{\partial(uv)}{\partial y} \\ \frac{\partial(uv)}{\partial x} + \frac{\partial(v^2)}{\partial y} \end{pmatrix} = \nabla \cdot F(\mathbf{u}).$$

A locally conservative DG discretization will be employed for the Navier-Stokes equation (9)-(11). We denote by $\mathcal{E}_h$ a shape-regular triangulation of the domain $\bar{\Omega}$ into triangles, where h is the maximum diameter of elements. Let Γ_h^I be the set of all interior edges of $\mathcal{E}_h$ and Γ_h^B be the set of all boundary edges. Set $\Gamma_h = \Gamma_h^I \cup \Gamma_h^B$. For any nonnegative integer r and $s \geq 1$, the classical Sobolev space on a domain $E \subset \mathbb{R}^2$ is

$$W^{r,s}(E) = \{v \in L^s(E) | \,\forall\, |m| \leq r,\ \partial^m v \in L^s(E)\}.$$

We define the spaces of discontinuous functions

$$\begin{aligned} V &= \{\mathbf{v} \in L^2(\Omega)^2 : \quad \forall E \in \mathcal{E}_h,\ \mathbf{v}|_E \in (W^{2,4/3}(E))^2\}, \\ M &= \{q \in L^2(\Omega) : \quad \forall E \in \mathcal{E}_h,\ q|_E \in W^{1,4/3}(E)\}. \end{aligned}$$

The jump and average of a function ϕ on an edge e are defined by:

$$\begin{aligned} [\phi] &= (\phi|_{E_k})|_e - (\phi|_{E_l})|_e, \\ \{\phi\} &= \frac{1}{2}\left((\phi|_{E_k})|_e + (\phi|_{E_l})|_e\right). \end{aligned}$$

Further, let $\mathbf{v}$ be a piecewise smooth vector-, or matrix-valued function at $\mathbf{x} \in e$ and denote its jump by

$$[\mathbf{v}] := \mathbf{v}^+ \cdot \mathbf{n}_{E^+} + \mathbf{v}^- \cdot \mathbf{n}_{E^-},$$

where e is shared by two elements E^+ and E^-, and an outward unit normal vector $\mathbf{n}_{E^+}$ (or $\mathbf{n}_{E^-}$) is associated with the edge e of an element E^+ (or E^-). The tensor product of two tensors $\mathbf{T}$ and $\mathbf{S}$ is defined as $\mathbf{T} : \mathbf{S} = \sum_{i,j} T_{ij} S_{ij}$.

Let $\mathbb{P}_N(E)$ be the set of polynomials on an element E with degree no more than N. Based on the triangulation, we introduce two approximate subspaces $\mathbf{V}_h (\subset V)$ and $M_h (\subset M)$ for integer $N \geq 1$:

$$\begin{aligned} \mathbf{V}_h &= \{\mathbf{v} \in L^2(\Omega)^2 : \quad \forall E \in \mathcal{E}_h,\ \mathbf{v}_h \in (\mathbb{P}_N(E))^2\}, \\ M_h &= \{q \in L^2(\Omega) : \quad \forall E \in \mathcal{E}_h,\ q \in \mathbb{P}_{N-1}(E)\}, \end{aligned}$$

The work was motivated by the work of Girault, Rivière and Wheeler on discontinuous finite elements for incompressible flows presented in a series of papers [8, 15]. Some projection methods [2, 19] have been developed to overcome the incompressibility constraints $\nabla \cdot \mathbf{u} = 0$. An implementation of the operator-splitting idea for discontinuous Galerkin elements was developed in [8]. We appreciate the advantages of the discontinuous Galerkin methods, such as local mass conservation, high order of approximation, robustness and stability. In this work, we will make use of the underlying physical nature of incompressible flows in the literature and extend the interior penalty discontinuous Galerkin methods to investigate dynamical behavior of vortex dominated lid-driven and cylinder flows.

The paper is organized as follows. In Section 2, a temporal discretization for the Navier-Stokes equation is presented by operator-splitting techniques, and subsequently, the nonlinearity is linearized. Both pressure and velocity can be solved successively from linear elliptic and Helmholtz-type problems, respectively. In Section 3, a local numerical flux will be given for the nonlinear convection term and an interior penalty discontinuous Galerkin scheme will be used in spacial discretization for those linear equations, and in Section 4, simulations of a lid-driven cavity flow up to $Re = 7500$, and a transient flow past a circular cylinder are presented, while a numerical investigation on the Strouhal-Reynolds-number has been done, comparable to the experimental results. Finally, Section 5 concludes with a brief summary.

2. Temporal splitting scheme

We consider here a third-order time-accurate discretization method at each time step by using the previous known velocity vectors. Let Δt be the time step, $M = \frac{T}{\Delta t}$, and $t_n = n\Delta t$. The semi-discrete forms of problem (6)-(8) at time t_{n+1} is

$$\begin{aligned}&\frac{\gamma_0\mathbf{u}^{n+1} - \alpha_0\mathbf{u}^n - \alpha_1\mathbf{u}^{n-1} - \alpha_2\mathbf{u}^{n-2}}{\Delta t} - \nu\Delta\mathbf{u}^{n+1} + \nabla p^{n+1}\\ &= -\beta_0\mathcal{N}(\mathbf{u}^n) - \beta_1\mathcal{N}(\mathbf{u}^{n-1}) - \beta_2\mathcal{N}(\mathbf{u}^{n-2}) + \mathbf{f}(t_{n+1}),\end{aligned} \tag{13}$$

$$\nabla\cdot\mathbf{u}^{n+1} = 0, \tag{14}$$

which has a timestep constraint based on the CFL condition (see [14]):

$$\Delta t \approx O\Big(\frac{\mathcal{L}}{\mathcal{U}N^2}\Big),$$

where $\mathcal{L}$ is an integral length scale (e.g. the mesh size) and $\mathcal{U}$ is a characteristic velocity. Because the semi-discrete system (13)-(14) is linearized, thus, a time-splitting scheme can be applied naturally, i.e., the semi-discretization in time (13)-(14) can be decomposed into three stages as follows.

- The first stage
  When $\mathbf{u}^n$ and $\mathbf{u}^{n-1}$ ($n \geq 1$) are known, the following linearized third-order formula can be used

$$\frac{\gamma_0\tilde{\mathbf{u}} - \alpha_0\mathbf{u}^n - \alpha_1\mathbf{u}^{n-1} - \alpha_2\mathbf{u}^{n-2}}{\Delta t} = -\beta_0\mathcal{N}(\mathbf{u}^n) - \beta_1\mathcal{N}(\mathbf{u}^{n-1}) + \mathbf{f}(t_{n+1}) \tag{15}$$

  with the following coefficients for the subsequent time levels ($n \geq 2$)

$$\gamma_0 = \frac{11}{6},\ \alpha_0 = 3,\ \alpha_1 = -\frac{3}{2},\ \alpha_2 = \frac{1}{3},\ \beta_0 = 2,\ \beta_1 = -1.$$

Especially, by using the Euler forward discretization at the first time step ($n = 0$), we can get a medium velocity field $\mathbf{u}^1$ by

$$\frac{\mathbf{u}^1 - \mathbf{u}^0}{\Delta t} = -\mathcal{N}(\mathbf{u}^0) + \mathbf{f}(t_1),$$

and $\mathbf{u}^2$ by

$$\frac{\gamma_0\mathbf{u}^2 - \alpha_0\mathbf{u}^1 - \alpha_1\mathbf{u}^0}{\Delta t} = -\beta_0\mathcal{N}(\mathbf{u}^1) - \beta_1\mathcal{N}(\mathbf{u}^0) + \mathbf{f}(t_2),$$

which adopts the following coefficients to construct a second-order difference scheme for the time level ($n = 2$) in (15)

$$\gamma_0 = \frac{3}{2},\ \alpha_0 = 2,\ \alpha_1 = -\frac{1}{2},\ \alpha_2 = 0,\ \beta_0 = 2,\ \beta_1 = -1.$$

- The second stage
  The pressure projection is as follows

$$\gamma_0\frac{\tilde{\tilde{\mathbf{u}}} - \tilde{\mathbf{u}}}{\Delta t} = -\nabla p^{n+1}. \tag{16}$$

To seek p^{n+1} such that $\nabla \cdot \tilde{\tilde{\mathbf{u}}} = 0$, we solve the system

$$-\Delta p^{n+1} = -\frac{\gamma_0}{\Delta t}\nabla \cdot \tilde{\mathbf{u}}, \tag{17}$$

with a Neumann boundary condition being implemented on the boundaries as

$$\begin{aligned}\frac{\partial p^{n+1}}{\partial \mathbf{n}} =& \mathbf{f}^{n+1} - \beta_0\mathbf{n}\cdot\Big(\frac{\partial \mathbf{u}^n}{\partial t} + \nabla\cdot F(\mathbf{u}^n) - \nu\Delta\mathbf{u}^n\Big)\\ &- \beta_1\mathbf{n}\cdot\Big(\frac{\partial \mathbf{u}^{n-1}}{\partial t} + \nabla\cdot F(\mathbf{u}^{n-1}) - \nu\Delta\mathbf{u}^{n-1}\Big)\\ &- \beta_2\mathbf{n}\cdot\Big(\frac{\partial \mathbf{u}^{n-2}}{\partial t} + \nabla\cdot F(\mathbf{u}^{n-2}) - \nu\Delta\mathbf{u}^{n-2}\Big) := G_n.\end{aligned}$$

One can compute the vorticity $\omega^n = \nabla \times \mathbf{u}^n$ at time $t^n = n \cdot \Delta t$. Then we use p^{n+1} to update the intermediate velocity $\tilde{\tilde{\mathbf{u}}}$ by (16).

- The third stage is completed by solving

$$\gamma_0\frac{\mathbf{u}^{n+1} - \tilde{\tilde{\mathbf{u}}}}{\Delta t} = \nu\Delta\mathbf{u}^{n+1},$$

which can be written as a Helmholtz equation for the velocity

$$-\Delta\mathbf{u}^{n+1} + \frac{\gamma_0}{\nu\Delta t}\mathbf{u}^{n+1} = \frac{\gamma_0}{\nu\Delta t}\tilde{\tilde{\mathbf{u}}}. \tag{18}$$

From the three stages given above, we notice that (15) in the semi-discrete systems is presented in a linearized and explicit process, moreover, (17) and (18) are obviously a type of elliptic and Helmholtz problems at each time step as $n \geq 2$. We decouple the incompressibility condition and the nonlinearity, then the pressure and velocity semi-discretizations (17)-(18) will be formulated by the interior penalty discontinuous Galerkin methods in spacial discretizations in the next section.

3. The spatial discretizations

For spacial approximations, assume that piecewise polynomials of order N are employed, then the approximation space can be rewritten as $\mathbf{V}_h = \bigoplus_{k=1}^{K} \mathbb{P}_N(E_k)^2$. In the approximating polynomial space for the velocity or pressure restricted to each element, a high-order nodal basis can be chosen, consisting of Lagrange interpolating polynomials defined on a reference simplex introduced in [11, 12]. We let $\mathbf{u}$ be approximated by $\mathbf{u}_h \in \mathbf{V}_h$ and adopt a suitable approximation for the term F, i.e., $F(\mathbf{u}) \approx F(\mathbf{u}_h)$, where $F(\mathbf{u}_h)$ also can be represented as the L^2-projection of $F(\mathbf{u}_h)$ on each element of $\mathcal{T}_h$. Multiplying the nonlinear term by a test function $\mathbf{v}_h \in \mathbf{V}_h$, integrating over the computational domain, and applying integration by parts, we have

$$\int_{\Omega} (\nabla \cdot F) \cdot \mathbf{v}_h d\mathbf{x} = -\sum_{E_k} \int_{E_k} (F \cdot \nabla) \cdot \mathbf{v}_h dx + \sum_{e \in \Gamma_h} \int_{e} \mathbf{n}_e \cdot [F \cdot \mathbf{v}_h] ds, \tag{19}$$

where the term $(F \cdot \nabla) \cdot \mathbf{v}_h$ equals to $F_{ij} \frac{\partial v_{hi}}{\partial x_j}$, for $i, j = 1, 2$ and the indexes i, j correspond to the components of the related vectors. On each edge $e \in \partial E_1 \cap \partial E_2$ shared by two elements, to ensure the flux Jacobian of purely real eigenvalues, we may define $\lambda^+_{E_1,e}$, $\lambda^-_{E_2,e}$ the largest eigenvalue of the Jacobians $\frac{\partial}{\partial \mathbf{u}}(F \cdot \mathbf{n}_e)\big|_{\bar{\mathbf{u}}_{E_1}}$ and $\frac{\partial}{\partial \mathbf{u}}(F \cdot \mathbf{n}_e)\big|_{\bar{\mathbf{u}}_{E_2}}$, respectively, where $\bar{\mathbf{u}}_{E_1}$ and $\bar{\mathbf{u}}_{E_2}$ are the mean values of u_h on the elements E_1 and E_2, respectively. The global Lax-Friedrichs flux is generally more dissipative than the local Lax-Friedrichs flux, therefore, we primarily consider the local flux on each edge. Although the Lax-Friedrichs flux is perhaps the simplest numerical flux and often the most efficient flux, it is not the most accurate scheme. A remedy of the problem is to employ high-order finite elements. By replacing the integrand in the surface integral as

$$\mathbf{n}_e \cdot [F \cdot \mathbf{v}_h] = \mathbf{n}_e \cdot \{F\} \cdot [\mathbf{v}_h] + \frac{\lambda_e}{2} [\mathbf{u}_h] \cdot [\mathbf{v}_h],$$

with $\lambda_e = \max(\lambda^+_{E_1,e}, \lambda^-_{E_2,e})$, one can get a DG discretization for the nonlinear term in (19) by the local Lax-Friedrichs flux.

For the pressure correction step (17) and the viscous correction step (18), we use the SIPG method to approximate the correction steps. Choosing the orthonormal Legendre basis and the Legendre-Gauss-Lobatto quadrature points gives a well-conditioned Vandermonde matrix and the resulting interpolation well behaved, which greatly simplifies the formulas. The C^0 continuity condition of the basis in the discontinuous Galerkin formulation is not required. Enforcing a weak continuity on the interior edges by a penalty term, we have for (17)

$$a(p_h^{n+1}, \phi_h) = L_p(\phi_h), \ \phi \in M_h,$$

where

$$\begin{aligned} a(p_h^{n+1}, \phi_h) = & \sum_{E_k \in \mathcal{E}_h} \int_{E_k} \nabla p_h^{n+1} \cdot \nabla \phi_h dx - \sum_{e_k \in \Gamma_h} \int_{e_k} \{\frac{\partial p_h^{n+1}}{\partial \mathbf{n}}\} [\phi_h] ds \\ & - \sum_{e_k \in \Gamma_h} \int_{e_k} \{\frac{\partial \phi_h}{\partial \mathbf{n}}\} [p_h^{n+1}] ds + \sum_{e_k \in \Gamma_h} \frac{\sigma_e}{|e_k|^\beta} \int_{e_k} [p_h^{n+1}][\phi_h] ds, \end{aligned}$$

$$\text{and} \quad L_p(\phi_h) = \sum_{E_k \in \mathcal{E}_h} \int_{E_k} \frac{\gamma_0}{\Delta t} \nabla \cdot \tilde{\mathbf{u}} \phi_h dx + \sum_{e_k \in \partial\Omega} \int_{e_k} \phi_h G_n.$$

In general, σ_e shall be chosen sufficiently large to guarantee coercivity, more accurately, the threshold values of σ_e in [4] are given for $\beta = 1$ in the above formula, which is referred to an SIPG scheme. Especially, as $\beta > 1$, the scheme is referred to an over-penalized scheme and the threshold values of σ_e are presented in [20, 21]. Analogously, the SIPG discretization for (18) is given by

$$a(\mathbf{u}_h^{n+1}, \mathbf{v}_h) + \frac{\gamma_0}{\nu \Delta t}(\mathbf{u}_h^{n+1}, \mathbf{v}_h)_\Omega = L_{\mathbf{u}}(\mathbf{v}_h), \quad \forall\, \mathbf{v}_h \in \mathbf{V}_h,$$

where

$$\begin{aligned} a(\mathbf{u}_h^{n+1}, \mathbf{v}_h) = & \sum_{E_k \in \mathcal{E}_h} \int_{E_k} \nabla \mathbf{u}_h^{n+1} : \nabla \mathbf{v}_h d\mathbf{x} - \sum_{e_k \in \Gamma_h} \int_{e_k} \{\nabla \mathbf{u}_h^{n+1}\}\mathbf{n} \cdot [\mathbf{v}_h] ds \\ & - \sum_{e_k \in \Gamma_h} \int_{e_k} \{\nabla \mathbf{v}_h\}\mathbf{n} \cdot [\mathbf{u}_h^{n+1}] ds + \sum_{e_k \in \Gamma_h} \frac{\sigma_e}{|e_k|^\beta} \int_{e_k} [\mathbf{u}_h^{n+1}] \cdot [\mathbf{v}_h] ds, \end{aligned}$$

and

$$L_{\mathbf{u}}(\mathbf{v}_h) = \Big(\frac{\gamma_0}{\nu \Delta t} \tilde{\tilde{\mathbf{u}}}_h, \mathbf{v}_h\Big)_\Omega - \sum_{e_k \in \partial\Omega} \int_{e_k} \Big(\nabla \mathbf{v}_h \cdot \mathbf{n}_e - \frac{\sigma_e}{|e_k|^\beta} \mathbf{v}_h\Big) \mathbf{u}_0.$$

where $\beta = 1$. As a DG method, these SIPG schemes have some attractive advantages of DG methods including high order hp-approximation, local mass conservation, robustness and accuracy of DG methods for models with discontinuous coefficients and easy implementation on unstructured grids, while the flexibility of p-adaptivity (different orders of polynomialsmight be used for different elements) in DG methods has become competitive for modeling a wide range of engineering problems.

4. Numerical results

We present a lid-driven flow problem to verify the efficiency and robustness of the interior penalty discontinuous Galerkin method, and then investigate a flow past a cylinder with walls or without a wall, as well as the relationship between the Strouhal number and the Reynold number. Throughout the section, time steps $\Delta t \leq 1E - 03$ are taken.

Example 1. The lid-driven boundary conditions are given by:

$$\begin{cases} u(x,1) = 1;\ u(1,y) = 0;\ u(0,y) = 0;\ u(x,0) = 0; \\ v(x,1) = 1;\ v(1,y) = 0;\ v(0,y) = 0;\ v(x,0) = 0. \end{cases}$$

Here the mesh size of the initial coarse grid is 0.2 and then it is uniformly refined three times with piecewise discontinuous elements being applied into the fully discrete SIPG approach.

Employing the uniform mesh (see the left profile in Figure 1) and approximation polynomials of order 3, we illustrate the velocity vector, pressure and vorticity profiles in Figure 2, observing that the main characterization of solutions can be captured by fine meshes well, except a few very small oscillations occurred close to the upper boundary along the lid for computing velocity in y−direction

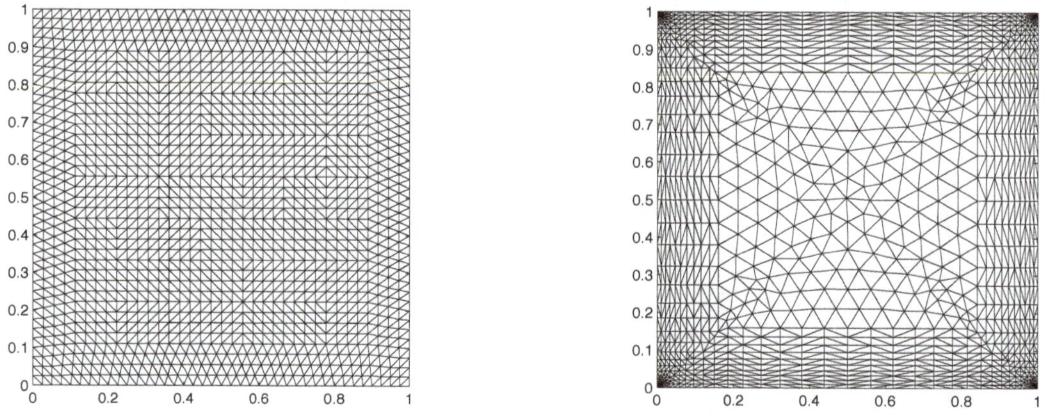

Figure 1. #1: A uniform mesh (left); #2: An initial locally refined mesh (right).

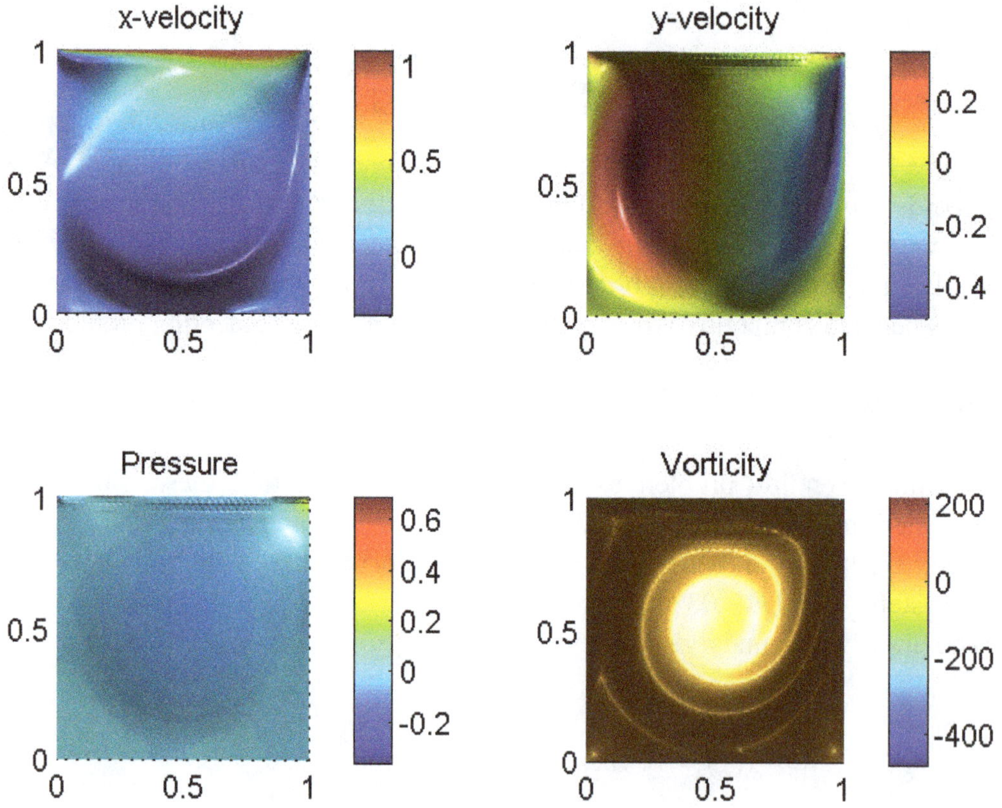

Figure 2. The profiles of velocity components (u, v), pressure and vorticity for Re=2000 by using mesh #1 and $N = 4$.

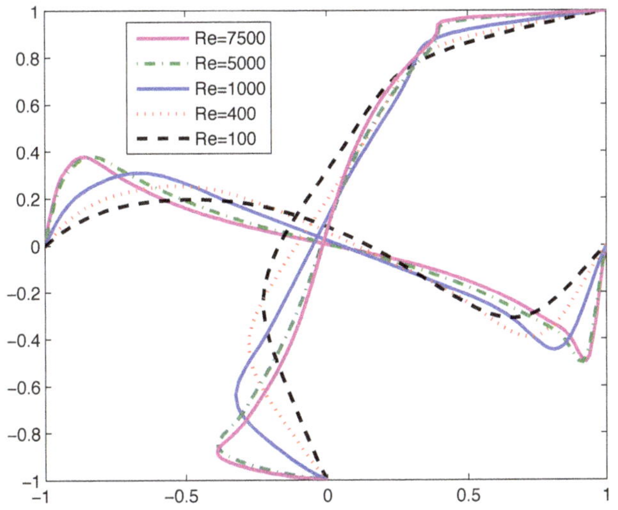

Figure 3. Velocity profiles (u, v) through geometric center of the cavity for Re = 100, 400, 1000, 5000, 7500.

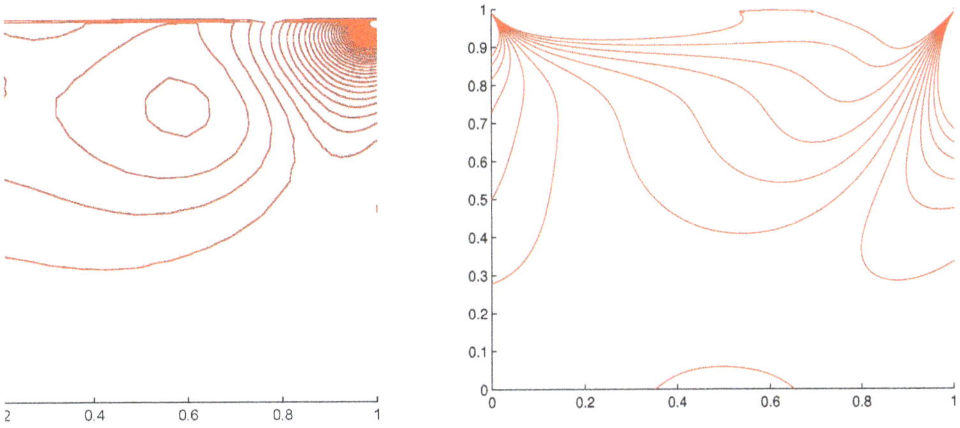

Figure 4. Re=100 and $N = 3$, mesh #2. Left: pressure contour; Right: vorticity contour.

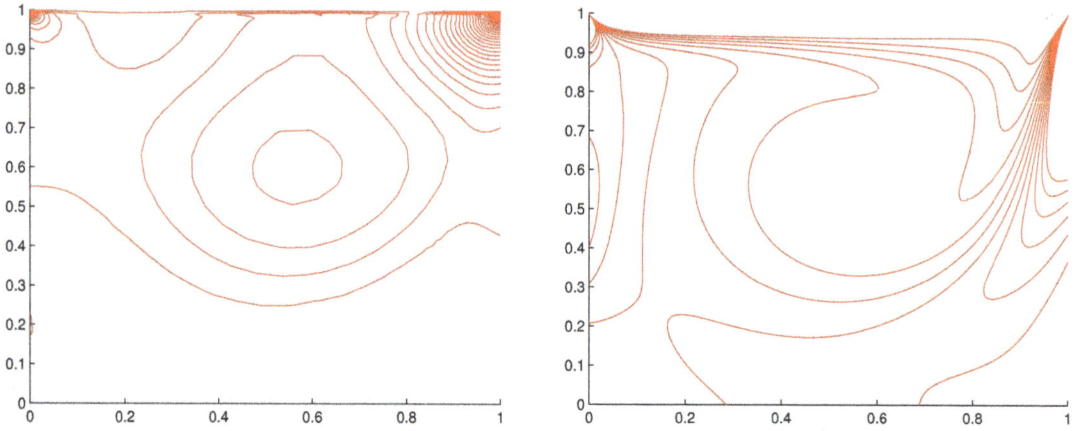

Figure 5. Re=400 and $N = 3$, mesh #2. Left: pressure contour; Right: vorticity contour.

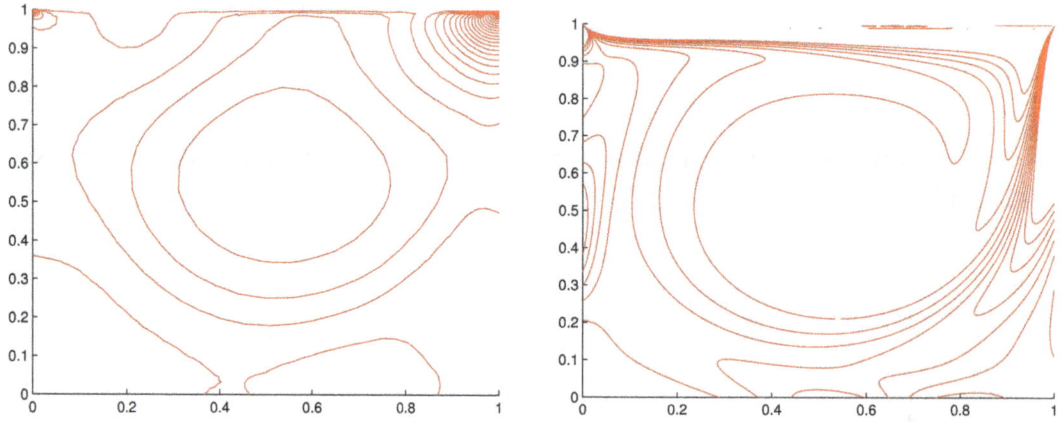

Figure 6. Re=1000, N=4, mesh #2. Left: pressure contour; Right: vorticity contour.

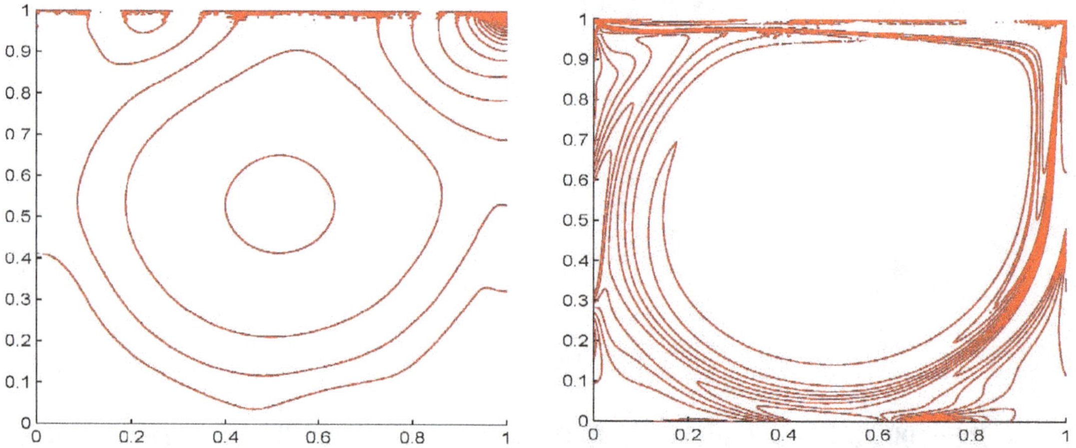

Figure 7. Re=5000, N=3, mesh #2 with a refinement once. Left: pressure contour; Right: vorticity contour.

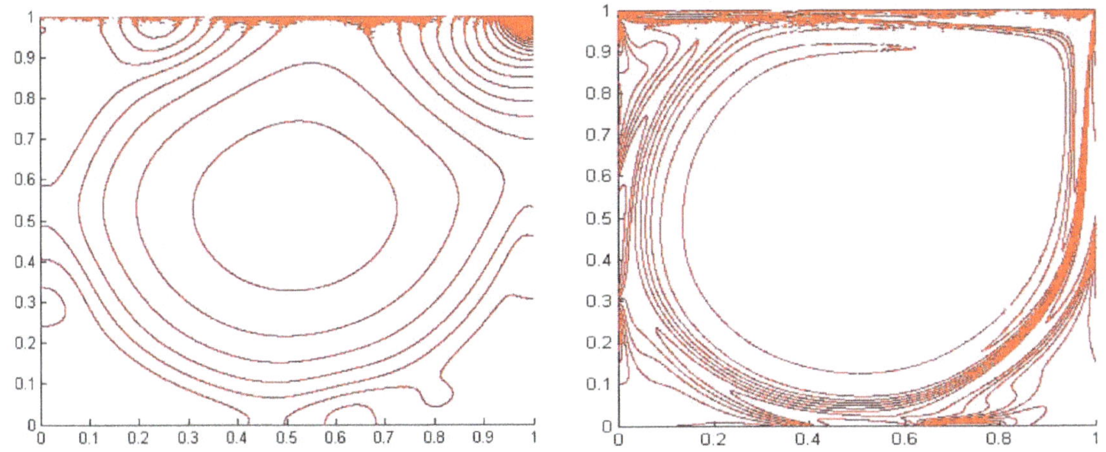

Figure 8. Re=7500, N=3, a refined mesh of #2. Left: pressure contour; Right: vorticity contour.

and pressure. Oscillations on the upper lid propagates from a velocity singularity that exists at the corners. The boundary condition at the vertex is a jump from zero velocity on the edge to a unit velocity on the upper edge. Nature prevents this singularity with a boundary layer forming along all walls, making the vertex velocity zero. It is adaptable to adopt an adaptive meshes for solving those singularity problems. Here, we apply the semi-implicit SIPG method in a locally refined mesh (see the right profile in Figure 1) to solve the incompressible flow. In Figure 3, the velocity profiles of (u, v) through the geometric center of the cavity are plotted with $Re = 100, 400, 1000, 5000, 7500$ taken. From Figures 4-8, with different Reynolds numbers taken up to 7500, the vorticity field exhibits the expected characteristics of a driven cavity flow consisting of a region of vortical flow centrally located. Energy enters the cavity through the viscous boundary later formed by velocity gradients on the upper driven edge. Convection distributes flow properties throughout the domain. Moreover, a video on the dynamical evolution of vorticity isolines ($Re = 1000$, $N = 4$) can be browsed through a website (Available from: `https://youtu.be/UfGWvnoiW58`). These numerical simulations are performed for the Navier-Stokes equations which illustrate the effectiveness of the DG method.

Example 2. We simulate a channel flow past a circular cylinder with a radius 0.05 at the origin $(0, 0)$ for $Re = 100$ by the discontinuous Galerkin method in the domain $(-1, 3) \times (-0.5, 0.5)$. The inflow boundary condition is $(u, v) = (1, 0)$, while the outflow boundary is $\frac{\partial \mathbf{u}}{\partial \mathbf{n}} = 0$. To the boundary conditions on the upper and lower sides, we present two different conditions for comparison (see Figure 9), which are wall ($u = 0$, $v = 0$) and homogeneous Neumann boundary conditions ($u = 1$, $\frac{\partial v}{\partial \mathbf{n}} = 0$), respectively. The homogeneous Neumann boundary condition is a special non-reflecting case, where the boundary flux is zero. For reference, the density of the fluid is given by $\rho = 1\ kg \cdot m^{-3}$ and a locally refined mesh ($\max h = 0.088$) will be used for the simulations.

Cylinder flow contains the fundamentals of unsteady fluid dynamics in a simplified geometry. The flow properties and unsteadiness are well defined through years of experimental measurements across a wide range of Reynolds numbers [13], making the cylinder an ideal validation testcase for unsteady numerical fluid dynamics simulations.

Verification in a numerical domain requires insights from physics for a proper comparison to ex-

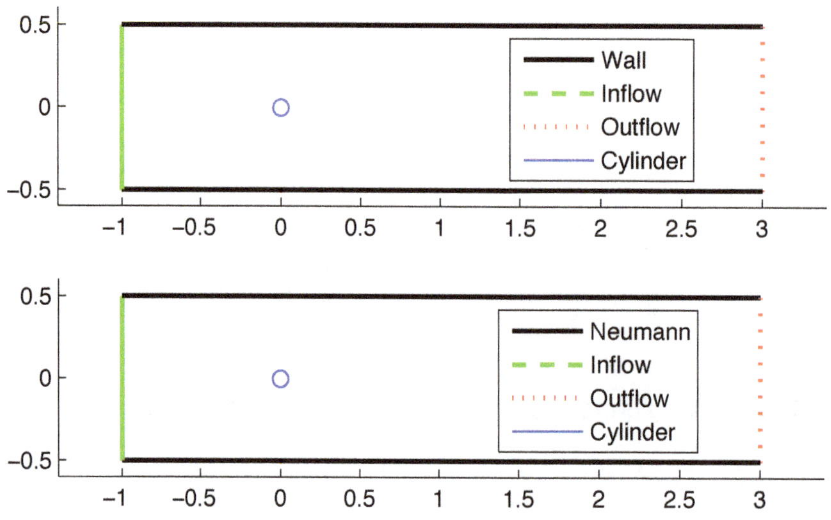

Figure 9. Flow past a cylinder and $N = 3$. Top: A channel with walls; Bottom: A channel without a wall. The free stream velocity on the inflow boundary is $\mathbf{u}_\infty = (1, 0)$.

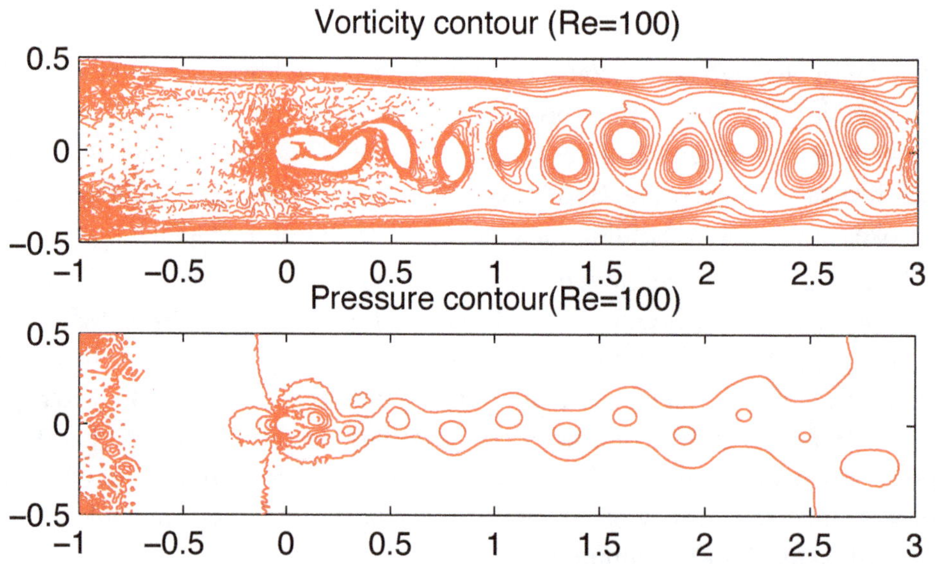

Figure 10. Flow past a cylinder in a channel with walls, Re=100. Top: Vorticity contour; Bottom: Pressure contour.

perimental and theoretical data. In Fig. 10, we observe that the boundary layers forms along the upper and lower walls. From continuity of mass, the presence of a boundary layer decreasing the flow velocity near the wall requires an increase in the centerline average flow velocity. The cylinder's wake provides a similar increase in centerline velocities. This implies a non-intuitive reality that drag can increase velocities within constrained domains. This effect is compensated for in wind tunnel test [1] environments topologically similar to Fig.10 with a constant mass flow rate and no-slip walls. Drela[3] develops an analysis for 2D wind tunnels resulting in an effective coefficient of drag of

$$C_d = \left(1 - \frac{1}{2}\frac{c}{H}C_d - \frac{\pi}{2}\frac{A}{H^2}\right)C_{d_{un}},$$

and an effective Reynolds number of

$$Re = \left(1 + \frac{1}{4}\frac{c}{H}C_d + \frac{\pi}{6}\frac{A}{H^2}\right)Re_{un},$$

where the *un* subscript represents the uncorrected value, H represents the domain height, A represents the cylinder area and c represents the cylinder radius. Drela's analysis does not specifically include the boundary layer forming on the upper and lower walls. The flow physics associated with wall boundary layer drag differs from cylinder drag in that the wall drag is a distributed effect of monotonically increasing drag with downstream distance rather than a conceptual point source of drag. The wall boundary layer tends to provide a steady acceleration of flow within the interior flow domain (i.e. non-boundary layer portion) leading to an effective buoyancy drag. A secondary feature of the wall boundary layer is that downstream flow features such as vortices are convected at a higher perturbation velocity compared to the initial upstream velocity. For numerical validation of raw experimental data, either the wind tunnel geometry should exactly match the numerical geometry, or the numerical geometry should be corrected using the concepts introduced above to match the actual wind tunnel geometry. Alternatively, the open-air corrected values should be used for validation. The above analysis provides insight into the domain height necessary to reduce volume blockage (c/H) and wake buoyancy (A/H^2) effects.

Alternatively, the flow without walls in Figure 11 has no interference of the boundary layers along the channel on the up and bottom boundaries, thus the pressure contours expend after flow passing through the cylinder. We also compared the components of the velocity profiles along the x direction in Figure 12, and observed that the boundary layers are produced in the top picture rather than in the bottom one. If the effect of the boundary layers disappeared, the velocity in the $x-$direction would reduce dispersively, in other words, the vortex lifespan is less than those produced in the channel with walls. The velocity profiles in the $y-$direction have been given in Figure 13.

We localize the domain around the cylinder and refine the mesh, then show the vorticity startup behavior in Figure 14 as well as the pressure Figure 15. Upon startup, two vortices of opposite direction are formed on the upper and lower aft portion of the cylinder. Given a low total simulation time, the flow field resembles the symmetrical low Reynolds number steady flow. As time progresses however, instabilities are magnified and the upper-lower symmetry increases. Given a total time of beyond $t = 10$, an autonomous and phased locked set of street vortices are generated. Surface pressures (Figure 16 at $Re = 80, 140$) generated reflect the process, including a steady startup portion and the eventual vortex shedding frequency. Validation at $Re > 41$ requires sufficient time to obtain the unsteady behavior.

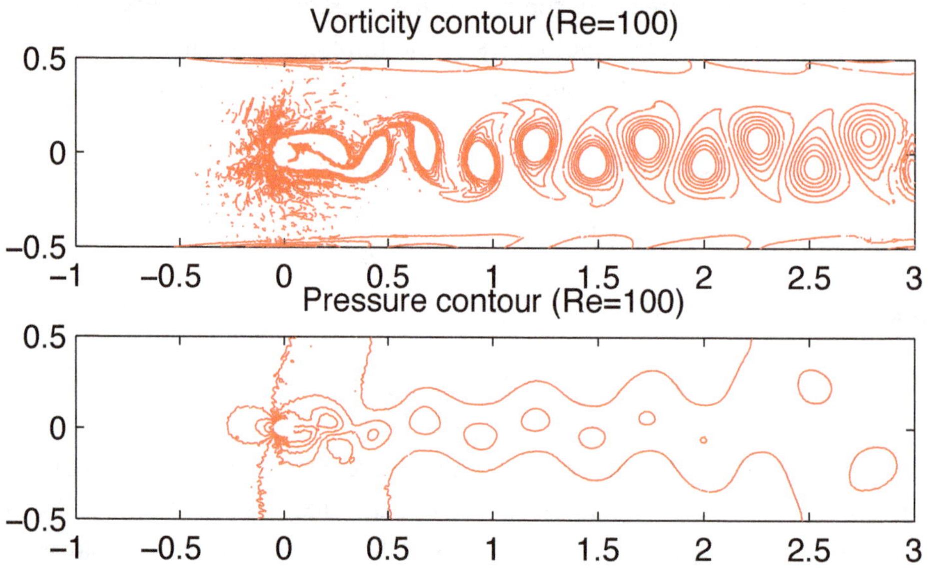

Figure 11. Flow past a cylinder in a channel without a wall, Re=100. Top: Vorticity contour; Bottom: Pressure contour.

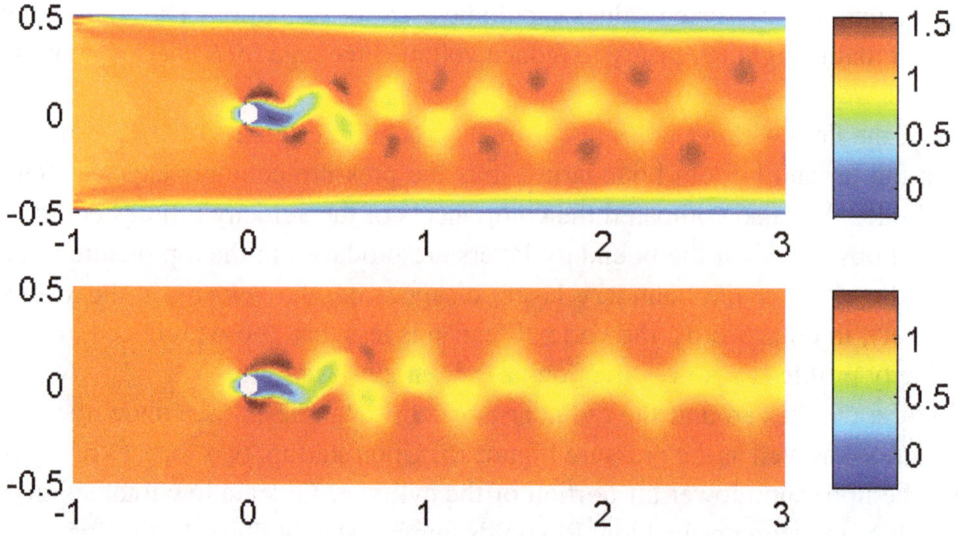

Figure 12. Flow past a cylinder in a channel, Re=100. Top: Velocity in x-direction with walls; Bottom: Velocity in x-direction without a wall.

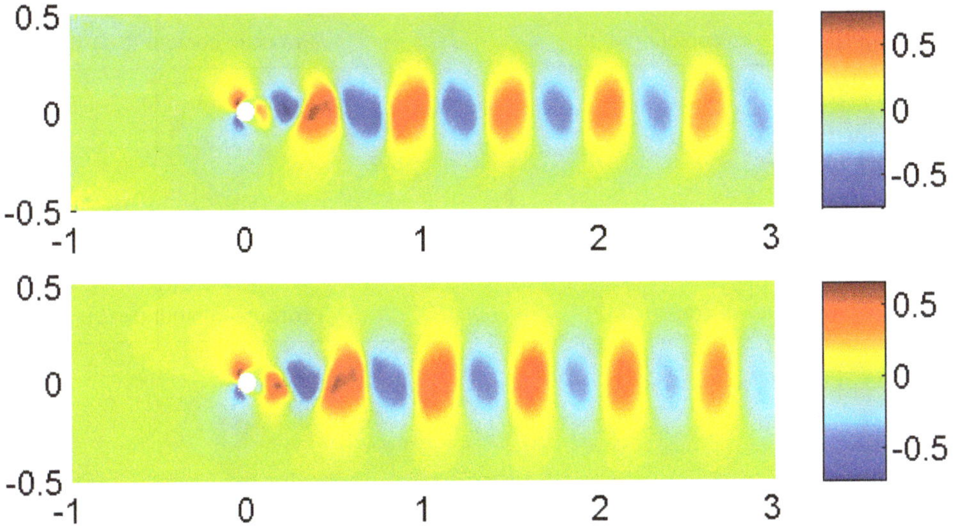

Figure 13. Flow past a cylinder in a channel, Re=100. Top: Velocity in y-direction with walls; Bottom: Velocity in y-direction without a wall.

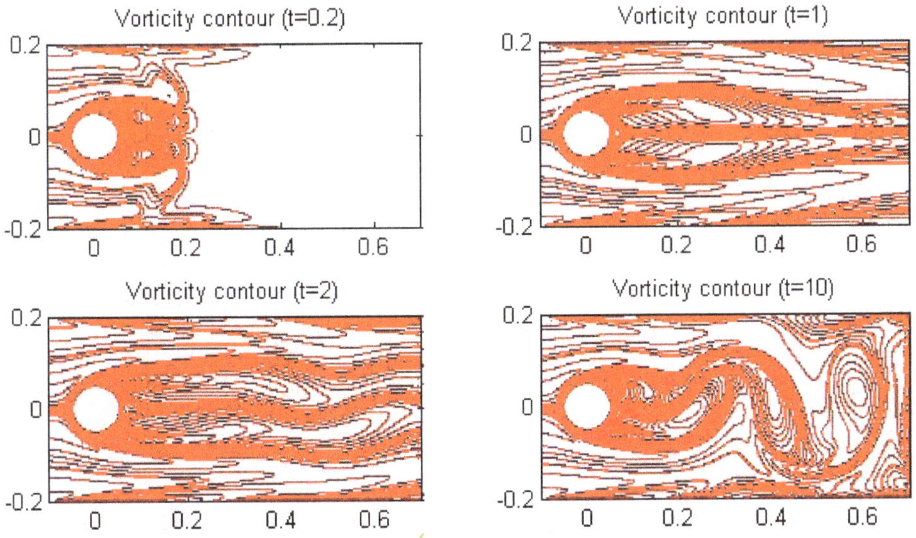

Figure 14. Vorticity contours of flow past a cylinder without a wall at different time.

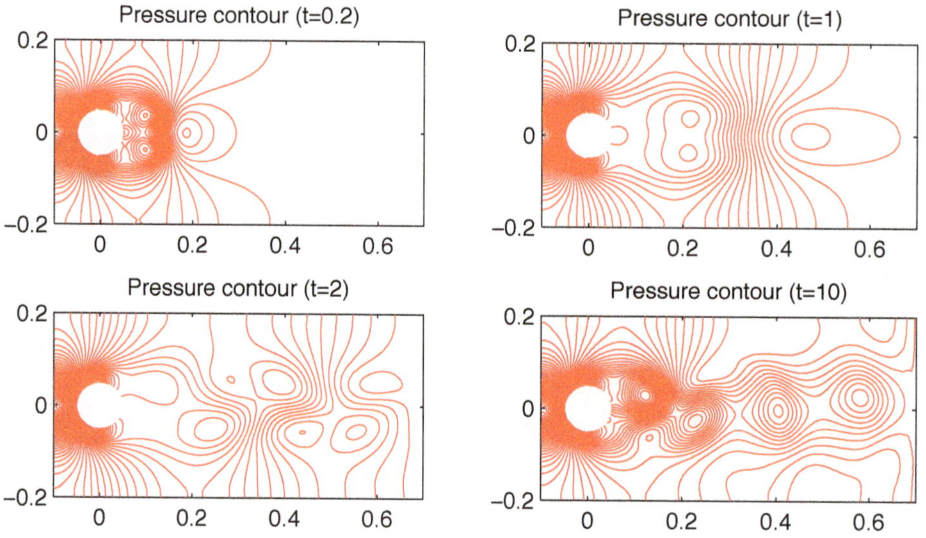

Figure 15. Pressure contours of flow past a cylinder without a wall at different time.

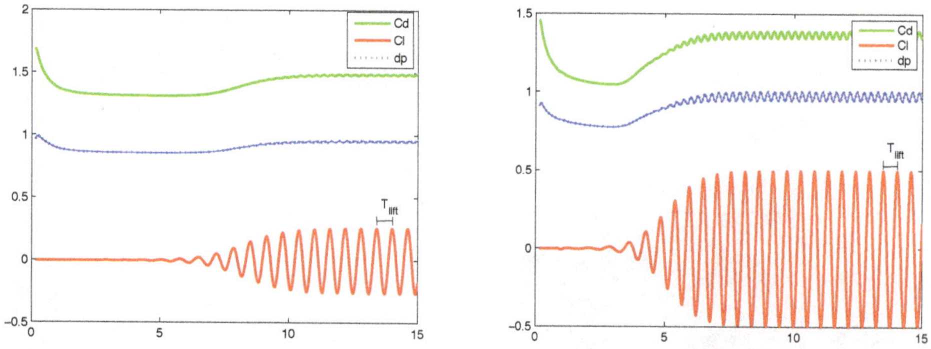

Figure 16. Drag coefficients, Lift coefficients and the variation of pressure with time, *Re* = 80 **(left),** *Re* = 140 **(right).**

To reduce the effect of the boundary layers along the walls, the coefficients of drag and lift as well as the difference of pressure between the leading edge and the trailing edge on the cylinder shall be computed in a larger domain. Then in a domain $\Omega := [-1, 5] \times [-1, 1]$, higher order DG finite elements have been investigated. Based on the velocity $\mathbf{u}_\infty$ and the diameter of the cylinder $D = 0.1$, we will chose different viscosity coefficients $\nu = 1e-3, 5e-4, 2.5e-4$ etc. to simulate flow with different Reynolds numbers, that is, the cases $Re = 100, 200, 400$, respectively. Our interest is the drag coefficient C_d, the lift coefficient C_l on the cylinder and the difference of the pressure between the front and the back of the cylinder

$$d_p = p(t; -0.05, 0) - p(t; 0.05, 0).$$

We use the definition of C_d and C_l given in [17] as follows:

$$C_d = \frac{2}{\rho D u_{\max}^2} \int_S \Big(\rho\nu \frac{\partial \mathbf{u}_{\mathbf{t_S}}(t)}{\partial \mathbf{n}} n_y - p(t) n_x\Big) ds,$$
$$\text{and} \quad C_l = -\frac{2}{\rho D u_{\max}^2} \int_S \Big(\rho\nu \frac{\partial \mathbf{u}_{\mathbf{t_S}}(t)}{\partial \mathbf{n}} n_x + p(t) n_y\Big) ds,$$

where $\mathbf{n} = (n_x, n_y)^T$ is the normal vector on the cylinder boundary S directing into Ω, $\mathbf{t_S} = (n_y, -n_x)^T$ the tangential vector and $\mathbf{u}_{\mathbf{t_S}}$ the tangential velocity along S. In the literature, Fey et al. in [6] propose the Strouhal number represented by piecewise linear relationships of the form

$$St = ST^* + \frac{m}{\sqrt{Re}}$$

with different values St^* and m in different shedding regimes of the 3D circular cylinder wake. We apply the periodic T_{lift} of the lift coefficients (see Figure 16) to express the periodic $T := \frac{1}{f}$ appearing in the definition of Strouhal number, i.e.,

$$St(Re) = \frac{D}{T_{lift} \mathbf{u}_\infty}, \tag{20}$$

which comes from the classical definition (5). From the evolution of C_d, C_l and d_p as in Figure 16, we may find a period T_{lift} of the lift coefficients for different values of Re to calculate Strouhal number by (20). In Figure 17, a comparison of Strouhal numbers between the experimental estimates in Fey etc. [6] and our estimates from (20) indicates a behavioral match between the unsteady onset at approximately $Re = 50$ and the beginning of the transition to turbulence at $Re = 180$. Beyond $Re = 180$, the onset of turbulence changes the flow physics by drastically increasing the energy spectrum of the shed vorticity. The transition appears as a marked decrease in the Strouhal number prior to $Re = 200$. As the present SIPG solver does not include a 3D turbulence model, our estimates follow the laminar results into the actual turbulent region.

5. Conclusion

This paper developed a Symmetric Interior Penalty discontinuous Galerkin numerical solver for the incompressible Navier Stokes equations of fluid flow, with the temporal splitting technique applied in

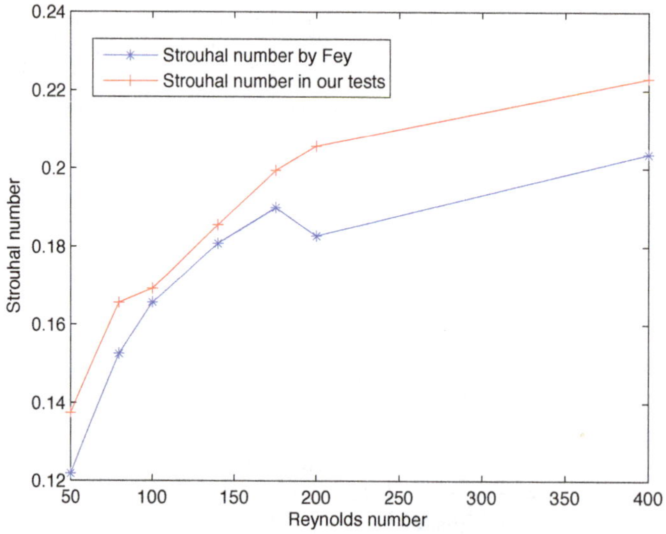

Figure 17. A comparison of Strouhal-Reynolds-number between our estimate and the linear fit in [6] for $50 \le Re \le 400$.

decoupling the diffusion and pressure terms and the local Lax-Friedrichs flux for the DG discretization of the nonlinear convection term. Two testcases are presented: a lid-driven cavity and a cylinder flow. The SIPG method produces stable discretizations of the convective operator for high order discretizations on unstructured meshes of simplices, as a requirement for real-world complex geometries.

Acknowledgments

The authors would like to thank the anonymous referee and editor very much for their valuable comments and suggestions, which greatly help us improve the presentation of this article. The work of the first author was partially supported by the Natural Science Foundation of Gansu Province, China (Grant 145RJZA046), and Special Program for Applied Research on Super Computation of the NSFC-Guangdong Joint Fund (the second phase).

Conflict of Interest

We declare no conflicts of interest in this paper.

References

1. J. B. Barlow, W. H. Rae, and A. Pope, Low-Speed Wind Tunnel Testing, John Wiley, 1999.
2. A. J. Chorin, *On the convergence of discrete approximations to the Navier-Stokes equations*, Math. Comp., **23** (1969), 341-353.
3. M. Drela, Flight Vehicle Aerodynamics, MIT Press, Boston, 2014.

4. Y. Epshteyn, B. Rivière, *Estimation of penalty parameters for symmetric interior penalty Galerkin methods*, J. Comput. Appl. Math., **206** (2007), 843-872.
5. C. Foias, O. Manley, R. Rosa and R. Temam, Turbulence and Navier-Stokes equations, Cambridge University Press, 2001.
6. U. Fey, M. König, and H. Eckelmann, *A new Strouhal-Reynolds-number relationship for the circular cylinder in the range* $47 \leq Re \leq 2 \times 10^5$, *Physics of Fluids*, **10(7)** (1998), 1547-1549.
7. U.Ghia, K. N. Ghia, C. T. Shin, *High-Re solutions for incompressible flow using the Navier-Stokes equations and a multigrid method.* J. Comput. Phys., **48** (1982), 387-411.
8. V. Girault, B. Rivière, and M. F. Wheeler, *A splitting method using discontinuous Galerkin for the transient incompressible Navier-Stokes equations*, ESAIM: Mathematical Modelling and Numerical Analysis, **39(6)** (2005), 1115-1147.
9. V. Girault, B. Rivière, and M. F. Wheeler, *A discontinuous Galerkin method with non-overlapping domain decomposition for the Stokes and Navier-Stokes problems*, Math. Comp.**74** (2005), 53-84.
10. O. Goyon, *High-Reynolds number solutions of Navier-Stokes equations using incremental unknowns*, Comput. Method. Appl. M.**130** (1996), 319-335.
11. J. S. Hesthaven, *From electrostatics to almost optimal nodal sets for polynomial interpolation in a simplex*, SIAM J. Numer. Anal., **35(2)** (1998), 655-676.
12. J. S. Hesthaven, C. H. Teng, *Stable spectral methods on tetrahedral elements*, SIAM J. Sci. Comput., **21** (2000), 2352-2380.
13. S. F. Hoerner, Fluid-Dynamic Drag, Hoerner Fluid Dynamics, Bakersfield, 1965.
14. G. Karniadakis, S. J. Sherwin, Spectral/hp element methods for CFD, Oxford University Press, New York, 2005.
15. S. Kaya, B. Rivière, *A discontinuous subgrid eddy viscosity method for the time-dependent Navier-Stokes equations*, SIAM J. Numer. Anal., **43(4)** (2005), 1572-1595.
16. B. Rivière, M. F. Wheeler, and V. Girault, *Improved energy estimates for interior penalty, constrained and discontinuous Galerkin methods for elliptic problems. Part I*, Comput. Geosci., **3** (1999), 337-360.
17. M. Schäfer, S. Turek, *The benchmark problem 'flow around a cylinder'*, In Flow Simulation with High-Performance Computers II, Hirschel, E.H.(ed.). Notes on Numerical Fluid Mechanics, vol. **52**, Vieweg, Braunschweig, (1996), 547-566.
18. J. Shen, *Hopf bifurcation of the unsteady regularized driven cavity flow*, J. Comput. Phys., **95** (1991), 228-245.
19. J. Shen, *On error estimates of the projection methods for the Navier-Stokes equations: Second-order schemes*, Math. Comp., **65(215)** (1996), 1039-1065.
20. L. Song, Z. Zhang, *Polynomial preserving recovery of an over-penalized symmetric interior penalty Galerkin method for elliptic problems*, Discrete Contin. Dyn. Syst. – Ser. B **20(5)** (2015), 1405-1426.
21. L. Song, Z. Zhang, *Superconvergence property of an over-penalized discontinuous Galerkin finite element gradient recovery method*, J. Comput. Phys., **299** (2015), 1004-1020.

22. R. Temam, Navier-Stokes Equations: Theory and Numerical Analysis, AMS Chelsea publishing, Providence, 2001.

23. R. Temam, Navier-Stokes Equations and Nonlinear Functional Analysis, Volume 66 of CBMS-NSF Regional Conference Series in Applied Mathematics, SIAM, Philadelphia, second edition, 1995.

24. M. F. Wheeler, *An elliptic collocation-finite element method with interior penalties,* SIAM J. Numer. Anal., **15** (1978), 152-161.

Permissions

All chapters in this book were first published in MATHEMATICS, by AIMS Press; hereby published with permission under the Creative Commons Attribution License or equivalent. Every chapter published in this book has been scrutinized by our experts. Their significance has been extensively debated. The topics covered herein carry significant findings which will fuel the growth of the discipline. They may even be implemented as practical applications or may be referred to as a beginning point for another development.

The contributors of this book come from diverse backgrounds, making this book a truly international effort. This book will bring forth new frontiers with its revolutionizing research information and detailed analysis of the nascent developments around the world.

We would like to thank all the contributing authors for lending their expertise to make the book truly unique. They have played a crucial role in the development of this book. Without their invaluable contributions this book wouldn't have been possible. They have made vital efforts to compile up to date information on the varied aspects of this subject to make this book a valuable addition to the collection of many professionals and students.

This book was conceptualized with the vision of imparting up-to-date information and advanced data in this field. To ensure the same, a matchless editorial board was set up. Every individual on the board went through rigorous rounds of assessment to prove their worth. After which they invested a large part of their time researching and compiling the most relevant data for our readers.

The editorial board has been involved in producing this book since its inception. They have spent rigorous hours researching and exploring the diverse topics which have resulted in the successful publishing of this book. They have passed on their knowledge of decades through this book. To expedite this challenging task, the publisher supported the team at every step. A small team of assistant editors was also appointed to further simplify the editing procedure and attain best results for the readers.

Apart from the editorial board, the designing team has also invested a significant amount of their time in understanding the subject and creating the most relevant covers. They scrutinized every image to scout for the most suitable representation of the subject and create an appropriate cover for the book.

The publishing team has been an ardent support to the editorial, designing and production team. Their endless efforts to recruit the best for this project, has resulted in the accomplishment of this book. They are a veteran in the field of academics and their pool of knowledge is as vast as their experience in printing. Their expertise and guidance has proved useful at every step. Their uncompromising quality standards have made this book an exceptional effort. Their encouragement from time to time has been an inspiration for everyone.

The publisher and the editorial board hope that this book will prove to be a valuable piece of knowledge for researchers, students, practitioners and scholars across the globe.

List of Contributors

Armel Andami Ovono
Université des Sciences et techniques de Masuku, Franceville, Gabon

Alain Miranville
Université de Poitiers, Laboratoire de Mathématiques et Applications, UMR CNRS 7348 - SP2MI, Boulevard Marie et Pierre Curie - Téléport 2, F-86962 Chasseneuil Futuroscope Cedex, France

Paul Bracken
Department of Mathematics, University of Texas, TX 78539-2999 Edinburg, USA

Morris W. Hirsch
Department of Mathematics, University of Wisconsin, Madison WI 53706, USA

Filippo Dell'Oro and Vittorino Pata
Dipartimento di Matematica, Politecnico di Milano, Via Bonardi 9, 20133 Milano, Italy

Claudio Giorgi
DICATAM, Università degli Studi di Brescia, Via Valotti 9, 25133 Brescia, Italy

Ken Shirakawa
Department of Mathematics, Faculty of Education, Chiba University, 1-33, Yayoi-cho, Inage-ku, Chiba, 263-8522, Japan

Hiroshi Watanabe
Department of Computer Science and Intelligent Systems, Faculty of Engineering, Oita University, 700 Dannoharu, Oita, 870-1192, Japan

Onur Alp Ilhan
Faculty of Education,University of Erciyes Melikgazi 38039, Kayseri, Turkey

Shakirbay G. Kasimov
Mechanics and Mathematics Faculty,National University of Uzbekistan ,Tashkent, Uzbekistan

Yanfang Li, Yanmin Liu, Xianghu Liu and He Jun
Mathematics Department, Zunyi Normal College, 563006, Guizhou, P. R. China

Yunjiao Wang, Kiran Chilakamarri and Demetrios Kazakos
Department of Mathematics, Texas Southern University, 3100 Cleburne, Houston, TX, 77004, USA

Maria C. Leite
Department of Mathematics, University of South Florida at St. Pete, 140 7th Avenue South St. Petersburg, Florida 33701, USA

Mario Ohlberger and Barbara Verfürth
Applied Mathematics, University of Münster, D-48149 Münster, Germany

Weiping Gao and Yanxia Hu
School of Mathematics and Physics, North China Electric Power University, Beijing, 102206, China

Lunji Song
School of Mathematics and Statistics, Lanzhou University, Lanzhou, 730000, China
Department of Mathematics, The University of Alabama, Tuscaloosa, AL 35487, USA

Charles O'Neill
Department of Aerospace Engineering and Mechanics, The University of Alabama, Tuscaloosa, AL 35487, USA

Index

www.ingramcontent.com/pod-product-compliance
Lightning Source LLC
Chambersburg PA
CBHW082032190326
41458CB00010B/3344
9781682855782